Clean Steel:
Superclean Steel

Clean Steel: Superclean Steel

Conference Proceedings
6–7 March 1995
London, UK

Edited by
J. Nutting and R. Viswanathan

Organised by
The Institute of Materials
on behalf of
The Electric Power Research Institute
Palo Alto, California, USA

THE INSTITUTE OF MATERIALS

Book 635
Published 1996 by
The Institute of Materials
1 Carlton House Terrace
London SW1Y 5DB

ISBN 0 901716 90 1

British Library Cataloguing-in-Publication Data
A catalogue record of this book
is available from The British Library

Cover picture courtesy of
Y. Ikeda, H. Yoshida, Y. Tanaka and T. Fukuda (see p.80)

Typeset by
Fakenham Photosetting Ltd
Fakenham, UK

Printed and bound at
The University Press
Cambridge, UK

Contents List

Preface vii

Application of Clean Steel/Superclean Steel Technology in the Electric Power Industry – Overview of EPRI Research and Products 1
R. Viswanathan

The EPRI Survey on 'Superclean Steels' 33
J. Nutting

Residual Contents in Purchased Scrap for Use in Basic Electric Arc Technology 53
P. A. Bralsford, E. Hydes and G. A. Honeyman

Manufacture of Superclean 3–3.5% NiCrMoV Steels for Gas Turbine Components 59
E. Potthast, K. Langner and F. Tince

Production and Properties of Superclean Monoblock LP Turbine Rotor Forgings 71
Y. Ikeda, H. Yoshida, Y. Tanaka and T. Fukuda

Development of Production Technology and Manufacturing Experiences with Superclean 3.5 NiCrMoV Steels 89
W. Meyer, R. Bauer and G. Zeiler

Production of a HP/LP Forging of 2CrMoV Steel 101
T. Tsuchiyama, H. Nakashima, M. Miyakawa, M. Okamura, P. Murpy and R. C. Schwant

A New Superclean High Pressure–Low Pressure Rotor Steel for Advanced Combined Cycle Power Plants 109
M. Yamada, Y. Tsuda, M. Miyazaki, Y. Tanaka, T. Azuma and Y. Ikeda

Development of Superclean 12Cr Steel Disc for Heavy Duty Gas Turbines 123
Y. Fukui, M. Shiga, S. Nakamura and K. Hidaka

Stress Corrosion in 3.5NiCrMoV Welded Steel 133
S. R. Holdsworth, M. Nougaret and B. Vittemant

Susceptibility to Embrittlement of 4% Ni–CrMo High Strength Steels and 11% Cr–NiMoV Heat-Resistant Steel – Roles of Melting Methods and Residual Elements 157
J. BOURRAT and H. SCHAFF

Pitting and Stress Corrosion Cracking of Conventional and High Purity LP Turbine Rotor Steel 167
J. DENK

100 000H Isothermal Ageing Test Results of NiCrMoV Steels for Low Pressure Steam Turbines 181
Y. TANAKA, T. AZUMA, N. YAEGASHI and Y. IKEDA

Influence of Steel Cleanness on the Operational Properties of Turbine Components 195
K. H. MAYER and C. BERGER

Low Residuals in Bolting Steels 213
H. EVERSON and J. ORR

Zero Inclusion Steels 223
J. H. LOWE and A. MITCHELL

Clean Steels for Fusion 233
D. S. GELLES

Temper Embrittlement in High Strength Pressure Vessel Steels for Polyethlene Manufacture 243
G. A. HONEYMAN and J. E. HEESOM

Improvements of Manufacturing and In Service Mechanical Properties of Nuclear Components in Mn Ni Mo Steel through the Lowering of Contents in Residual Elements 255
L. DUNARD-ROUX, B. HOUSSIN and T. BERGER

Recent Developments in High Pressure Vessels for the Petrochemical Industry: Materials and Products 265
P. BOCQUET, A. CHEVIET and R. DUMONT

Low Residual HY80 Steel for Offshore Applications 285
G. A. HONEYMAN, M. A. WALSH and M. G. KEARNEY

Clean Steel/Superclean Steel Workshop – Questions and Answers 295

Report on the Round Table Discussion 307
J. NUTTING

Author Index 313

Index 315

Preface

The improved understanding of the factors leading to the long term in-service embrittlement of L.P. rotors which developed from basic research carried out in the 1970s enabled the late Bob Jaffee to formulate his ideas for the production of superclean steels. A history of the progress from experimental studies to actual rotors has been recorded in the proceedings of four previous workshops organised by the Electric Power Research Institute of Palo Alto, California, U.S.A.

The proceedings of the Fifth Workshop are recorded in the present volume which appears some three years after that of the Fourth Workshop.

The progress during the intervening period, as outlined in the twenty one papers presented, has been extensive. Not only have the properties of superclean 3.5Ni–Cr–Mo–V steels been thoroughly established, the economies of their production has been assessed and the extent of their use in the power generation industry has been recorded.

The concepts involved with the already developed superclean steels are applicable to a wider range of steels than the conventional 3.5Ni–Cr–Mo–V varieties. Already some progress is now being made with the development of single shaft HP–LP rotor steels to a superclean specification and the same comment applies to the more highly alloyed 9–12% Cr steel, where, once again, low residual contents are found to alleviate the long-term embrittlement problems.

It is also clear that the superclean steels can be used for other than conventional power generation equipment. The use of these improved materials for general pressure vessels, nuclear components and the petro-chemical industry's pressure vessels is now becoming more widespread, where the requirement is to minimise in-service embrittlement at elevated temperatures. But, superclean steels also have excellent fracture toughness, at around ambient temperatures, with good stress corrosion resistance and these characteristics have led to the application in off-shore structures.

It is not surprising, therefore, that there is considerable interest in the future development and applications of superclean steels. Many views were expressed on these topics at a round table discussion, a summary of which is given in the proceedings.

The Editors of the proceedings are grateful to the other members of the organising committee for their unstinted help and to Miss Lisa Davies of The Institute of Materials, for her assistance as General Organiser of the Conference. Many others helped, but in particular, our thanks are due to Monsieur André Coulon, for his continued interest and advice.

We hope that the present volume will be looked upon as a further fitting tribute to the memory of the late Bob Jaffee.

J. Nutting and R. Viswanathan.

Application of Clean Steel/Superclean Steel Technology in the Electric Power Industry – Overview of EPRI Research and Products

R. VISWANATHAN
Electric Power Research Institute
3412 Hillview Avenue, Palo Alto, CA

1. INTRODUCTION

The adverse effects of impurity elements phosphorus (P), antimony (Sb), tin (Sn), arsenic (As), sulphur (S), oxygen (O), and deoxidants aluminum (Al) and silicon (Si) on the mechanical properties of steels have been known for many decades. P, Sb, Sn and As interactively with Si and Mn cause temper embrittlement and lead to reduction in the fracture toughness (K_{IC}) and increase in the ductile–brittle transition temperature (FATT). Presence of sulphide inclusions, and nonmetallic inclusions containing Al and Si can facilitate cavity nucleation at the grain boundaries and in the grains, thus facilitating creep fracture at high temperatures and ductile fractures in the upper shelf region. These changes result in reduced creep ductility at high temperatures and reduced fracture toughness at lower temperatures. Advancements in the steel making technology during the last two decades have enabled reduction of these impurities and deoxidants to levels as low as 20 ppm leading to what might be called 'clean steels'. Further realisation that in view of the low sulphur levels achievable, even Mn is no longer necessary to 'fix' the sulphur and can, therefore, be reduced to levels as low as 0.01 to 0.02% has resulted in 'superclean steels'.

The development that has made 'clean' and 'superclean' steels possible was secondary steel refining via ladle furnaces in conjunction with vacuum degassing in the ladle and during casting. Ladle treatment of molten steel for the purpose of desulphurisation and vacuum carbon deoxidation were developed only in 1975. These techniques provided a production means of manufacturing high purity steels that previously could only be made in the laboratory as a control for temper embrittlement research studies. Thus, the practical solution to the problem in alloy steels was to refine the steels from Mn, Si, P, Sn, As, and Sb during steel making operations. Manganese, silicon, and phosphorus are easily removed during the oxidising stage of steel making, because they oxidise preferentially to iron, and enter the oxidising slag. This generally is done in the electric arc furnace. Tin, arsenic, and antimony are controlled by scrap selection, with basic oxygen furnace steel scrap used as the starting material for electric arc furnace melting if possible. After separation of the oxidised steel from the oxidising slag, it is transferred to a reducing slag in a ladle refining furnace, which removes the sulphur. Vacuum treatment of the desulphurised steel in the ladle furnace accompanied by argon bubbling provides a means for deoxidation via the vacuum carbon deoxidation (VCD) process. This leaves no oxide particles dispersed in the steel as would occur if deoxidation were done with silicon. The alloying elements are added at this stage. Casting into molds contained in a vacuum chamber, called vacuum stream degassing (VSD) completes the double degassing treatment. The extensive degassing combined with argon bubbling removes nitrogen and hydrogen as well as oxygen. The result is a superclean steel free to the maximum extent possible of grain boundary segregates and of nonmetallic oxides and sulphides. Alternate routes to produce clean/superclean steels also include use of Al deoxidation in lieu of VCD and conventional pouring in air with argon shrouding.

The primary driver in cleaning up steels has been the desire to achieve improved fracture toughness, although benefits in other properties such as resistance to creep, stress corrosion, etc. have accrued incidentally and serendipitously, making cleanliness even more desirable. In general, any modifications to alloy content or heat treatment to improve toughness of the steels have an opposite effect on the creep strength. Getting rid of the impurities (tweaking) provides one of the few viable ways of achieving the best results with respect to all the mechanical properties.

Good fracture toughness and freedom from impurity induced temper embrittlement are crucial requirements for rotors and discs. Regardless of the operating temperature, the final failure scenario for all catastrophic rotor/disc bursts involves either failure due to peak thermal stresses during a start-up/shut-down transient, during overspeed testing or during steady operation at low temperatures. Hence FATT and fracture toughness considerations at low/intermediate temperatures are important.[1,2]

Reduced fracture toughness of rotors and discs arising either inherently due to sulphides and inclusions or due to impurity-induced embrittlement results in a decrease of the component life under base load and cyclic operation due to increased risk of brittle fracture, and is one of the major causes of early component retirement. Embrittlement can also have a major constraining effect on the operating procedure. To keep transient stresses low, stringent controls have to be exercised with respect to the start–stop cycles. For instance, during each cold start, the rotors may need to be pre-warmed to a temperature above the ductile–to–brittle fracture appearance transition temperature (FATT) over a period of several hours prior to imposition of full load. These requirements lead to increased operational costs for the plant and decreased flexibility and availability. Reduced toughness also increases inspection costs due to the more frequent inspections required. In the case of low-pressure (LP) rotors, the LP cross-over temperature in fossil plant turbines is kept below about 370°C (700°F) to avoid embrittlement, even at the risk of reduced efficiency. Increase of the LP cross-over temperature to 400 to 427°C (750–800°F) would result in major efficiency gains. Improving the toughness of LP rotor steels also means that the increasing of the strength such as the tensile strength and the proof strength can be obtained along with maintaining the conventional toughness level by the modification of the tempering condition. Accordingly, the length of the last stage blading can be increased to improve the turbine thermal efficiency. Toughness of steels thus has a significant effect on the reliability, availability, cycling ability, efficiency and operating and maintenance costs (O&M) of turbines.

Table 1 shows the principal 'driver' requiring advancements in the different types of turbine rotors. For a more complete summary of rotor developments the reader is referred to Ref. 3. In the case of HP/IP rotors operating at elevated temperatures, improved creep strength permitting operating at higher temperatures and providing greater efficiency and improved toughness providing greater reliability, longevity and cycling ability have been the drivers. For LP rotors, increasing the LP cross-over temperature from 370°C (700°F) to 400°C (750°F) without the risk of temper embrittlement has been the principal driver since such a change would increase the turbine efficiency substantially. Increased longevity under stress corrosion conditions has also been an incidental benefit. For HP/LP combined rotors, reduced installed cost/kwh by housing the entire rotor inside a single casing has been the driver. This becomes possible if the creep strength requirements in the high temperature end and toughness requirements in the low temperature end can both be satisfied in a single forging. In all cases, improved reliability of the rotor leads to decreased inspection costs since the inspection intervals can be extended. An order of magnitude estimate the cost impact of various improvements is provided in Appendix 1, strictly for illustration purposes.

EPRI has been active in sponsoring and catalysing research pertaining to improved HP, LP and HP/LP rotor steels primarily through improvements in cleanliness since the mid 70s. Although the deleterious effects of impurities were known prior to that, there were no systematic studies of the synergistic effects of impurities. Although it was known that Mn contributed to embrittlement, high Mn levels were routinely maintained although no longer required in view of the steady decrease in the S content of steels.[3b] In order to make clean steels a reality, the steel makers, especially in the U.S. needed to be convinced of three things:

1) That impurity elements do in fact adversely affect the toughness, creep ductility, stress corrosion and key properties of steel.

2) That clean steel components in the sizes needed can be successfully made with available technologies, without driving up the costs to unrealistic levels.

Table 1 Principal Drivers for Development of Cleaner Rotor Steels

Rotor	Primary Requirement	Current Material	Advanced Material	Advantages	Principal Drivers
HP and IP <538°C (<1000°F)	Creep strength with adequate toughness	1Cr 1 Mo 0.25V	Clean 1Cr 1 Mo 0.25V	• Improved toughness • Incidental improvement in creep strength ductility	• Reliability • Longevity • Cycling ability
HP and IP <565°C (<1050°F)	Creep strength with adequate toughness	12CrMoV 11CrMoVNbN 11CrMoVTaN	Further modifications with • Nb, W, Ta • Cleanliness • High N	• Improved creep strength	• Efficiency gain due to operability up to 620°C (1150°F) in USC plants. • Cycling ability.
LP <370°C (<700°F)	Toughness	3.5%NiCrMoV	Superclean 3–3.5%NiCrMoV	• Improved toughness (freedom from embrittlement). • Incidental improvements in SCC resistance. • Incidental improvements in creep strength/ ductility	• Efficiency gain due to increased LP cross-over temperature to 400°C (750°F) in USC plants. • SCC resistance
HP/LP single shaft	Creep strength in HP end and toughness in LP end	• 1Cr1Mo0.25V differentially heat treated. • 1Cr1Mo0.25V and 3.5NiCrMoV bolted. • 2CrMoNiWV	• Superclean 2.5NiCrMov • Clean 2CrMoNiWV • Clean 2CrNiMoVNbW • Clean 2Cr1Mo2Mn 1.5NiV • Clean 1.8NiCrMoV	Improved toughness maintaining the creep strength of CrMoV at HP and maintaining the toughness of 3.5NiCrMoV at the LP.	• Reduction in installed cost. • Cycling ability.
Combustion turbine discs	Creep strength with adequate toughness	1Cr1Mo0.25V 1%CrNiMoV	Superclean 11%CrNiMoV	• Improved toughness • Freedom from embrittlement	Efficiency gain due to higher temperature capabilities.

3) That superclean steels with a very low Mn could be made and forged into product shape.

All of these concerns were systematically addressed through a series of research and development and demonstration projects. A summary of these projects, the principal outcome of each project leading to the next step and the EPRI product relating to each of these developments is provided in Table 2. The objective of this paper is to review the various EPRI-sponsored developments relating to clean/superclean steel technology.

2. HP AND IP ROTORS

These rotors typically operate at a maximum temperature of either 540°C (1000°F) or 565°C (1050°F) at the steam inlet end with temperature falling to about 345°C (650°F) at the steam exit end of the high pressure (HP) or the Intermediate Pressure (IP) turbines. For temperature up to 540°C (1000°F), a 1CrlMo 1/4V steel is generally used and for temperatures up to 565°C (1050°F) variations of a 12%Cr martensitic steel are used.

Figure 1 shows the reduction in the concentration of impurity elements in 1Cr1Mo 1/4V steels over the last few decades, due to advancements in steel making and refining technologies.

Numerous improvements have been made since the late 1940s in terms of rotor compositions, fabrication practices and cleanliness, as illustrated in Fig. 2 for high temperature forgings, based on the extrapolation of results cited in Refs 4 and 5. The toughness and rupture ductility have seen corresponding improvements as shown in Fig. 2.[4,5] The key events in this evolution include:

1) Switching to higher creep strength CrMoV rotors from the NiMoV rotors in 1948.

2) Introduction of D-grade rotors austenitised at 954°C (1750°F) in place of the earlier C-grade rotors austenitised at 1010°C (1850°F) to eliminate notch sensitivity and low ductility problems around 1953.

3) Use of electric furnace melting and vacuum pouring introduced around 1958 to reduce bore segregation of nonmetallic inclusions and sulphide streaks.

4) Introduction of 12 percent Cr rotors for 1050°C service in 1960.

5) Application of vacuum arc remelting process in 1970.

6) Demonstration of clean steel CrMoV rotors about 1984.

7) Demonstration of higher creep strength in modified 12 percent Cr rotor compositions about 1987.

The integrity of rotors has also been improved by advances in welding procedures for the initial fabrication or repair incorporating smaller forgings of higher quality.

The first EPRI project dealing with clean steels, RP1343 was started in the

Table 2 EPRI Resources Pertaining to Clean/Superclean Steels

Project	Performed By	Outcome	Product
RP559	University of Pennsylvania	Embrittling effects of impurities and synergisms in pressure vessel and rotor steels were demonstrated and the need for steel cleanliness was established.	EPRI Report NP-1501, September 1980.
TPS82–628	Engineering Materials and Processes	A survey of advanced steel making processes was completed.	EPRI RD3336, December 1983.
RP2741–1	University of Leeds	Demonstrated that low Mn superclean steels could be forged.	Report on RP2741-01, 1987.
RP1343	Westinghouse Electric	Improved toughness due to reduced S content was demonstrated in three trial HP rotors of 1Cr1MoV composition using VCD, VSD and low S steel making.	EPRI CS-4516, May 1986.
RP2060–1,-2	Japan Steel Works/ Bethlehem Steel Company	Four structural steels were evaluated in high purity form with low levels of Si and Mn. The basis for superclean 3.5NiCrMoV LP rotor development was established. In 1CrMoV HP rotors, the need for compensating for loss of bainitic hardenability due to removal of Mn by adding Ni and the impossibility of developing superclean versions based on straight 1Cr1Mo1/4V were identified.	EPRI NP-5399, September 19897.
RP1403–8	Vereinigte Edelstahlwerke	A trial rotor (VEW1) of superclean LP 3.5%NiCrMoV was made and improved mechanical properties were demonstrated.	Paper in Ironmaking Steelmaking, Vol. 13, p 322, 1986.

Table 2 *(continued)* EPRI Resources Pertaining to Clean/Superclean Steels

Project	Performed By	Outcome	Product
RP2741–3	Vereinigte Edelstahlwerke	A superclean LP 3.5NiCrMoV rotor (VEW2) with relaxed specification was evaluated. Concluded that the more stringent specs of VEW1 were needed.	RP2741–3 Final Report, unpublished.
RP1403–15	Toshiba/JSW/GE	A superclean LP 3.5NiCrMoV trial rotor was evaluated. Concluded that the steel was superior to conventional steel.	Draft EPRI Report RP1403–15, December 1991.
RP1403–18	ASEA Brown Boveri	Stress corrosion evaluations were completed on superclean 3.5NiCrMoV LP material from VEW1. Superiority in crack initiation was demonstrated.	EPRI GS-6907, 1990.
RP1403– 32	Daedalus Associates	All developments on superclean 3.5NiCrMoV LP rotor steels were summarised and reviewed.	EPRI GS-6610, December 1989.
RP1403–32	Daedalus Associates	Developments on superclean 3.5NiCrMoV LP rotors were reviewed and a concise guide for utility use including chemistry specification was prepared.	EPRI GS-6612, December 1989.
RP2426–4, 2741–4	Bethlehem Steel Corporation	Following up on the work in RP2060–1,-2, a superclean version of 1Cr1MoV HP rotor in which loss of Mn was compensated by adding 2.5%Ni and Nb was developed based on evaluation of lab heats. It was found the alloy could match the creep strength of 1CrMoV HP and the toughness of 3.5NiCrMoV LP. This paved the way for evaluation of 2.5NiCrMoV for HP/LP single shaft application.	EPRI ER-6887, July 1990.

Table 2 *(continued)* EPRI Resources Pertaining to Clean/Superclean Steels

Project	Performed By	Outcome	Product
RP1403–55	General Electric/Toshiba Corp./Japan Steel Works	A trial rotor of 2.5NiCrMoV for HP/LP was made and evaluated. The alloy had FATT below room temperature and creep strength comparable to CrMoV, but was found to be notch weak in creep.	EPRI TR-103689, February 1994.
RP1403–21	KWU-Siemens Corp. Saarstahl. Asea Brown Boveri. MAN Energie Inc. GEC Alsthom Inc. Parsons Inc.	Laboratory heats of a 2%CrMoNiWV European alloy was evaluated. A trial rotor of the optimised chemistry was made. This material is currently under evaluation.	Reference 29.
RP1403–7	Engineering Materials and Processes	A survey of open literature provided information on the capabilities of current 12Cr steels up to 565°C (1050°F) and modified 12Cr steels up to 593°C (1100°F).	EPRI CS-5277, July 1987.

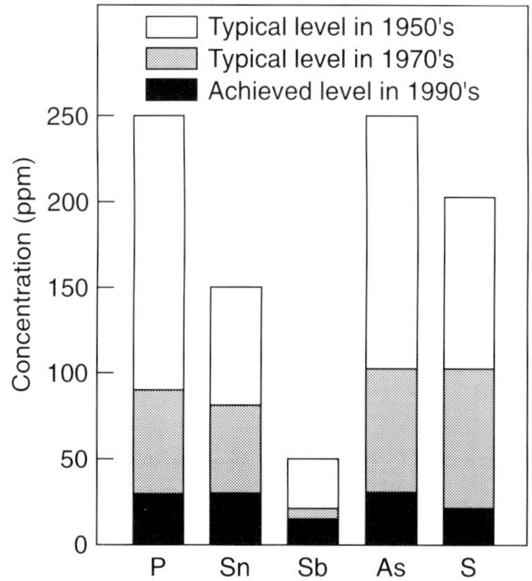

Fig. 1 Trends in impurity levels in Cr–Mo–V rotor steels

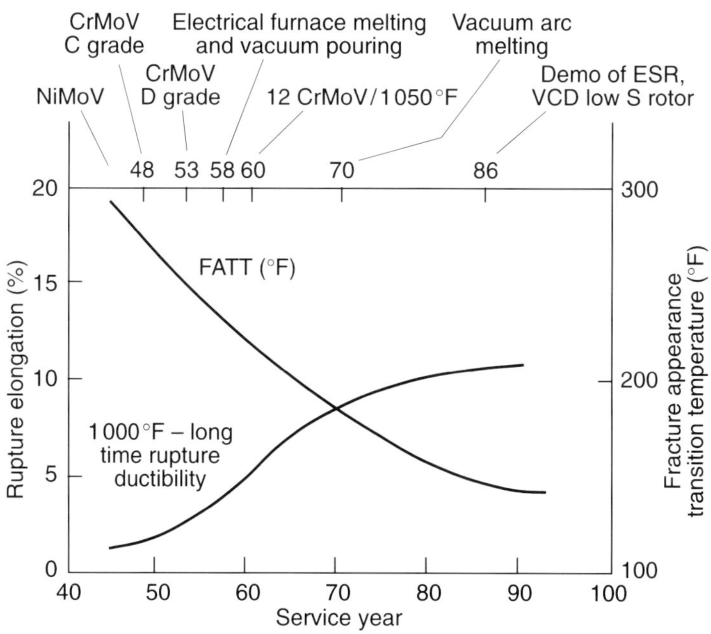

Fig. 2 Evolution of HP rotor properties in relation to changes in steel making technology

wake of the catastrophic failure of the Gallatin HP/IP rotor, wherein presence of sulphide stringers at the bore had been implicated in a major way.[6] To reduce the sulphur level in HP/IP 1Cr1Mo 1/4V type (ASTM A470 – Class 8) rotors, Swaminathan and Jaffee demonstrated the effectiveness of three advanced steel making technologies.[7,8] Three full-size 30 ton Cr–Mo–V rotor forgings were produced from 100 ton ingots each by vacuum carbon deoxidation (VCD), electroslag remelting and low sulphur vacuum silicon deoxidation (low S) processes. Ladle refining was applied to the liquid metal in the low S and VCD processes, whereas in the ESR process refining of the steel was achieved during electroslag remelting. All the processes reduced sulphur to very low levels in the range 10–20 ppm resulting in excellent bore quality and cleanliness. In addition, the VCD process also resulted in low P (30 ppm) and low Si levels. Table 3 provides a comparison of the composition of the advanced rotors with the conventional rotors. The centreline cleanliness of the rotors were appreciably better than conventional rotors.

Table 3. Chemical Compositions of Advanced Technology and Conventional 1CrMoV Forgings Evaluated by Jaffee and Swaminathan[(7,8)]

Element	Low S	VCD Low S	ESR[2]	Conventional[3]	ASTM A470, Class 8
Carbon	0.31	0.28	0.31	0.32	0.25–0.35
Manganese	0.78	0.76	0.78	0.83	1.0 max
Phosphorus	0.007	0.004	0.009	0.009	0.015 max
Sulphur	<0.001	0.001	0.002	0.009	0.018 max
Silicon	0.23	0.05	0.19	0.27	0.15–0.35
Nickel	0.33	0.40	0.27	0.23	0.75 max
Chromium	1.13	1.18	1.18	1.07	0.9–1.5
Molybdenum	1.15	1.21	1.18	1.17	1.0–1.5
Vanadium	0.23	0.26	0.26	0.25	0.2–0.3
Tin	0.002	0.010	0.003	0.006	–
Antimony	0.0012	0.0015	0.001	0.001	–
Copper	0.04	0.04	0.04	0.09	–
Aluminum	0.004	0.005	0.009	0.003	–
Arsenic	0.003	0.006	0.011	0.010	–
Nitrogen[4]	84	96	105	–	–
Hydrogen	1.5	0.37	1.6	–	–
Oxygen	16	38	33	–	–

1. Elements in wt%, gases in ppm. 2. Average of top and bottom analyses. 3. Mean value of 29 Westinghouse conventional rotors (1971–77). 4. Gas analysis from hot top.

Compared with conventional forgings, the fracture toughness, creep–rupture strength, rupture ductility, and LCF strength were found to be superior, as shown in Fig. 3. Considering the fact that, in general, improvements in rupture strength and ductility and in rupture strength and fracture

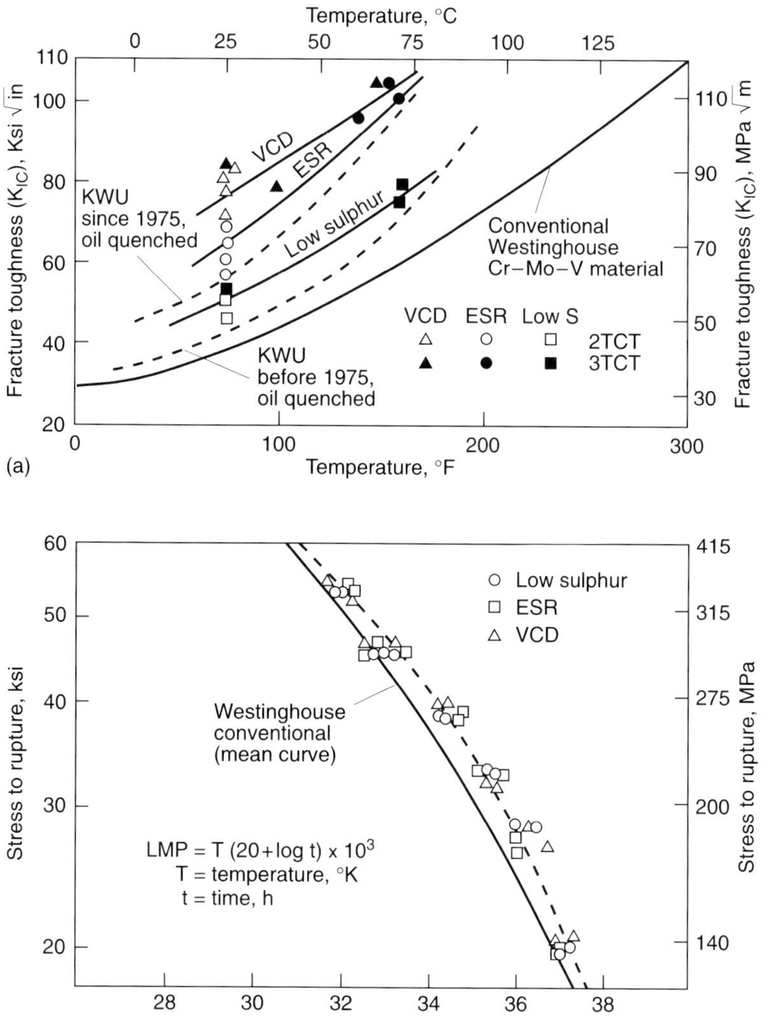

Figs 3a and **b** Improved mechanical properties achieved by advanced melting technologies and oil quenching compared with air-cooled conventional Cr–Mo–V steel rotor forgings

toughness are mutually opposed, the simultaneous improvements achieved with respect to all of these properties must be considered a major milestone in rotor technology. The three trial rotors made in this project were installed at three operating plants in the U.S. rated at 520 MW each. Rotors with low levels of Si produced by the VCD process have been used in several commercial plants recently.[9] In Germany, on the other hand, fracture toughnesses comparable to the low sulphur steels have been achieved since 1973 even in commercial cleanliness CrMoV steels by oil quenching, in contrast to the practice of air quenching in the U.S.[9a] However, the extreme reduction of

Mn content required in 'superclean' steels has not been performed in 1Cr1Mo 1/4V rotor steels. The application of superclean steel technology to this steel was tried in EPRI project RP2060[10] with a view of increasing the toughness, but the results were not promising. The 1Cr1MoV steel lost hardenability as a result of the absence of Mn and compensation to restore hardenability was necessary in order to achieve a fully bainitic structure at the slow cooling rates

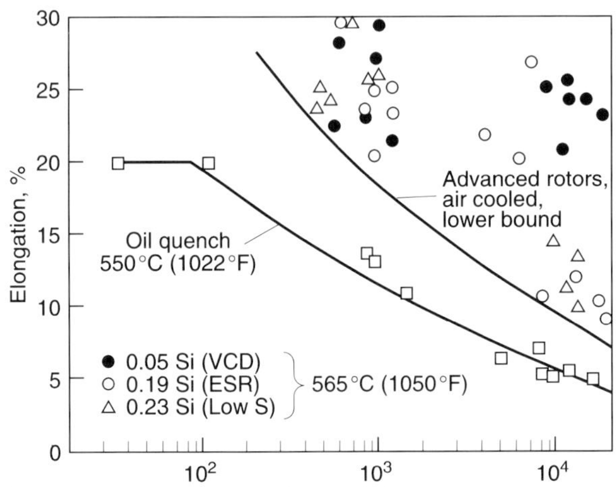

(c)

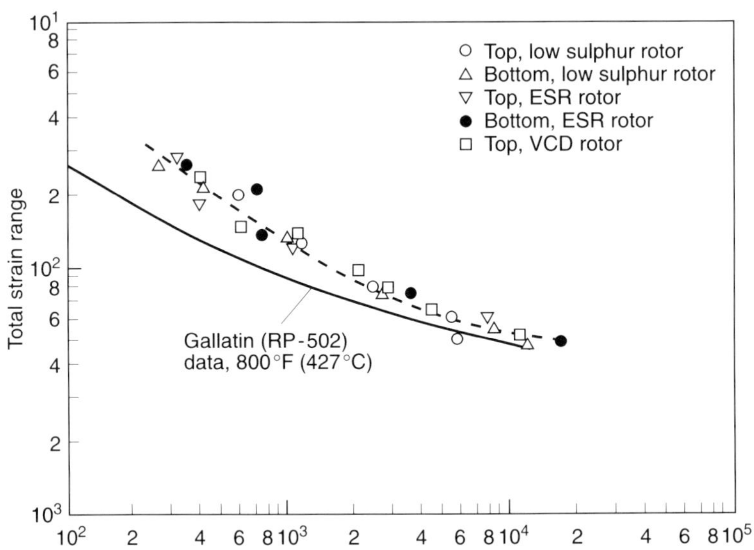

(d)

Figs 3c and **d** Improved mechanical properties achieved by advanced melting technologies and oil quenching compared with air-cooled conventional Cr–Mo–V steel rotor forgings

characteristic of the center of rotors. The alloy modifications made for compensation eventually led to the 2.5% NiCrMoV steel discussed later as a candidate for HP/LP combined shaft.

With respect to 12%Cr steels for applications at and above 565°C (1050°F), EPRI has compiled and summarised all the available information.[11] There is limited data on what appear to be clean steel versions of 12%Cr steels in the literature.[12–15] Several modified versions of 12%Cr steels for use above 593°C (1100°F) have been investigated. The most recent compilation of these compositions may be found elsewhere.[13,16] Many of these compositions investigated tend naturally to be low in Si, S and P since many of the rotors are produced by VCD, ESR or other refined processes. For all practical purposes these may be called 'clean' steels. Since alloying element modifications and not specifically improved cleanliness is the focus of these studies, they are not reviewed here. There are, however, no reported studies pertaining to superclean versions of 12% steel for steam turbine applications. Two recent studies have reported results of evaluation of superclean trial discs of 12%Cr steels for combustion turbine disc applications. The superclean steels were shown to have significantly improved toughness and freedom from long term embrittlement compared to conventional 12%Cr steels.[17,18] This development would clearly permit use of superclean steel discs at higher temperatures with attendant gains in efficiency.

3. LP ROTORS

LP rotors are typically made of a 3.5%Ni, 1.5%Cr, 0.5%Mo, 0.1%V steel. The maximum permissible temperature for these rotors has been limited to about 370°C (700°F) due to the possibility of in-service temper embrittlement. Cleanliness therefore has been a focus primarily to eliminate temper embrittlement so that increased efficiencies could be realised by operating at higher temperatures.

Figure 4 illustrates the decrease in FATT as a function of the year of manufacture for LP rotor materials. This is the core FATT in the as-fabricated condition. The plot shows that the FATTs have been quite low and that toughness has not been much of a concern since the early 60s. The concern has been mainly in-service temper embrittlement and post exposure FATT. With the introduction of VCD and reduction of Si in the early 60s, the embrittlement problem was reduced. With the advent of the superclean steels, where both Mn and Si are reduced using a number of alternative techniques, the embrittlement problem has been virtually eliminated.

The application of clean steel and superclean steel technologies with respect to 3.5Ni 1.5Cr 0.5Mo 0.1V steels typically used for rotor steels was first investigated in EPRI project RP2060.[10] The superclean steels contained approximately 0.001%S, 0.002%P, 0.02%Si and 0.02%Mn, 25 ppm oxygen and

50 ppm N. Laboratory ingots 100–200 Kg were made, forged and evaluated. These evaluations showed the superclean version of NiCrMoV LP rotor steels had superior ductility and toughness properties and high temperature creep strength compared to the conventional NiCrMoV rotor steel. The next step was to make and evaluate trial rotors of the 3.5NiCrMoV steel used for low pressure steam turbine rotors. EPRI sponsored two trial rotors (34 tons 1186 mm dia.) cast by Vereinigte Edelstahlwerke A.G. (now Boehler A.G.) in Kapfenberg, Austria in 1985.[19–21] Japan Steel Works and Toshiba sponsored a third trial rotor (106 tons 1800 mm dia.) made by JSW in Muroran, Japan. Sections of these trial rotors were widely distributed throughout the world under EPRI aegis. Some of the results from these studies conducted under EPRI project RP1403–15 have been reported by Watanabe, Yoshioka, Schwant and others.[22,23] On the basis of the successful trial 3.5NiCrMoV rotor demonstrations, superclean 3.5NiCrMoV was selected for the Chubu Electric Company's 700 MW units for the Kawagoe and Hikenen plants. Other trial and production 3.5NiCrMoV rotors were manufactured by Kobe Steel and Japan Casting and Forging. Two recent retrofits for Duke Power Company's River Bend Station were manufactured by Kobe Steel. One Japanese steel manu-

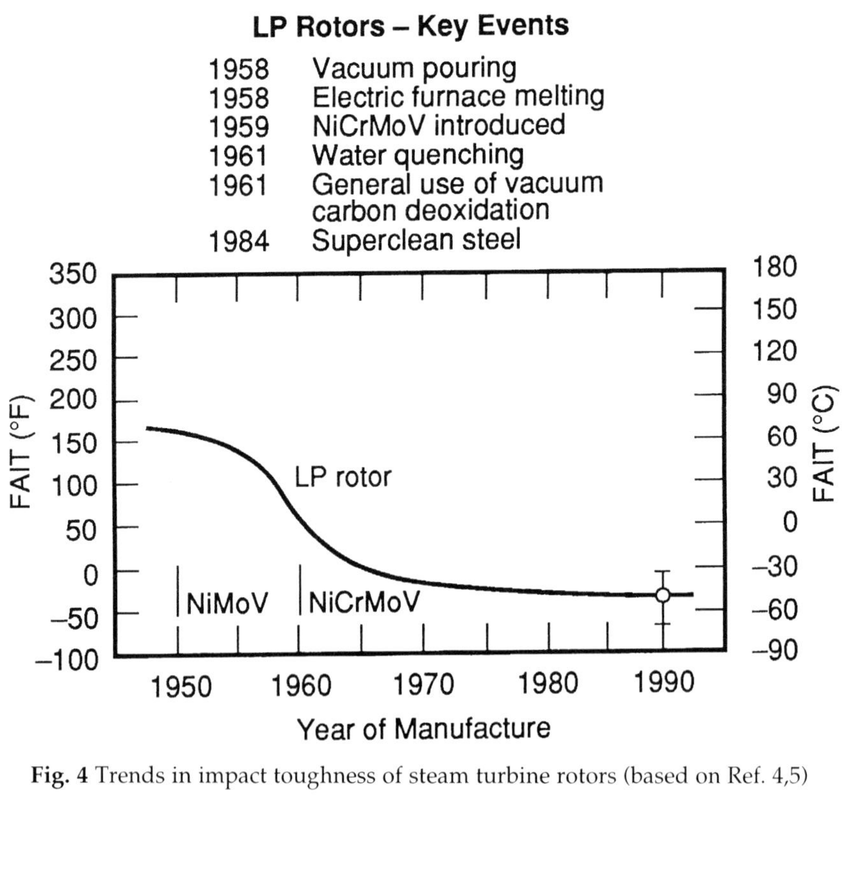

Fig. 4 Trends in impact toughness of steam turbine rotors (based on Ref. 4,5)

facturer reported production and delivery of 28 superclean LP rotors in the 700 MW class for improved efficiency and SCC resistance.

In the case of the Kawagoe USC (Ultra Super Critical) plant of Chubu Electric Company in Japan, the increase in the inlet steam temperature of LP turbine was applied without the cooling by the extracted steam in order to improve the thermal efficiency. In this USC plant, the superclean LP rotors were used for the first time in the world. The LP inlet temperature of this plant was about 30°C (54°F) higher than that of the conventional plants and therefore the thermal efficiency was improved by about 0.1% with no change in the turbine operation and the inspection interval, etc.[9]

The superclean 3.5NiCrMoV steels have lower FATT, higher fracture toughness and higher upper shelf energy compared to conventional grade as shown in Figs 5 and 6. The FATT ranges from −20°C (−4°F) to −70°C (−94°F)

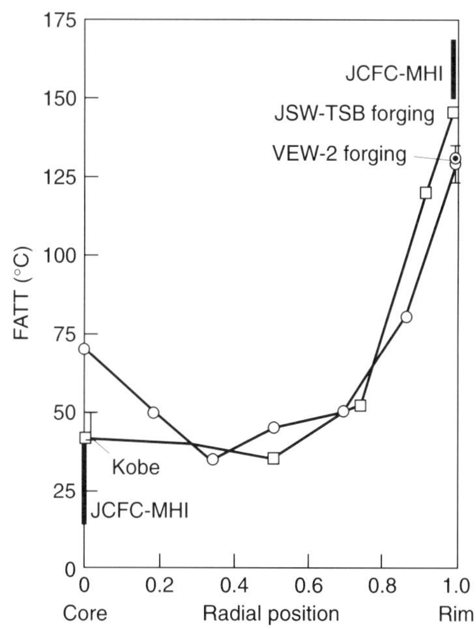

Fig. 5 Variation of FATT as a function of radial position (various rotor forgings)

at the core and mostly below −110°C (−166°F) at the rim. These values are somewhat below those achievable for conventional grade. Most importantly, the superclean steels are immune to temper embrittlement (see Fig. 7). The higher fracture toughness and lower FATTs translate into larger critical sizes and hence increased longevity under steady and cycling conditions.

The increased creep strength of superclean 3.5NiCrMoV steels compared to conventional grade steel is shown in Fig. 8. The increased strength was derived without compromising the creep ductility. This is thought to derive

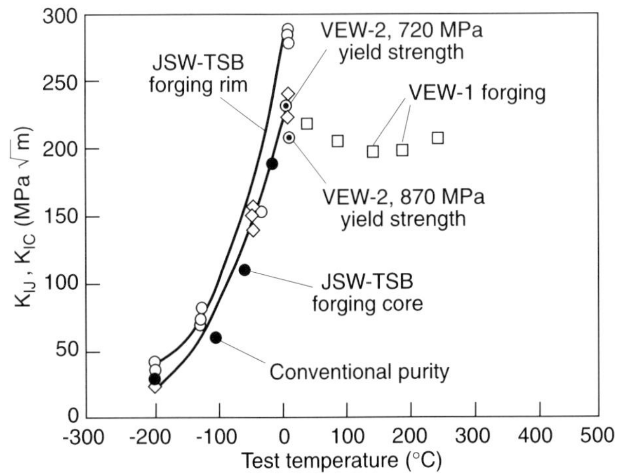

Fig. 6 Fracture toughnesses for superclean and conventional purity 3.5NiCrMoV steel

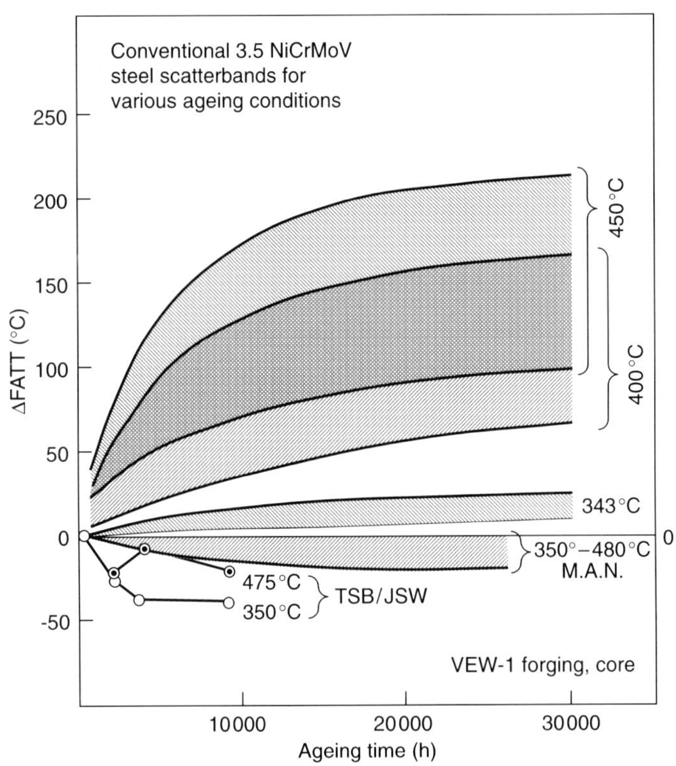

Fig. 7 Change in FATT versus time (temper embrittlement) for conventional material and for the superclean VEW-1 forging

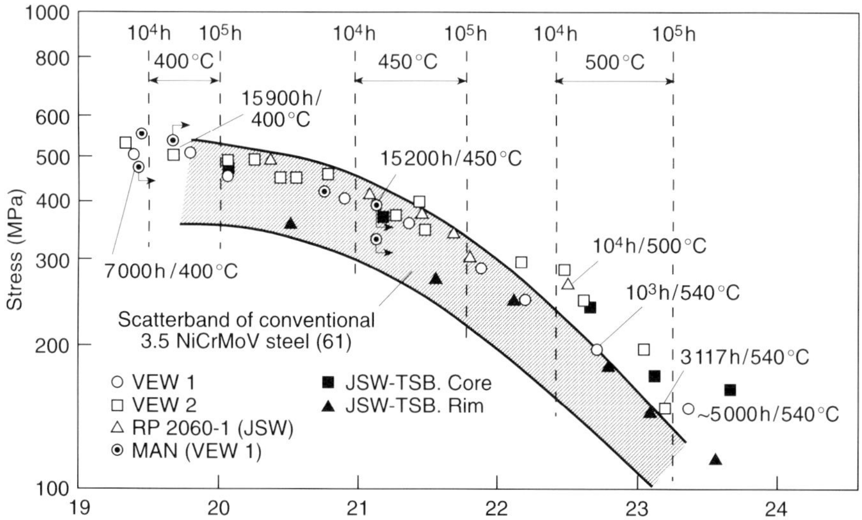

Fig. 8 Stress rupture of superclean and conventional 3.5NiCrMoV steel

from the absence of MnS particles at the prior austenite grain boundaries which might cavitate and initiate grain boundary creep fractures. The service temperature ceiling for superclean grade 3.5NiCrMoV is limited by creep strength to a maximum of 500°C instead of to a maximum of 350°C limited by susceptibility to temper embrittlement. This makes it a good candidate for use in single shaft HP/LP configurations for steam temperatures up to 500°C. Superclean steels have also been used extensively for Combustion Turbine Disk applications.

An incidental but nevertheless key benefit of superclean 3.5NiCrMoV steels is their greater resistance to SCC initiation. The results are clear that crack propagation rates are the same for superclean and conventional 3.5NiCrMoV. However, data from many sources indicate that crack initiation is much more difficult in superclean grade steel, probably because of its greater cleanliness. In applications where rotors and discs must operate in the presence of condensed steam, this characteristic could spell the difference between success and failure, particularly if the steam were contaminated by salt impurities or oxygen. Figure 9 presents initiation data on 3.5NiCrMoV heat treated to 860 MPa yield from Scarlin and Denk.[24]

EPRI has prepared a superclean steel guide for utility use.[25] This guide summarises a very large body of available literature and provides a sample compositional specification as shown in Table 4.

Superclean 3.5NiCrMoV has become widely produced and accepted for LP as well as single shaft HP–LP rotors with steam temperatures up to 500°C (930°F), the presently accepted temperature limit for adequate creep strength. Superclean 3.5NiCrMoV steels have also found a 'niche' market for combustion turbine parts. There seems to be considerable interest in this application

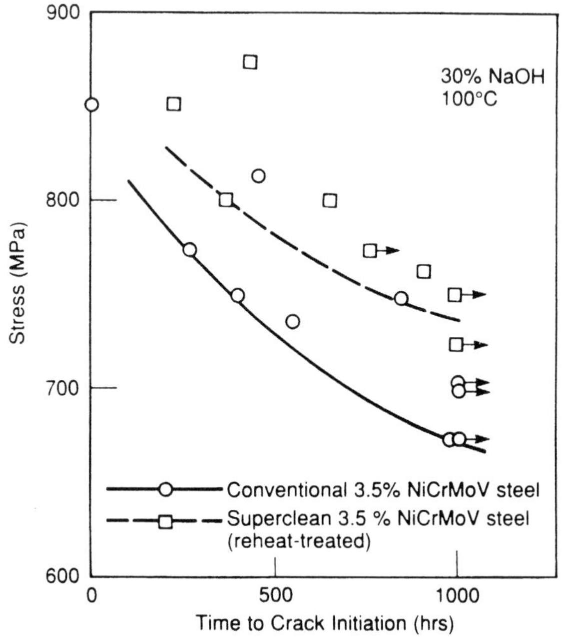

Fig. 9a Load level versus crack initiation time for conventional and superclean 3.5NiCrMoV steel

Source: Scarlin and Denk (53).

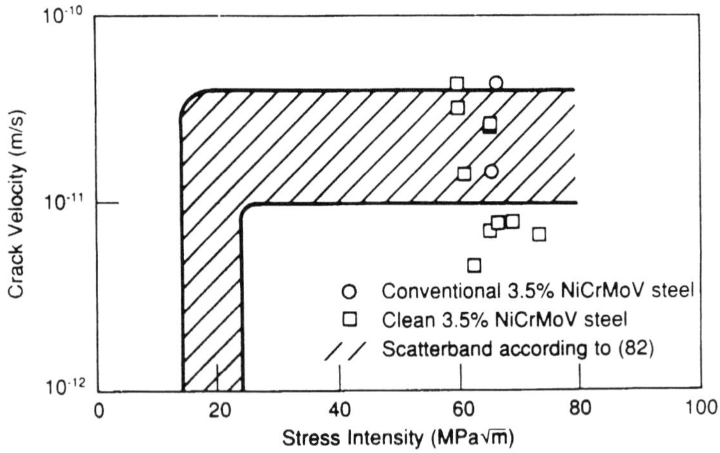

Fig. 9b Stress corrosion crack velocity versus stress intensity factor (in aerated and deaerated pure water) for conventional and superclean 3.5NiCrMoV steels at 97°C

Table 4 Chemical Analyses and Associated *J* Factor of Superclean 3.5NiCrMoV Steel LP Rotor Forgings

Source	Weight	C	Mn	Si	S	Ni	Cr	No	V	P	Al	Sn	As	Sb	$J \times 10^4$
EPRI Super Clean Specification															
Aim		0.25	0.02	0.02	0.001	3.50	1.65	0.45	0.10	0.002		0.002	0.002	0.001	1.6
Max.		0.30	0.05	0.05	0.002	3.75	2.00	0.50	0.15	0.005	0.005	0.005	0.005	0.002	10
Conventional Rotor Chemistry															
GE (80 heat average)		0.26	0.30	0.042	0.012	3.47	1.77	0.39	0.12	0.008		0.006	0.006	0.001	47.9
Production Sized Trial Rotors															
EPRI/VEW-1	34 ton	0.30	0.02	0.03	0.001	3.44	1.68	0.41	0.09	0.0024	0.005	0.0032	0.0024	0.0005	2.8
EPRI/VEW-2	34 ton	0.27	0.02	0.01	0.001	3.73	1.71	0.43	0.10	0.004	0.006	0.005	0.006	0.0015	2.7
TSB-JSW	105 ton	0.25	0.03	0.01	0.001	3.66	1.77	0.41	0.12	0.003	<0.005	0.003	0.006	0.0017	4.2
Hitachi-Kobe	50 ton	0.25	0.04	0.03	0.0012	3.68	1.75	0.36	0.10	0.002		0.002			
Production Rotors															
Kawagoe 1(A)	120 ton	0.25	0.02	0.03	0.001	3.64	1.75	0.43	0.13	0.002	<0.005	0.003	0.003	0.0012	2.5
Kawagoe 1(B)	120 ton	0.25	0.02	0.03	0.001	3.65	1.75	0.42	0.13	0.002	<0.005	0.003	0.003	0.0012	2.5
Kawagoe 2(A)	120 ton	0.25	0.02	0.02	0.001	3.61	1.75	0.42	0.13	0.002	<0.005	0.003	0.003	0.0009	2.5
Kawagoe 2(B)	120 ton	0.25	0.02	0.03	0.001	3.64	1.76	0.42	0.13	0.002	<0.005	0.003	0.003	0.0008	2.5
MHI-JCFC	76 ton	0.25	0.03	0.01	0.0012	3.43	1.78	0.38	0.11	0.003	0.001	<0.001	0.003	0.0008	2.4

J = (Mn + Si) (P + Sn) [expressed as %].

especially in Europe. Numerous papers describing production and properties of superclean steels are contained in reference Volume 12, 15 and 16. A keynote paper reviewing most of these developments by Jaffee is contained in Ref. 12 (Sapporo workshop).

3.1 HP/LP Single Shaft Rotors

In many combined cycle plants, it is cost effective to have a single shaft HP/LP rotor configuration for the steam turbine. The need for multiple casings is eliminated, the unit is more compact and simpler to install. The rotor used in this fashion should combine in a single forging high creep strength at one end with high toughness at the other end. EPRI has sponsored two major development efforts, one involving a superclean version of a 2.5%NiCrMoV steel jointly with the General Electric Company, Toshiba Corporation and Japan Steel Works under project RP1403–55 and a second project involving KWU–Siemens in the lead with Saarstahl, and several European turbine manufacturers (Asea Brown Boveri, GEC-Alsthom, MAN Energie and Parsons) as partners under RP1403–21 to evaluate a 'clean steel' version of a 2%CrMoNiWV steel. A superclean version of the European steel was not attempted due to hardenability reasons.

In the GEC/Toshiba/JSW project, based on the successful results of laboratory work by Bodnar and Hansen of BethForge and Jaffee of EPRI on an improved CrMoV rotor steel,[26] a superclean 2.5%Ni–CrMoV trial high pressure–low pressure rotor shaft was produced by the Japan Steel Works Ltd. (JSW).[27] The shaft consisted of a 1270 mm (50.0 in) diameter high pressure (HP) section and a 1750 mm (68.9 in) diameter low pressure (LP) section. The chemistry and dimensions of the ingot were optimised based on a study at JSW with special attention being directed to the reduction of chemical segregation in the ingot. The steel was refined in a basic electric arc furnace and a ladle refining furnace, and then cast into a 90 metric ton (99 ton) ingot using the vacuum carbon deoxidation (VCD) process. The final chemistry was 0.23% C, 0.03% Si, 0.03% Mn, 0.003% P, 0.001% S, 2.5% Ni, 1.6% Cr, 1.2% Mo and 0.25% V. After the preliminary heat treatment by double normalising, the HP and LP portions were quenched differentially. The LP portion was water spray quenched from 935°C (1715°F) and the HP section was forced air cooled from 950°C (1742°F). The LP portion of the rotor shaft was tempered to two tensile strength levels of 860 MPa (125 ksi) and 790 MPa (115 ksi) by two stage tempering for evaluation. The FATT at the center of the LP and HP sections were approximately 22°C (72°F) and 3°C (37°F), respectively, for a strength level of 790 MPa (115 ksi). Increasing the tensile strength of the LP section to 860 MPa (125 ksi) increased the FATT to 49°C (120°F). Creep rupture tests revealed that the creep rupture strength of the HP section was equivalent to the strength of a conventional CrMoV rotor steel. Generally, other characteristics such as fatigue properties and fracture toughness were equal to or in excess of those of the standard CrMoV turbine rotor steel levels. However, notch bar creep rupture tests showed a tendency for notch weakening and the notch rupture ductility was somewhat marginal and could interfere with the use of this alloy in steam turbine applications. Figure 10 illustrates the

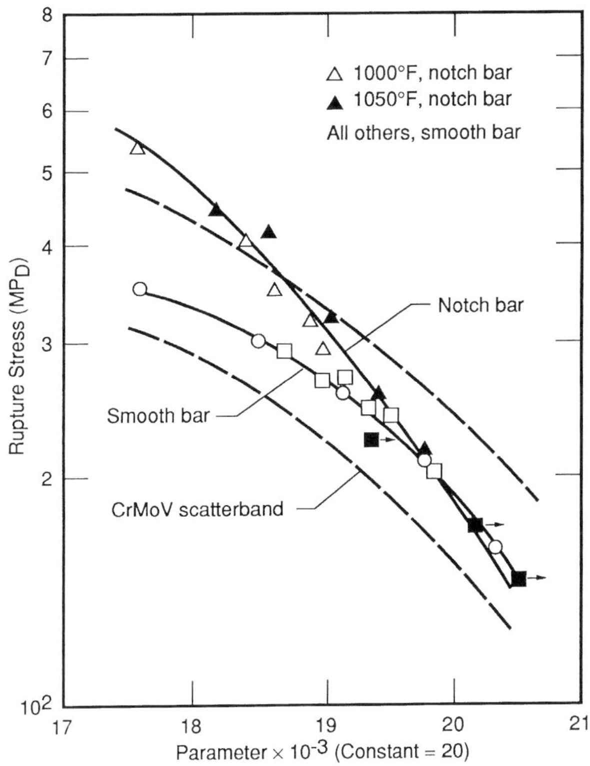

Fig. 10 Creep rupture data on the trial rotor material from the EPRI/GE/JSW/ Toshiba 2.5%NiCrMoV HP/LP rotor alloy

crossover of the notch bar stress rupture curve with respect to the smooth bar curve. The apparent cause of this difficulty was the removal of niobium from the original proposed chemistry because of the concern about the potential adverse effects of chemical segregation. The proposed solution to this problem would involve the restoration of the niobium in conjunction with electroslag remelting (ESR) to minimise the segregation tendency.

The EPRI–Europe rotor project RP1403–21 involves evaluation of a 2%CrMoNiWV steel – which was proposed by Saarschmiede in 1980.[28] The alloy has been meanwhile subjected to extensive testing in collaboration with turbine manufacturers. In the past 6 years, 25 rotors for HP/LP and 20 rotors for HP/IP combination turbines have been manufactured. The purpose of the EPRI project was to optimise a clean steel version of this alloy using lab heats, followed by production and evaluation of a trial rotor using the optimised composition. A combination rotor of 1250 mm (50") diameter in the HP section and 1750 mm (69") diameter in the LP section has been produced from a 95 ton ESR ingot of 2300 mm (91") diameter.[29] The rotor was first quenched and tempered to an 0.2% yield strength of approx. 703 MPa (102 ksi) by differentially cooling the HP section (air/water quenching simulating oil quenching) and the LP (water quenching) section from 950°C (1745°F) austenitising temperature. This resulted in a FATT of approx. −15°C (−10°F) in tangential rim locations and of +55°C (131°F) in the core of the HP section as well as of +57°C (135°F) in the core of the LP section. Following this testing, a through-section radial core of 395 mm (15.5") diameter was taken from the HP section and from the LP section for further testing by the turbine manufacturers. Subsequently, the rotor was tempered to a lower 0.2% yield strength of approx. 628 MPa (91 ksi) and tested in the same way as the higher-strength material. The FATT values in this case were from −28 to −40°C (−18 to −40°F) in tangential rim locations, +14°C (57°F) in the core of the HP section and +19°C)66°F) in the core of the LP section. The results available so far have met the assumptions with respect to this steel. They need, however, to be substantiated by the further investigations, particularly the long-term testing, to be conducted by the turbine manufacturers.

In the absence of long term creep data on the clean steel version of the 2%CrMoNiWV trial rotor, we could still use the data on the various production rotors compiled previously for purposes of comparison[30] with the assumption that the clean steel version will exhibit higher rupture strength. Figure 11 shows a comparison of the Larson–Miller rupture data for the 2.5%Ni superclean CrMoV and the production version of the 2%CrMoNiWV alloys. The data for both fall within the General Electric Company's scatterband of conventional 1Cr1Mo0.25V HP/IP rotor steels. Slightly higher long term rupture strength is indicated for the 2.5%NiCrMoV alloy compared to the 2%CrMoNiWV alloy. Figure 12 compares the rupture ductilities of the two alloys. The tendency for the decrease in rupture ductility with time is evident for the 2%NiCrMoV alloy consistent with the notch weakening observed to this alloy.

Several parallel developments have been taking place in the meantime with respect to HP/LP single shaft development. With the exception of the EPRI-European 2%CrMoNiWV alloy, the experience base and long term performance data is insufficient at this time. For the record, Tables 5 to 7 provide a compilation of available data relating the various contending alloys for the single shaft HP/LP application based on information summarised in Ref. 27. Among the various alloys listed, the EPRI-European alloy 2CrMoNiWV has the longest accumulated rupture data as well as service experience. While the 10^5h rupture strength values reported in Table 7 for the 2%CrMoNiWV alloy are based on tests up to 40,000 hr, the rupture strength values for the other alloys are based on test duration of less than 5,000 h. All of the alloys have HP bore FATTs that are much lower than conventional HP rotor alloy 1Cr1Mo1/

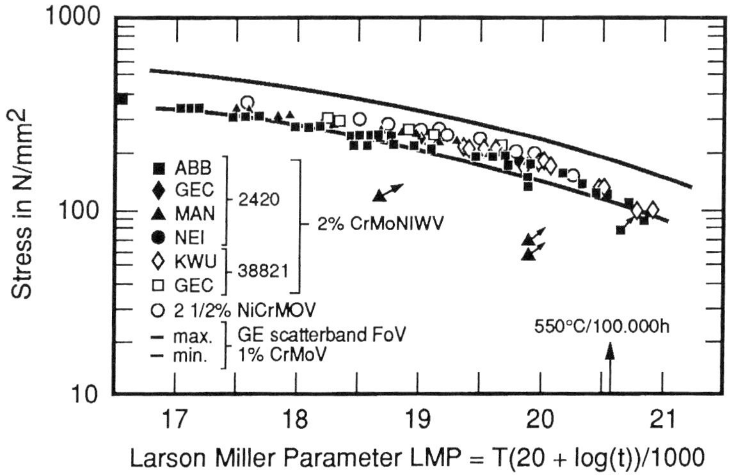

Fig. 11 Larson-Miller plot of the creep rupture properties for the 2%CrMoNiWV rotors and the 2 1/2 NiCrMoV trial rotor (based on Ref. 30)

4V (80 to 100°C). None of the steels other than the EPRI/GE/JSW/Toshiba steel call for a 'superclean' specification. They do however qualify to be classed as 'clean' steels.

4. FUTURE ISSUES

In summary, the general benefits of cleaning up steels have been amply demonstrated in laboratory studies. Significant improvements in fracture

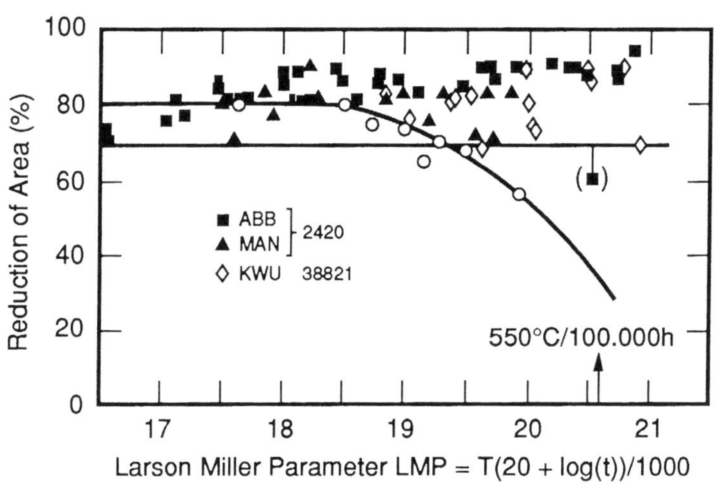

Fig. 12 Larson-Miller plot of the creep rupture ductility for the 2%CrMoNiWV rotors in terms of reduction of area (%) (based on Ref. 30)

toughness, FATT and freedom from temper embrittlement have been achieved while maintaining or improving creep strength and ductility of steels. In the case of superclean LP rotor steels improved SCC resistance has also been achieved.

While the general trends towards lower impurity levels in commercial rotors have continued, industry has not rushed to embrace specifications calling for 'clean' or 'superclean' standards except in niche markets. In Europe, clean and superclean 3.5NiCrMoV and 12%Cr steels are finding extensive application in Combustion Turbine Disk applications. Use of nearly 30 superclean 3.5NiCrMoV LP rotors has been reported in Japan for increased efficiency and SCC resistance. There is no industry standard per se that can be used as a benchmark for each of these categories. Several technical and commercial issues remain to be resolved before widespread use of these steels is likely to occur. These issues are listed below:

Table 5 Summary of Single Piece HP/LP Turbine Rotor Papers

		2.5Ni-CrMoV EPRI/GE/JSW/ Toshiba	CrNiMoVNbW JSW/Toshiba	2Cr1Mo0.2Mn1.5NiV Kobe	2CrMoNiWV EPRI/Saarstahl/ Europe Rotor	1.8Ni-CrMoV Hitachi
Ingot Wt. (MT)		90	75	80	95	75
LP Dia. (mm)		1750	1750	1380	1750	1720
HP Dia. (mm)		1270	1000	935	1250	1320
wt%						
C		0.23	0.24	0.24	0.22	0.23
Si		0.03	0.02	0.02	0.06	0.01
Mn		0.03	0.45	0.20	0.70	0.20
P		0.003	0.004	0.004	0.005	0.004
S		0.001	0.0009	0.004	0.001	0.002
Ni		2.50	1.69	1.48	0.76	1.74
Cr		1.60	2.22	1.99	2.13	2.03
Mo		1.20	1.08	1.69	0.86	1.17
V		0.24	0.19	0.23	0.32	0.26
Nb			0.015			
As		0.002	0.004		0.007	0.003
Sn		0.003	0.004		0.005	0.003
Sb		0.0010	0.0010		0.0007	0.0010
W			0.19		0.65	
		(°C)	(°C)	(°C)	(°C)	(°C)
Aust.	LP	935	900	9 10	954	950
	HP	950	970	950	954	970
Quench	LP	WS	WS	WQ	WQ	
	HP	FAC	FAC	FAC	Air/water simulating OQ	

WS = Water Spray, WQ = Water Quench, FAC = Forced Air Cool, OQ = Oil Quench.

Table 6 Summary of Single Piece HP/LP Turbine Rotor Papers – LP Section

		2.5Ni-CrMoV EPRI/GE/ JSW/Toshiba	2.25CrNiMoVNbW JSW/Toshiba	2Cr1Mo0.2Mn1.5NiV Kobe	2CrMoNiWV EPRI/Saarstahl/ Europe Rotor		1.8Ni-CrMoV Hitachi
		Metric (MPa) (Joules) (mm) (°C)	Metric (MPa) (Joules) (mm) (°C)	Metric (MPa) (Joules) (mm) (°C)	Metric (MPa) (Joules) (mm) (°C)		Metric (MPa) (Joules) (mm) (°C)
LP	TS	772		774	736	810	832
Surf.	0.2% YS	640		635	628	703	
	El%	22.7		20	23	22	21.6
	RA	73		70	75	72	71.9
	FATT	−62		−68	−28	−15	0, 48
	RTE	210		130	212	188	144, 183
LP	TS	805	839	779	768	825	864
Bore	0.2% YS	651		620	636	716	
	El%	22	22.2	18	20	20	20.9
	RA	67.5	61.2	60	73	70	58.7
	FATT	+22	+23	−5	+19	+57	46
	RTE	165	71	72	114	52	48

1 ksi = 6.89 MPa; 1 ft-lb = 1.355 Joules

Table 7. Summary of Single Piece HP/LP Turbine Rotor Papers – HP Section

		2.5Ni-CrMoV EPRI/GE/ JSW/Toshiba	2.25CrNiMoVNbW JSW/Toshiba	2Cr1Mo0.?Mn1.5NiV Kobe	2CrMoNiWV EPRI/Saarstahl/ Europe		1.8Ni-CrMoV Hitachi
		Metric (MPa) (Joules) (mm) (°C)	Metric (MPa) (Joules) (mm) (°C)	Metric (MPa) (Joules) (mm) (°C)	Metric (MPa) (Joules) (mm) (°C)		Metric (MPa) (Joules) (mm) (°C)
HP	TS	789		768	743	795	
Surf.	0.2% YS	656		617	636	686	
	El%	23		19	23	21	
	RA	70		67	74	71	
	FATT	−3		+5	−40	−16	
	RTE	157		94	213	193	
HP	TS	792	871		780	840	
Bore	0.2% YS	653			653	726	
	El%	21.7	23.1		20	19	
	RA	66.8	63.6		72	66	
	FATT	+3	+40		+14	+55	
	RTE	157	46		70	41	
RS ($10^{5\,HRS}$ @ 1000°F)		161	177	166	145		171
RA (1000°F)		66	80	85	70		
RA (1050°F)		58	80				
CDR		0.46		1.09			

1 ksi = 6.89 MPa; 1 ft-lb = 1.355 Joules

- Clearer definitions of what constitutes 'clean' and superclean steels and broad standards conforming to these definitions need to be evolved. Sample specifications for HP, LP and HP/LP rotors should be developed.
- The need for clean/superclean steels must be placed in the context of design and operation requirements. It is not our objective to make steels as clean as they can get, but as clean as we need. For instance, what FATT, K_{IC}, etc. are needed (rather than the lowest we can achieve at all costs) for different applications and what cleanliness is adequate to meet these requirements will have to be decided by the turbine manufacturers based on customer needs on a case-by-case basis. Some broad classifications of various applications and the corresponding cleanliness required would however be very useful in helping the utility customer understand his needs better.
- The question of whether there are threshold levels for impurity contents below which no further benefits are derived needs to be examined. We need a clearer understanding of the synergism between the impurities P, Sb, Sn, etc. and S. Some impurities are known to be more deleterious in some steels than in others. For instance, P is the key impurity in CrMoV HP/IP rotors, while Sn becomes important in NiCrMoV LP rotors and discs. It may, therefore, be economical to clean steels selectively for specific impurities rather than to attempt en mass cleaning. The effect of nominal relaxation to the specifications e.g. Mn in superclean steels from 0.05 to 0.1% could significantly decrease the costs. The effect of such relaxation on the mechanical and other properties needs to be investigated
- The superclean 3%NiCrMoV is most likely to find wider application as disks in nuclear LPs where SCC has been plaguing the industry; also as monoblock rotors as more orders are placed. In the fossil LP, the primary benefits of raising the LP cross-over temperature and that of designing longer last stage blade containing rotor, both involve design changes and can, therefore, apply only to new plants and not for retrofit. Most plants being built today are combustion turbine/combined cycle plants where the steam turbine is small/medium. Development of superclean/clean HP/LP single shaft would be more the focus than improving the superclean 3.5NiCrMoV steels. The superclean steels are also finding wide applications in LP rotor combustion turbine parts.
- For nuclear LP rotors and discs, the key concern is SCC. Documentation of the benefits of clean steels, particularly clarification of interactive effects between yield strength, cleanliness, and K_{ISCC} and da/dt will be very useful. In the case of monoblock LP rotors, weld repairs would require availability of suitable filler metals.
- The benefits of extending superclean steel technology to 12%Cr HP rotors steels remain to be investigated. Extension of both clean and superclean

steel technologies to other applications such as blading, casings, combustion turbine (CT) discs and pressure vessels needs to be evaluated. Significant interest already exists with respect to CT disc application of superclean 12%Cr steels.

- It would be desirable to document and quantify the benefits from these steels in terms of 1) improved life, 2) avoidance of forced outrages, 3) increasing inspection interval, 4) more rapid startup and cycling ability, 5) avoidance of need to prewarm before loading, 6) increased efficency due to higher LP rotor cross over temperature. Some kind of cost/benefit calculations involving specific 'numbers' for one or more typical applications would be most useful in moving this research into the market place.
- A list of turbine manufacturers who commercially offer clean steel/ superclean rotors/discs for HP/IP/LP applications and an estimate of price differentials compared to conventional rotors would be useful. A list of steel makers who have the capability to make these components would also be useful.
- The views of equipment manufacturers and forgemasters on what are currently the impediments to market penetration and how these can be overcome would be very valuable. In addition, an estimate of orders booked to date for clean steel versions and the specific application is also desirable.

The purpose of this workshop is primarily to identify, document and propose solutions to the technological and commercial barriers to the widespread use of clean/superclean steels. It is hoped that the workshop participants will actively confront these issues during the next two days. An initial attempt to address some of these issues has been made by Professor Jack Nutting under EPRI sponsorship (RP8046–2). He has just completed an industry survey, the results from which he will be presenting later during the workshop.

Acknowledgments and thanks are due to many individuals and organisations. I would like to acknowledge my gratefulness to Professor Jack Nutting for acting as the Chairman of the organising committee and pulling together this entire program. My sincere thanks are also due to Drs G. Honeyman, G. Neate, E. Potthast, D. Thornton and P. A. Coulon who generously contributed their valuable time to serve as the members of the organising committee. I am also grateful to Lisa Davis, who on behalf of the Institute of Materials has provided efficient logistics support to this workshop.

This is the fifth workshop on clean/superclean steels. The previous workshops were held in July 1986, July 1987, June 1988 and August 1989 in Kapfenberg, Dusseldorf and Sapporo, respectively, under the able stewardship of Dr R. I. Jaffee. Any publication on the subject of clean/superclean steels would be incomplete without an acknowledgment of the pioneering technical contributions and leadership of Dr Jaffee. His greatest achievement was in bringing together manufacturers and steel makers from across the globe in a relationship of mutual trust and shared goals, transcending narrow national and commercial interests. This workshop will provide continuity to this relationship. It is in this spirit of humble dedication to Bob Jaffee's contributions that I am now opening this workshop.

ACKNOWLEDGMENTS

The author is grateful to Dr E. Potthast, Mr W. Wiemann, Professor J. Nutting and Mr Asa Kaplan for reviewing this manuscript and suggesting numerous useful revisions.

Appendix 1: Estimated Cost Benefits of Various Improvements To A 500 MW Turbine

	Million $
• Avoid replacement of a rotor	5
• Avoidance of 1 month of forced outage	21.6
• Elimination of 1 inspection during life	1.1
• Elimination of 4h of pre-warming/more rapid start up	0.6/per year
• 1% improvement in turbine efficiency	0.5/per year

REFERENCES

1. R. Viswanathan and R. I. Jaffee: 'Toughness of Cr-Mo-V Steels for Steam Turbine Rotors', *ASME, Journal of Eng. Materials and Technology*, December 1983, **V. 105**,286.
2. R. Viswanathan and R. I. Jaffee: 'Metallurgical Factors Affecting the Reliability of Fossil Steam Turbine Rotors', *ASME, Journal of Engineering for Gas Turbines and Power*, July 1987, **V. 107**, 642.
3. R. Visanathan: *Damage Mechanisms and Life Assessment of High Temperature Components*, ASM International, Metals Park, OH, 1989, 303–306 and 400–410.

3b. J. Nutting: 'Overheating Characteristics of Clean 3.5NiCrMoV', *Final Report RP2741–01*, EPRI, Palo Alto, CA, 1987.

4. D. P. Timo, R. M. Curran and R. J. Plazec: 'Rotor Forgings for Steam Turbine Generators', R. I. Jaffee ed. *EPRI Report WS79–235*, 1979.
5. R. M. Curran, D. L. Newhouse and J. C. Newman: 'The Development of Improved Rotor Forgings for Modern Large Steam Turbines', ASME Paper 82JPGCPWR- Vol.25, Joint Power Generation Conference, Denver, 1982.

6. L. D. Kramer, D. D. Randolph and D. A. Weisz: 'Analysis of The Tennessee Valley Authority, Gallatin Unit No. 2, Turbine Rotor Burst', *Winter Meeting of ASME*, New York, December 5–10, 1976.

7. V. P. Swaminathan, J. E. Steiner and A. Mitchell: 'Advanced Rotor Forgings for High Temperature Steam Turbines', *Report CS–4516*, Electric Power Research Institute, Palo Alto, CA, May 1986.

8. V. P. Swaminathan and R. I. Jaffee: 'Significant Improvements in Properties of CrMoV HP Rotors by Advanced Steel Making', *Life Assessment and Improvement of Turbogenerator Rotors for Fossil Power Plants*, R. Viswanathan, ed., Pergamon Press, New York, 1985, 657.

9. M. Yamada: Toshiba Corporation Personal Communication to J. Nutting, September 2, 1994.

9a. J. Ewald, C. Berger, K. H. Keinburg and W. Wiemann: 'Present Quality Level of Heat Resistant Forgings Made of 1%CrMoV Steels', *Steel Research 57* (1986), 83–92 and 172–177.

10. T. Ohhashi, R. L. Bodnar and others: 'High Purity Steels for Utility Components', *Report NP–5399, RP2060–1, –2*, Electric Power Research Institute, Palo Alto, CA, September 1987.

11. D. L. Newhouse: 'Guide to 12CR Steels for High and Intermediate Pressure Turbine Rotors for the Advanced Coal-Fired Steam Plant', *Report CS–5277, RP1403–7*, Electric Power Research Institute, Palo Alto, CA, July 1987.

12. P. A. Coulon: 'Improved Superclean 11%Cr Steel for High Temperature Turbine Rotors', *Superclean Rotor Steels*, Proc. of a Workshop at Sapporo, Japan (EPRI GS–6921), R. I. Jaffee, ed., Pergamon Press, New York, 1989.

13. K. Sato et al: 'High Purity 12%Cr Martensitic Steel', *Superclean Rotor Steels*, Proc. of a Workshop at Sapporo, Japan (EPRI GS–6921), R. I. Jaffee, ed., Pergamon Press, New York, 1989.

14. P. A. Coulon: 'Influence of Impurities in Improved 12% Chromium Steel for High Temperature Steam Turbine Rotor', *Symposium on Residual and Unspecified Elements in Steel*, ASTM A1 Proceedings, Bal Harbor, Florida, November 11, 1987.

15. K. Sato et al: 'Manufacturing of High Purity 12% Cr Rotor Forgings', *Clean Steels Technology*, R. Viswanathan, ed., ASM International, Metals Park, OH, 1992.

16. C. Berger, K. H. Mayer and S. M. Beech: 'High Temperature Rotor Forgings of High Strength 10% CrMoV Steels', *12th International Forgemasters Meeting*, Forging Industry Education and Research Foundation, Cleveland, OH, 1994.

17. J. Taira et al: 'Development and Production of High Purity 12%Cr Steels for Gas Turbine Disc Material', *Proceedings of the 12th International Forgemasters Meeting*, Chicago, published by Forging Industry Education and Research Foundation, Cleveland, OH, 1994.

18. M. Shiga et al: 'Development of 12%Cr Steel for Gas Turbine Disc', *12th*

International Forgemasters Meeting, Forging Industry Education and Research Foundation, Cleveland, OH, 1994.

19. R. I. Jaffee, P. Machner, W. Meyer and J. E. Steiner: 'Production and Properties of a Superclean 3.5NiCrMoV LP Rotor Forging', EPRI Project RP1403–8, *Ironmaking Steelmaking*, 1986, **13**, 322–326.
20. 'Production and Properties of a Low Pressure Model Rotor with Relaxed Specification', *VEW Final Report to EPRI on RP2471–3*, 1986.
21. R. H. Richman and W. P. McNaughton: 'Superclean Steel Development: Status Report', *GS–6610, RP1403–32*, December 1989.
22. O. Watanabe, Y. Yoshioka and R. C. Schwant: 'Evaluation of a Superclean NiCrMoV Low Pressure Turbine Rotor Forging for Advanced Steam Conditions', *Report of RP1403–15*, Electric Power Research Institute, December 1991.
23. Y. Yoshioka, O. Watanabe and M. Miyazake: 'Production and Evaluation of Superclean LP Rotor Forgings', *Proc. of the Sapporo Workshop*, Electric Power Research Institute, GS–6921, 1989.
24. R. B. Scarlin and J. Denk: 'Stress Corrosion Cracking Behavior of a Clean 3.5% NiCrMoV Steel', *Report GS–6907, RP1403–18*, Electric Power Research Institute, Palo Alto, CA, July 1987.
25. R. H. Richman and W. P. McNaughton: 'Superclean Steel Development: A Guide to Utility Use', *Report GS–6612, RP1403–32*, Electric Power Research Institute, December 1989.
26. R. L. Bodnar, S. S. Hansen and R. I. Jaffee: 'Improved Superclean NiCrMoV Rotor Steel', *Report ER–6887, RP2426–4*, Electric Power Research Institute, Palo Alto, CA, July 1990.
27. Y. Tanaka, O. Watanabe, A. Kaplan and others: 'Production and Properties of a Superclean 2.5%NiCrMoV HP/LP Rotor Shaft', *Report TR–103689, RP1403–55*, Electric Power Research Institute, Palo Alto, CA, February 1994.
28. H. Finkler and E. Potthast: 'Presentation of a New Steel for High Pressure Shafts', *Rotor Forgings for Turbines and Generators*, R. I. Jaffee, ed., Pergamon Press, New York, 1980.
29. E. Potthast, R. Viswanathan and W. Wiemann: 'Advanced 2%CrMoNiWV Steel for Combination Rotors', *Proceedings of the 12th International Forgemasters Meeting*, Chicago. Published by Forging Industry Education and Research Foundation, Cleveland, OH, 1994.
30. W. Wiemann: 'The Development of an Improved 2%CrMoNiWV HP/IP Rotor', *COST Project Report EVR 13452 EN*, J. B. Marriott, ed.; 1991.

The EPRI Survey on 'Superclean Steels'

J. NUTTING
Consultant Metallurgist

Following the successful development and introduction of superclean steels for the manufacture of rotors for the power generation industry, The Electric Power Research Institute (EPRI) wished to determine the impact of 'Superclean' technology on the power generation industry and to assess some of the wider issues involved. As part of this programme, Professor J. Nutting, a consultant metallurgist, was commissioned by EPRI to carry out a survey of the above topics based on the replies to questionnaires which had been submitted to:

(1) STEELMAKERS AND FORGEMASTERS,
(2) BOILERMAKERS,
(3) POWER GENERATION PLANT BUILDERS, and
(4) POWER GENERATION PLANT USERS.

In framing the questionnaires, Professor Nutting sought advice from the following:

Dr G. A. Honeyman – Forgemasters Steels and Engineering Ltd.
Mr D. V. Thornton – G.E.C. Alsthom
Dr G. J. Neate – PowerGen

and Professor Nutting is grateful for the time and effort these gentlemen spent on these tasks. The draft questionnaires were then assessed by Dr R. Viswanathan of EPRI, who gave final approval for their circulation. Some forty four questionnaires in the four categories given above were sent to organisations in Europe, the Far East, Japan and the USA. In total some seventeen replies were received, these being predominantly from Steelmakers and Plant Builders. In some cases, replies were sent saying that it was not company or national policy to reply to such questionnaires. Where such replies had not been given, repeated requests for the return of the questionnaire failed to elicit a response. Where answers to the questions were given, there was a general consistency which would indicate that some significance can be given to the views expressed in this report. Where consistency was not found, the divergent views are presented and a brief comment has been given on possible reasons for the divergence.

A copy of the questionnaire sent to Steelmakers and Forgemasters is appended to this report.

GENERAL INTRODUCTION TO THE DEVELOPMENT OF SUPER CLEAN STEELS

It had been known for six or seven decades that the steels used for both HP and LP rotors were susceptible to temper embrittlement partly developing as a result of the initial heat treatment and particularly, with LP rotors, as a result of the in-service temperature conditions. In order to overcome such difficulties elaborate precautions needed to be taken during the shutting down and the restarting of turbines from ambient conditions and this greatly hindered flexibility in the use of the plant.

In spite of much academic research the real causes of temper embrittlement were imperfectly understood until with the development of Auger spectroscopy in about the 1970s it became possible to identify the embrittling species which had segregated to the austenite grain boundaries and so lowered intergranular cohesion, particularly at relatively high strain rate conditions.

The segregating elements were identified as phosphorus, arsenic, antimony and tin but the severity of the segregation was enhanced by two other elements normally present in power generation steels, silicon and manganese. This led the late Bob Jaffee of EPRI to contemplate the development of a new class of steel which would not be susceptible to temper embrittlement. The difficulty was the requirement to specify a low manganese content. Following discussions between Dr R. Jaffee and Professor J. Nutting it was agreed that if the sulphur content of the appropriate steels was reduced to the order of 0.001% then the manganese content could be reduced to 0.02% without risk of hot shortness developing during forging. Professor Nutting also pointed out that with such low sulphur and manganese contents the risk of overheating during forging would be eliminated as all the sulphur present in the steel would remain in solid solution at temperatures between 900 and 1400°C. In order to test the validity of these arguments small experimental casts with low sulphur and manganese contents, together with a low residual content, were made by Japan Steel Works and tested in Great Britain and the USA. It was found that the steels could readily be forged, that they were not susceptible to temper embrittlement and that no overheating occurred even after heating to 1400°C, i.e. well above the normal forging temperature. In the almost complete absence of phosphorus it was also found that the steels did not burn. These steels contained virtually no non-metallic inclusions either of oxides or sulphides.

The way was now open to the production of trial melts and their subsequent fabrication into simulated rotor forgings. This task was carried out by Vereinigte Edelstahlwerke AG, Austria (VEW) and the subsequent property testing by a variety of plant makers in a number of countries. Following the production of the first cast [VEW (1)] it became apparent that in the almost complete absence of manganese and with a low silicon content, the hardenability of the standard 3.5% nickel steel was too low and for this reason the nickel specification was raised to 3.5–3.75%. Thus VEW 2 was produced, which apart from increased nickel content had a lower oxide inclusion content, and which on testing gave extremely promising results. A further trial forging was produced by Japan Steel Works which was sectioned and tested with support from EPRI.

On the basis of the results a superclean 3.5 NiCrMoV rotor was installed at the Kawago plant in Japan and was visited by those attending the Super Clean Workshop at Sapporo in 1989.

Further work has since been carried out to assess the need for a superclean specification for the standard chromium–molybdenum–vanadium steel used for HP rotors and develop steels for combined HP–LP rotors having the good creep properties of Cr–Mo–V steels with the inherent toughness of 3.5% Cr–Mo–V materials.

Apart from the absence of temper embrittlement and overheating superclean steels have been found to have a lower susceptibility to stress corrosion cracking, improved fatigue properties due to the general reduction of the inclusion content and improved creep ductility due to the absence of MnS inclusions which normally act as void nucleators at the prior austenite grain boundaries. The steels have also been found to have a higher strength to ductility ratio after quenching and then tempering in the range 200–650°C.

Following the identification of the elements responsible for temper embrittlements various semi empirical relationships have been developed and reported in the literature so enabling the steel maker and the steel user to have a 'figure of merit' in relation to temper embrittlement. These relationships are of the form:

(1) 'J' $= (\mathrm{Mn} + \mathrm{Si})(\mathrm{P} + \mathrm{Sn})$

(2) 'K' $(\mathrm{Mn} + \mathrm{Si})(10\mathrm{P} + 5\mathrm{Sb} + 4\mathrm{Sn} + \mathrm{As})$

(3) 'X' (Bruscato) $\dfrac{10\mathrm{P} + 5\mathrm{Sb} + 4\mathrm{Sn} + \mathrm{As}}{10}$

In each case the element content is expressed as a wt%

These values are now frequently quoted by the steel maker together with the chemical analysis. The lower the value the more resistant the steel is to temper embrittlement.

STEELMAKERS AND FORGEMASTERS

Some fourteen companies were approached and seven replied. The replies were equally distributed between Europe and the Far East; there were no replies from the USA.

The basic electric arc furnace appears to be almost universally used as the primary method of steel making with a furnace capacity of 100–125 tonnes.

The other methods used when available were small vacuum induction melting units with capacities up to 5 tonnes. Secondary steel refining was chiefly of the vacuum arc degassing type with some usage of conventional ladle furnaces. The ASEA–SKF system was not widely used. Pouring methods used for the steel varied from simple air pouring, inert gas shielding, vacuum stream degassing, vacuum ingot casting, with the emphasis being on vacuum stream degassing. Where remelting techniques were available, the electro-slag refining method was chiefly used, although some companies also had vacuum arc remelting units. The capacities of these remelting units were consistent with them being used to produce ingots from which disc forgings would be manufactured. Where large ingots were produced, the maximum sizes ranged from 100 to 500 tonnes. The total steel melting capacity available of the seven major companies that replied was about 1 million tonnes per annum.

The companies replying all produced a wide range of products for the power generation industries with some specialising in mono-bloc rotors for turbines and generators, and others in discs and shafts. The non-power generation products covered a wide range of applications for the mechanical and chemical engineering industries, aircraft and aerospace applications, rolls and foundry products. In spite of the fluctuation in the trade cycle over the last four years, the shipped weight of forgings and castings to the power generation industries has increased over the last four years by about 15%, with a small trough occurring during 1992, which links with the low point of the trade cycle.

During the four year period (1990–1993) it would appear that a reasonable estimate of the total tonnage of alloy steel supplied to the electric power generation industries would amount to about 200,000 tonnes. That is to say on average about 50,000 tonnes/annum. From the answers given it is difficult because of some ambiguity to assess the total quantity of steel produced during the last four years to an EPRI superclean specification, but it could be between 7,000–10,000 tonnes.

The data obtained would indicate that of the total installed steel making capacity of the major producers of electric power generation steels, over the last four years on average only 5% of that capacity has been used for power generation applications and of these steels only 3.5–5% were produced to an EPRI superclean specification.

Of the rotors of various forms provided by the steelmakers to the plant makers over the last four years, by far the largest proportion (about 55–60%) were of the 3.5 Ni–Cr–Mo–V grade. Some companies had not been approached to supply superclean materials. Others had been approached, but had refused to tender; the reasons for the refusal were not given. It would appear that all the steelmakers had plant capable of refining the steel to give low phosphorus, low sulphur and low manganese contents required to meet the EPRI specification.

The steelmakers, whether they supplied superclean steel or not, reported values of the residual element contents of their conventional steels which were not widely different from those required by the EPRI superclean specification. The European producers had greatest divergence in relation to tin and arsenic whilst the Japanese producers had greatest divergence in relation to arsenic with only one producer having problems with tin. Antimony levels were all very close to the EPRI specification. All the steel makers produced power generation steels to meet their own internal standards of 'J', 'K' or 'X' values with the most popular criteria being the 'J' or 'K' values (see Introduction).

Those involved with providing steels for fabrication into pipework headers etc. had produced no superclean material for these applications. One steelmaker was surprised that this was the case in view of the many advantages which would accrue from the use of superclean steels in the manufacture of superheater tubes. Of those involved with producing small forgings for discs, only one company had produced ingots to a superclean specification, and no difficulties were found in their subsequent fabrication.

Questions were asked about joining problems and it was expected that the answers would come from those steelmakers involved with the production of discs for subsequent fabrication into rotors. As mentioned above the disc producers did not appear to have joining problems but two companies involved with the production of mono block rotors were interested in transition joint pieces which could be used for joining 3.5 Ni–Cr–Mo–V steel to Cr–Mo–V steel, hence to the production of single cylinder configurations.

There was some demand from the steelmakers to see a relaxation of the EPRI specification. One company wished to have the Mn specification raised from 0.2–0.05% to 0.08–0.12%, whilst another company expressed a similar view with a Mn relaxation to 0.1%. These requests would indicate an imperfect understanding of the reasons for establishing the initial EPRI specification. As outlined in the introduction, the susceptibility to temper embrittlement will be doubled and there will be serious risk of overheating during forging.

The European steelmakers favoured a relaxation in the residual element specification particularly for Sn, As and Sb. Such a request can clearly be related to the problems of obtaining low residual scrap in Europe, but the request in relation to antimony is strange in view of the fact that their normal production steels meet the required EPRI limit as mentioned above.

Another interesting view was that the specification was the concern of the plant builder, or plant user, and that the present specification could be met without too much difficulty. It must be remembered, however, that the specification does influence the price. There was surprising unanimity amongst the steelmakers that had supplied superclean steels that a premium of 15–20% would be charged in relation to conventional grades for ingot weights up to 100–150 tonnes – a somewhat higher premium of 40% was quoted for ingot weights in the 300–600 tonnes range. Companies that had not supplied superclean steels anticipated a premium of 30% – a value widely quoted in the literature.

Views were expressed about the influence of a relaxed specification on the expected premium. The same company that suggested a relaxation in the manganese content to 0.08 to 0.12% believed that this would reduce the premium to about 5%. A European view was that if the residual content could be relaxed but keeping the existing silicon, sulphur and manganese superclean levels, the premium would be reduced to 5%. The steelmakers were unanimous in their views that any increase in demand for superclean steels coming either from the power generation or other industries, would not influence the premium.

There was a surprising divergence of views as to whether a demand for superclean steels would develop from other than the power generation industries. About half anticipated some demand – about half did not. This divergence could be due to differences in the customer base of the different companies and from the details given in the replies to the questionnaire it was impossible to assess this accurately. But where positive answers were given, the increased demand was seen as likely to come from the chemical industry, and for pressure vessels and for offshore applications.

Even if the EPRI specification were relaxed, those companies that believed there would be an increased demand for a superclean steel were somewhat divided in their views about marketing such steels. Some said they would, others were doubtful. Those that gave a positive response seemed to be strongly in favour of developing new superclean steels.

In reply to a question about the accuracy of analytical procedures, there were some remarkable differences in views. The answers are presented in a tabular form (see Table 1). In some cases the answers were given specifically in relation to 3.5 Ni–Cr–Mo–V steel and in others they covered the whole steel product range. The steelmakers whose answers covered the whole range of their steel products in general gave replies which showed lower tolerances than those who gave specific tolerances.

The analytical tolerances were in general lower (that is to say more accurate) for the Japanese steelmakers than the others. In replying to the question about the need for improved analytical methods, all the steelmakers felt that these were accurate enough for dealing with the specification limits. There was one request for standardised methods to be established and another for improved analytical methods for hydrogen.

Further assessment of the data given in the questionnaires is presented in Table 2. This shows the accuracy of the methods used by the majority and by the minority in relation to the EPRI specification. Dividing the upper limit of the EPRI specification by the tolerance gives a value which if it is in excess of say 10 would indicate reasonably reliable results. From the table it can be seen that difficulties arise even with companies reporting highly accurate ana-

Table 1 Accuracy Ranges of the Analytical Procedures used by Various Steelmakers

Element	Majority Range wt%	Minority Range wt%
C	0.001–0.0025	0.007–0.01
Si	0.0008–0.003	0.01
Mn	0.001–0.005	0.05
S	0.0003–0.0005	–
P	0.00007–0. 0005	0.001
Ni	0.01–0.03	0.05
Cr	0.008–0.01	0.027–0.04
Mo	0.01– 0.013	0.0006–0.004
V	0.001–0.005	0.008–0.01
Sn	0.0003–0.0006	0.0008–0.001
As	0.0002–0.0005	0.001–0.0015
Sb	0.0002–0.0005	–
Cu	0.0007–0.0015	0.006–0.008
H	0.05–1 ppm	only 4 replies
O	1–2 ppm	5–6 ppm Some answers were not clear
N_2	1–5 ppm	–

Table 2 Accuracy Tolerances in Relation to EPRI Specification quoted by Various Steel Makers

	EPRI	Majority tolerance av.	Minority tolerance av.	Ratio EPRI upper limit Tolerance (majority)	Ratio EPRI upper limit Tolerance
C	0.25–0.30	0.002	0.009	35	30
Si	0.02–0.05	0.002	0.01	20	4
Mn	0.02–0.05	0.003	0.05	10	0.8
S	0.001–0.002	0.0004	–	5	5
P	0.002–0.005	0.0003	0.001	7	4
Ni	3.5–3.75	0.02	0.05	180	75
Cr	1.65–2.00	0.009	0.035	220	57
Mo	0.45–0.50	0.0012	0.00023	400	2000
V	0.10–0.15	0.003	0.009	50	16
As	0.002–0.005	0.0004	0.0013	12	4
Sb	0.001–0.002	0.0004	–	5	5
Sn	0.002–0.005	0.0005	0.0009	10	5
H	0.5–10 ppm	1 pm	–	10	10
O	10–25 ppm	1.5 pm	5.5 ppm	16	5
N	25–50 ppm	3 ppm	–	16	16

lytical procedures in relation to the elements sulphur, phosphorus and antimony where the ratios are 5–7.

For the steel companies reporting less accurate analytical procedures, difficulties arise in relation to silicon, manganese, sulphur, phosphorus, arsenic, antimony, tin and oxygen.

In commenting on the need for future work, all the steelmakers supported the views put forward in the questionnaire about the possible study of the synergism between the normal alloying elements and the impurities such as S, P, As, Sb and Sn, and an assessment of the level of impurity levels below which no further improvement of in service performance occurs.

It was also thought that further studies should be carried out on rotors coming out of service to give realistic data on in service embrittlement. Two steelmakers requested further work on establishing the 'safe' upper limit for manganese. This is linked to the view that a relaxation of the manganese specification would lower the superclean premium.

Further views were linked with the development of a superclean specification for Cr–Mo–V steel and for a standard for HP–LP combined rotor. Whilst papers have already been written on the development of superclean 12% Cr steels, there were a number of requests for further work in this area with a view to establishing a superclean standard for this class of material.

Questionnaire prepared
as part of the

E P R I

'CLEAN STEEL PROGRAMME'

for submission to

STEELMAKERS

and

FORGEMASTERS

please note different questionnaires will be submitted to Plantmakers, Boilermakers and the major Electric Power Generation companies

1 Name

Address

Telephone number and Fax number of company

2. Name and title of responsible person replying to this questionnaire.

Telephone number and Fax number if different from above

3. Available capacity for steel production (tonnes/annum)

4. Steelmaking facilities available

(a) Primary steelmaking

	Type	Tonnes/Annum
i	ACH	
ii	BOF	
iii	Basic Electric Arc	
iv	Other	

(b) Secondary refining available

	Type	Capacity
i	Vacuum Arc degassing	
ii	ASEA - SKF	
iii	Other	

(c) Pouring Techniques

	Type	Capacity
i	Air Pouring	
ii	Inert Gas Shielding	
iii	Vacuum Stream degassing	

(d) Remelting Techniques

	Type	Capacity
i	ESR	
ii	VAR	
iii	Other	

5. Are the ingots produced forged by your own or an associated company?

YES/NO

(i) If no, could you give the name of the forgemasters you use.

(ii) If yes, could you give details of:-

(a) Maximum size/weight of ingot you could forge

6. Brief description of major products range,
   (a) in relation to Power Generation

   (b) in other sectors

7. If involved with supplying steels for service in the electric power industries, could you provide information on the following:-
   (i) Total tonnage of steel supplied to Electric Power Generation Industries during each of the following years,

1990,	1991,	1992,	1993

   (ii) If you supply monobloc rotors, please give details of the numbers of the following components supplied over the last five years.

   Please indicate the normal size range and maximum size range for each item (a-e) below

		No's	Maximum Size
(a)	High pressure rotors (HP)		
(b)	Intermediate pressure rotors (IP)		
(c)	HP//IP rotors		
(d)	LP		
(e)	Generators		

   (f) What proportion of the items (a-e) above were in 3.5 Ni-Cr-Mo-V Steel?

   (g) What proportion of the items (a-e) above were in steel to an EPRI super clean composition?

   (h) Have you been asked to tender for superclean steel to an EPRI composition, but have declined to tender?

   YES/NO

   If no could you give reasons for declining to tender?

   (iii) If you supply discs for subsequent fabrication into rotors, please give details:-
   (a) Number of disc forgings supplied over the last five years.

   (b) Usual size of disc forgings in relation to diameter width and weight

   (c) What proportion of items in (a) above were in 3.5% Ni-Cr-Mo-V steel?

   (d) What proportion of the above items were in steel to EPRI superclean composition?

(iv) Are you involved with welding of superclean discs? YES/NO

If yes

(a) are suitable welding rods available? YES/NO

(b) are suitable weld procedures available? YES/NO

(v) Do you supply steel for fabrication into pipework, Headers, etc. YES/NO

(a) Give the approximate annual tonnage

(b) Was any of this steel supplied to an EPRI superclean specification? YES/NO

If yes could you give the approximate annual tonnage

(vi) Do you fabricate, within your own organisation, the products in (v) above? YES/NO

(vii) If so do you believe there is a need for transition pieces for making welded joints between differing types of steel in conventional and superclean grades? YES/NO

(a) If yes, what applications would you envisage for such transition pieces?

(b) What type of transition piece design would you favour?

(c) Would you be interested in further development work on such items? YES/NO

STEEL COMPOSITIONS

8. Do you supply steel of the following type to the Electric Power industries

Steel			YES/NO	Specification Numbers
A	1%	Cr-Mo-V		
B	2.5%	Cr-Mo variants		
C	3.5%	Ni-Cr-Mo-V		
D	9-12%	Cr-Mo-V		
E	Other			

If yes,

(a) please provide details of the normal residual element contents for, Al As Cu H N Sb Sn, in each of the above steels.

9. Do you produce steel to an internal standard in relation to,

	Yes	No
(a) 'J' value (Mn + Si) (P + Sn)		
(b) 'K' value (Mn + Si) (10P + 5Sb + 4Sn + As)		
(c) " X " value $\frac{10P + 5SB + 4 Sn + As}{100}$ (Bruscato)		
(d) Other values (please specify)		

10 Would you like to see the present EPRI preferred composition modified?

YES/NO

If yes, in what way?

11 The original EPRI super clean steel composition was drawn up to avoid difficulties with temper embrittlement and overheating. Have you carried out any research on the above topics which support your views on the influence of the composition changes you would propose?

YES/NO

If yes, please attach any results you have obtained to support your views.

12 What premium would you expect to charge for steels made to the present EPRI preferred composition above that for steels to a conventional specification? (An answer expressed as a % of the two values in relation to the finished forgeing price would be helpful.)

Do you adjust this premium in relation to the component size/weight (tonnage effect).

13 If there were to be a relaxed composition particularly in relation to S, P, Mn, As, Sb, and Sn, it is to be expected that this would enable you to offer such 'clean' steels at a lower premium than for super clean steels.

Could you give an indication of the premium you would charge for steels to the modified specification in relation to the conventional specification. (An answer expressed as a % of the two would be helpful.)

If the relaxed composition were adopted, do you anticipate that there would be a difference in the premium in relation to the component size/weight (tonnage effect)?

YES/NO

14 Do you envisage a demand developing for steels to an EPRI super clean steels to meet engineering needs in other than the electric power industries. If so, could you give an indication of the product types and specific niche markets.

15 If there were a growth in demand for the EPRI type super clean steels for non-electric power industry applications, what percentage increase in your total output of super clean steel would be required to bring about a significant reduction in cost?

16 If the present EPRI super clean steel composition were relaxed as outlined above, would you envisage a further increased demand for such clean steels for engineering applications, other than for the electric power industries?
For example,

(a) Would you publicise that you were able to produce such steels? YES/NO

(b) Could you give a brief indication of the product types that you think your customers would be interested in.

(c) Do you think the EPRI clean steel philosophy could be extended to the production of other grades of engineering steels supplied to the Electric Power industries? If so, could you indicate the compositions and product types that could be of interest.

17 (a) What tolerances do you work to in the laboratory analytical procedures you use in relation to the following elements.

Element	Tolerance	Element	Tolerance
C		Sn	
Si		As	
Mn		Sb	
S		Cu	
P		H	
Ni		O	
Cr		N	
Mo			
V			

(b) Do you think there is a need for improved analytical methods for any of the elements listed above. Please specify.

18 Are there any specific topics, or general directions in which you would wish research to be carried out for improving the types and qualities of steels supplied to the Electric Power industries? Examples of topics which could be tackled on a generally sponsored programme would be:-

(i) The Synergism between the impurities such as P, S, Sn, As, Sb, and the normal alloying elements such as Ni, Cr, Mo, V, Mn and Si.

(i i) An Assessment of the limits to the levels of impurity contents below which no further improvement of in-service properties occurs.

19 Could you give a realistic and cost effective composition for "clean steel" in relation to the elements normally specific.

20 Are there any other issues you would like to raise in connection with the EPRI clean steel programme which have not been included above?

BOILER MAKERS

Some six questionnaires were sent to boiler makers or organisations serving the needs of boiler makers. Only two replies were received and one reply was more concerned with answering the questions as a steel maker. The answers in this questionnaire have been incorporated therefore in the survey of the Steelmakers and Forgemasters.

The other reply was almost completely negative in the sense that no superclean steel had been used in the fabrication of pipe work and that in general there was little awareness of the advantages that superclean steels would have for power generation equipment.

POWER PLANT MAKERS

Twelve questionnaires were sent to power plant makers and associated organisations in Europe, Japan and the USA. Six replies were received; in two of these from European plant builders it was stated that all matters relevant to their needs would be met by the answers from steelmakers. Of the other four replies received, two came from Europe, one from Japan and one from the USA. Repeated requests for further replies failed to elicit a response. At least geographically the answers relate to the power plant producers in the three chief manufacturing regions.

Those companies replying to the questionnaire were heavily committed to supplying steam turbines and generator plant but in some cases they were also involved with gas turbines. Three of the companies have made offers to supply superclean steel rotors in IP, LP or single cylinder configurations but only two have been successful in obtaining orders. One company has not offered to supply rotors in any form to a superclean specification. However it can be inferred from the answers given that two European suppliers of power generation equipment have placed orders for superclean steels for service in power generation plant from a European steel maker and presumably the finished components have gone or are going into service.

When the plant builders wished to incorporate superclean steels into rotors, they appeared to have no difficulty in finding steelmakers willing to supply suitable forgings. These components could be supplied without greatly affecting the lead time for delivery and there was no tonnage related premium linked to the supply of the steel.

There would appear to have been no great difficulties associated with the further fabrication of the superclean rotors from the as machined forgings supplied by the steelmakers.

None of the companies supplying or offering to supply superclean rotors appear to have carried out creep or fatigue tests on EPRI specification steel although tests have been carried out on some steels which were near to the

EPRI specification. Similarly no tests had been carried out to assess the susceptibility of the steel to temper embrittlement. All the suppliers were involved with stress corrosion testing of superclean steels.

It was generally thought that it would be possible to relax the EPRI specification without bringing about a deterioration of the service performance, but none of the suppliers were able to put forward results to support this view. It was also generally felt that relaxing the specification would lead to a wider demand for power plant applications with possibly the wider use of 3.5% Ni–Cr–Mo–V steels in preference to Cr–Mo–V compositions.

As opposed to the steelmakers the power plant producers believed that no great advantages would ensue from the application of superclean steels to pipe work and headers or other power generating equipment.

One of the topics raised on the questionnaire was the advantages to be gained from the use of superclean steels in power generation plant. Some rather discursive replies were received.

The major quantifiable advantage identified by all plant builders was the ability to use superclean rotors with higher crossover or LP inlet temperatures than conventional rotors which then influences turbine efficiency. Such a change can also influence the general design of the turbine by making it possible to increase the length of the last stage blading which also increases the efficiency. It was stated that with a 30°C higher LP inlet temperature the thermal efficiency was improved by 0.1% with no change in the turbine operation and inspection interval.

It would appear that with the hope of developing single cylinder turbine design the creep properties of the existing EPRI superclean steel limit the temperature of operation. In order to raise the maximum temperature perhaps a new EPRI superclean specification is required for a higher chromium lower nickel version strengthened further by the strong carbide formers Nb and W. Work on such materials was thought to be more important than studying synergistic impurity effects in the existing EPRI superclean steel with the hope of 'fine tuning' the composition.

POWER PLANT USERS

Questionnaires were sent to the main power generation companies in Great Britain, and central organisations dealing with the needs of the power generation industries in Europe, Japan and the United States. The general tenor of the replies was that it was not the policy of the central organisations to reply on behalf of their members, or that they were specifically prohibited by national policy from replying to questionnaires of the type submitted. After much exchange of correspondence no progress was made and only one fully completed and a partially completed questionnaire were received from power generation companies in Great Britain. As neither of these companies had plant using superclean steel, the outcome of this part of the exercise was disappointing. Attempts were made to get informal views from 'friends in the trade' at casual meetings and conferences but anecdotal evidence of this type is not widely accepted as being a substitute for formal statements.

It would appear that when an electric power generating company outlines its requirements for new plant to the plant makers with a request to tender, the power generating company defines the size of plant required and the specific duty but at no stage are requests made that specific materials be used in the construction of the plant. This does not mean that the plant operator is not interested in what materials the plant is made from, but he leaves it to the plant maker to choose what is thought to be suitable. The plant operator then accepts after due consideration what is thought to be the most suitable tender. This really puts the onus on the plant builders to offer superclean steels if they think they have a cost effective advantage.

In relation to the advantages of superclean steels as seen by the user, the most important aspects would seem to be those associated with rapid start-up and avoidance of the need to pre-warm before loading. Increases in the intervals between inspection and the avoidance of forced outages are the next most important issues. Improved life is not looked upon as being of great importance and surprisingly in view of the plant makers views outlined above increased efficiency due to higher LP crossover temperatures is thought to be of only minor importance. Perhaps other plant users might have different orders of priority.

GENERAL COMMENTS AND CONCLUSIONS

Following the EPRI workshop on 'Super Clean Rotor Steels' held in 1989 in Sapporo, Japan, it became clear that such materials were not just laboratory curiosities but they were being used for the manufacture of rotors that were actually being used to generate electricity.

The major alloy steelmakers seem to have accepted that a further demand for such steels would arise and they have developed the appropriate steel-making techniques together with ancillary analytical procedures for dealing with alloying and impurity levels that were lower than previously specified. The global demand for superclean steel has now risen to 7,000–10,000 tonnes per annum but it is clear that there is installed capacity for production of much greater quantities.

It was part of accepted wisdom before this survey was carried out that an increase in demand for superclean steels would lead to a reduction in the premium charged for these steels in relation to a conventional grade (widely quoted as about 30%). The results of this survey show that any increase in demand is unlikely to lead to a lowering of the premium which now seems to be about 20% in relation to the basic steel cost. The surprising views have been expressed that if the manganese upper composition limit was raised to about 0.1% the premium would only be 5% and a similar premium was quoted if the manganese limit was kept to 0.02% but the residual impurity limits of As, Sb and Sn were slightly raised. Who should support such a study?

In spite of the clear advantages of using superclean 3.5 Ni–Cr–Mo–V steel for LP rotors, only a small proportion of the recently installed rotors are in the new material. It is difficult to understand why. It could be that wider usage will develop from their application to discs and it may be that as more LP rotors are used in nuclear plant where stress corrosion cracking is more prevalent, the advantages of superclean steels will become more apparent. Whatever should develop, there would appear to be no great difficulty in obtaining suitable forgings, the initial cost premium is not high in relation to the finished cost and there appears to be no difficulty in fabrication. Is it the fault of the plant makers for not stressing the advantages? Or should the plant users take a more active role in specifying their needs?

A wider dissemination of the characteristics of superclean steel may help. This certainly seems to be the case with the boiler makers who may be forgiven for believing that their products do not suffer from long term temper embrittlement, but a superclean type 2.25% Cr–Mo superheater tube steel would have a much better creep life than a conventional grade, due to the absence of MnS inclusions to act as grain boundary void nucleators.

The development of super critical power plant has led to the development of the 9–12% Cr type steels for high temperature rotors and the evidence suggests that perhaps further development of superclean varieties of these materials is one of the ways ahead. A further possibility is to produce a new grade of steel with the toughness of 3.5% Ni–Cr–Mo–V and the creep properties of 2.25% Cr–Mo–V which could then be used for single cylinder rotors – some progress has been made. It seems clear that such a steel would have to have a superclean specification, and would be of particular benefit for gas fired turbine combined cycle power plant.

Further work on such materials would seem to be preferable to a more detailed study of the existing superclean 3.5% Ni–Cr–Mo–V. Widening the superclean product range could be beneficial to the electric power industry and also to the steel producers who have done so much to make the late Dr. Bob Jaffee's dreams become a commercial reality.

Residual Contents in Purchased Scrap for Use in Basic Electric Arc Technology

P. A. BRALSFORD, E. HYDES and G. A. HONEYMAN

Forgemasters Steel & Engineering Ltd, P.O. Box 287, Brightside Lane, Sheffield, United Kingdom

1. INTRODUCTION

Low residual steels are being increasingly demanded by various industries in order to improve mechanical, environmental and elevated temperature properties. Power Generation require rotor forging steels with a low propensity to temper embrittle as designs are modified to increase thermal efficiency. Nuclear components use low residual steels to improve fracture toughness, and to minimise embrittlement from neutron irradiation. Some petrochemical plants, which operate at elevated temperatures, seek low residual steels to minimise in-service embrittlement and the Offshore Industry are developing low residual steels to improve fracture toughness in structures operating in hazardous environments.

Clean steel processing has been developed by the Forgemasters to meet the requirements of the various industries. Some residual elements are controlled through the steelmaking process, however, others are restricted by careful control and selection of scrap.

2. STEELMAKING

Typical compositions for a standard 3.5% NiCrMoV steel and its superclean counterpart is shown in Table 1. Residual elements defined for the standard grade include phosphorus, sulphur, tin, arsenic and antimony whereas

Table 1. Typical compositions for a standard and superclean 3.5% NiCrMoV

Composition Type	Chemical Composition wt%											
	C	Si	Mn	P	S	Cr	Mo	Ni	V	Sn	As	Sb
Standard	0.26	0.05	0.20	0.005	0.004	1.50	0.36	3.55	0.11	0.005	0.006	0.001
Superclean	0.26	0.01	0.03	0.004	0.001	1.71	0.47	3.66	0.12	0.004	0.004	0.001

superclean steel further restricts silicon and manganese. The latter two elements are known to exercise a considerable effect on grain boundary embrittlement.

The control of residual elements during steelmaking can be divided into two main areas:

- scrap control
- steelmaking process technology

Table 2 summarises the three stages of steelmaking and at what stage various residual elements are controlled. Phosphorus and manganese are removed in

Table 2. Steelmaking control of residual elements

		Influence on Residual Control		
Stage	Technology	Steelmaking	Additions	
Stage 1	Basic Electric Arc	Manganese Phosphorus Silicon	Tin Arsenic Antimony Copper	Scrap
Stage 2	Secondary Steelmaking (VAD)	Sulphur Hydrogen Oxygen	Phosphorus	Ferro Alloys
Stage 3	Teeming	Hydrogen Oxygen Nitrogen		

the Basic Electric Arc Furnace using oxidising slags whereas sulphur is lowered in the secondary steelmaking ladle using reducing slags and other processing techniques. Gaseous compounds are minimised by slag technology and degassing under vacuum. However, tin, arsenic, antimony and copper can only be kept low by careful selection of scrap since these elements remain constant throughout the whole steelmaking process. The importance of scrap and ferroalloy controls in ensuring optimum steel compositions is often under-rated by those technologists outside the Steel Industry and it is particularly critical during the manufacture of superclean steels.

3. ECONOMIC FACTORS

Scrap is either generated by the manufacturing and processing industries and is termed new scrap or from demolition or discard which is called old scrap. Old scrap availability is increasing due to heavy consumption and recycling whereas new scrap is decreasing and it is the latter which is particularly important for the manufacture of low residual clean steel. BOS plants consume primarily internal scrap so that most scrap purchases in the steel industry are to feed electric arc furnaces. The increasing scrap deficit in the Pacific Rim is a fundamental element in the current position on scrap prices. The deficit in the area has moved from 2 million tonnes in the 1970s to 12.5 million tonnes in 1992. The major countries with the largest imports of scrap are Korea, Taiwan, India and Indonesia followed more recently by an upsurge in demand by China. Due to the world market changes, the international trade in scrap has increased from 8% in 1983 to 13% in 1992.

The variation in scrap prices, however, is not only related to a supply and demand situation. Various policies operated by certain countries can influence the security of supply. Furthermore, speculation due to scrap being an international traded commodity, based on the US dollar, and currency exchange fluctuations cause market distortions. Figure 1 gives the variation in scrap prices from January 1994 to January 1995. In 1993, scrap prices reached an all time high but during the course of 1994 fell to previously stable levels. However, due to the recovery in the Steel Industry, prices again have increased dramatically and have exceeded the previous high levels of 1993.

In manufacturing superclean steels, there are four key elements which add premiums to the cost over conventional or standard grades:

- price of low residual scrap
- cost of alloying additions
- residence time in steelmaking – cost of electricity
- the use of dolomite lined ladles to prevent contamination

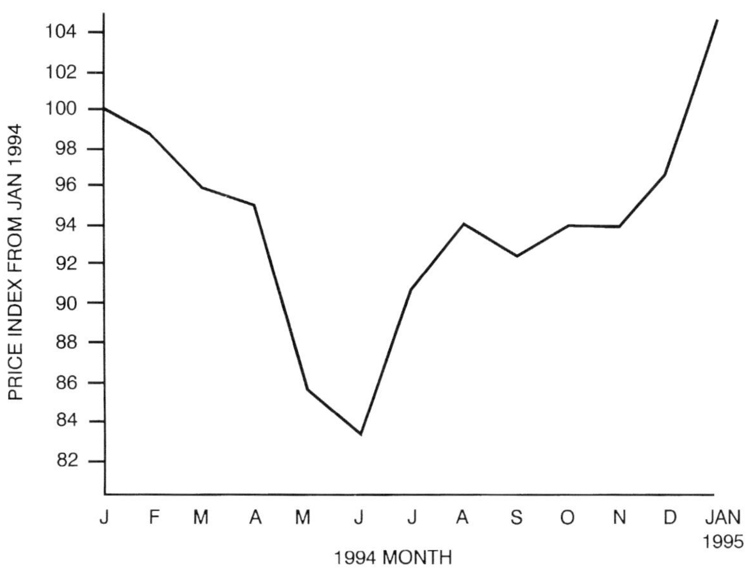

Fig. 1 Purchased ferrous scrap price index by month from Jan 1994

Whilst the cost of the actual steelmaking process and the ladle relining procedure can be accurately assessed, the fluctuations in raw material prices makes it impossible to give a definitive cost premium for superclean steels. Figure 2 shows the purchased ferro alloy price index for the three major

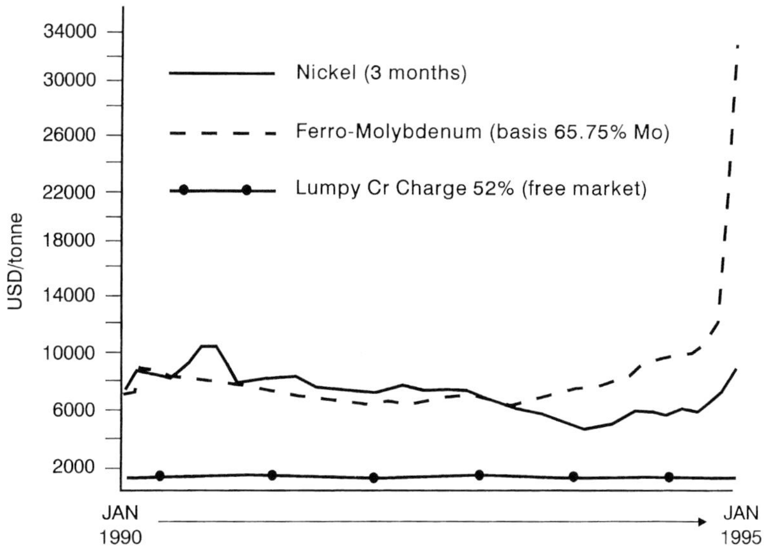

Fig. 2 Purchased Ferro Alloy price index

alloying elements of nickel, ferromolybdenum and free market chromium. Whilst chromium has remained fairly constant in price over the last five years, nickel has shown a gradual increase over the last 12 months. Ferromolybdenum, however, has shown a huge increase in recent months, being valued at over four times the price of 1994. The main reason for this change is that, due to the recession when many mines and reprocessing plants were mothballed, the price was kept steady, however, because of the recovery in the Steel Industry, the demand for molybdenum has substantially increased and currently there are not the stocks or the processing capabilities to meet this demand. Since more pure alloying additions are made in the secondary steelmaking process to produce superclean steels than for conventional steels then this also adds to the difficulty of comparing premiums for superclean steels in a sensible way.

4. SCRAP SELECTION

In order to meet the residual requirements of superclean steel, the selection of scrap is vital and significantly different to standard compositions. Table 3

Table 3. A comparison of the types of scrap used for a superclean and standard 3.5 NiCrMoV

Scrap %	Superclean	Standard
Turnings (External)	–	10
Bale	80	60
BOS	20	–
Turnings (Internal NiCrMoV)	–	30

gives a typical scrap mix for 3.5% NiCrMoV steel for a superclean and standard LP rotor forging. Turnings purchased from outside the Company as well as those produced internally can be used to charge the Basic Electric Arc Furnace for standard grades as, in general, the residual specifications for most Power Generation companies are not too restrictive. However, for superclean grades, typically specified by the EPRI standard, then turnings are not used due to their higher levels of tin, arsenic, antimony and copper. Furthermore, BOS scrap is required to levels of about 20% for superclean steel in order to maintain the low levels demanded by the EPRI. Table 5 gives typical compositions for various grades of scrap and it is readily apparent the importance of BOS scrap to the manufacture of low residual steels.

Table 4 compares the percentage ferroalloys added to a typical standard

Table 4. A comparison of the percentage of Ferro Alloys added to a typical standard and superclean cast

Alloy Addition	% Added	
	Superclean	Standard
High Carbon Ferrochrome	2.5	1.6
Molybdenum	1.2	0.4
Nickel	4.1	2.4

and superclean heat during secondary steelmaking in the Vacuum Arc Degassing unit. Higher percentages of pure alloying additions are added to superclean steel to maintain the low residual standard and to offset the lack of turnings added in the Basic Electric Arc Furnace which typically have higher chromium, molybdenum and nickel contents (Table 5).

Finally, the grades of ferroalloying additions must also be considered in order to ensure low residual steel is manufactured. Table 6 compares the phosphorus contents of low and high carbon ferrochrome. With the typical percentages of ferrochrome added to a superclean cast shown in Table 4, then the difference in phosphorus content could add as much as 0.001% phos-

Table 5. Typical compositions of various scrap types

Scrap Type	Element wt%						
	Cr	Mo	Ni	Cu	Sn	As	Sb
Carbon Turnings (Internal)	.30	.10	.35	.20	.020	.020	.004
Carbon Turnings (Bought)	.40	.15	.50	.30	.030	.0205	.004
Bale	.08	.015	.05	.04	.006	.007	.0035
Bos	.08	.015	.05	.04	.004	.005	.002

Table 6. A comparison of the Phosphorus content in Ferrochrome

Ferrochrome	%P
High Carbon	.010
Low Carbon	.025

phorus to the final heat analysis if the lower grade, carbon content version is used.

5. SUMMARY

The premium for superclean steels comes from a combination of steelmaking process technology and raw material costs. Whilst many companies are seeking to define this premium against standard steel compositions, the volatility of the scrap market pricing and the variable cost of ferroalloying additions makes it impossible to define accurately this premium. It is certain, however, that the Forgemasters have the capability and the technology to produce the low residual steels to the requirements of the various industrial sectors.

Manufacture of Superclean 3–3.5% NiCrMoV Steels for Gas Turbine Components

E. POTTHAST, K. LANGNER and F. TINCE
Saarschmiede GmbH Freiformschmiede, 66330 Völklingen, Germany

ABSTRACT

Facilities for the manufacture of superclean steels • Melting of 42 heats in a 125–t electric arc furnace • Secondary metallurgy in the ladle • Casting of forging ingots of 30–105 t weight • Mn, Si, Al, P, S, Sn, As and Sb contents during steelmaking • Embrittlement factors J, J', $\bar{X}$ and K • Embrittlement factors and toughness properties of superclean steel and of conventional steel at room temperature

1. INTRODUCTION

The economy of power stations can be improved by increasing the steam temperature. To protect the 3–3.5% NiCrMoV steels employed in the low-pressure section of turbines from long-term embrittlement due to the temperature increase to over 350°C, extremely low levels of Mn, Si, P, S, Al, Sn, As and Sb are necessary. Steels exhibiting those properties are called superclean steels.

2. SAARSCHMIEDE'S FACILITIES FOR CONVENTIONAL FORGING INGOTS

Saarschmiede has available one 125–t electric arc furnace with eccentric bottom tapping. Facilities for secondary metallurgy consist of 1 ladle furnace, 1 straight ladle degassing facility and 2 vacuum ingot casting facilities. The forging ingots can be top- or bottom-poured in air with argon shrouding or top-poured under vacuum.

3. MANUFACTURE OF SUPERCLEAN HEATS

Extremely low levels of Sn, As and Sb are obtained by the use of scrap with low contents of these elements. The elements Si, Mn and P, which can be influenced by metallurgical measures, are brought down to very low levels by several refining operations including slag changing. To prevent a significant degree of phosphorus pick-up during the refining process, the refining is accomplished at low temperatures. By the use of the residual steel pool method in combination with eccentric bottom tapping (EBT), slag carry-over during tapping is practically excluded. Tapping of the steel into a clean ladle is accompanied by Al-deoxidation and the addition of a basic slag. The slag forming during the tapping operation is largely removed and replaced by a fresh ladle slag. Final alloying and the major part of the desulphurisation process take place in the ladle furnace. The alloying and desulphurisation processes being terminated and after reaching the melt super-heat required for further ladle treatment, the ladle is transferred to the degassing facility, where the heat is thoroughly degassed and desulphurised to extremely low levels by intensive mixing of steel and slag. Ladle treatment of the steel is terminated by a gentle rinsing, which is called rinsing for cleanliness. This flushing promotes the float-out of oxidic inclusions and of slag droplets in the liquid steel. Casting of superclean heats into forging ingot molds is preferably done by the top-pouring technique using argon shrouding as the best oxidic cleanliness ratings have been obtained in our works by this technique.

4. METALLURGICAL RESULTS

4.1 Elements Al, Si, Mn, P and S

The oxidisable elements Al, Si, Mn and P are brought down to very low levels during the refining process in the furnace, namely:

Al to < 0.005%
Si to < 0.005%
Mn from 0.06 to 0.02%
P from 0.0025 to 0.0008%.

The deoxidation elements Al and Si oxidise already when the scrap is melted down. Their contents prior to tapping are below the threshold level detectable by spectrochemical analysis.

After Al-deoxidation of the steel and practically slag-free tapping, the average aluminium content is of the order of approx. 0.070% (Fig. 1a). During ladle treatment, the Al burn-off is controlled so as to result in 0.005 to 0.010% Al in cast samples.

As shown in Fig. 1b, the Si content in the steel rises only slightly during tapping and somewhat more strongly during the reduction of the ladle slag in the ladle furnace. Vacuum treatment and flushing for cleanliness cause only a slight decrease of the Si content.

The elements Mn and P partly oxidise and rise into the slag already during the melt-down phase (Fig. 2). Achieving extremely low contents of Mn and P requires several refining phases with slag changing at relatively low temperatures of the molten steel.

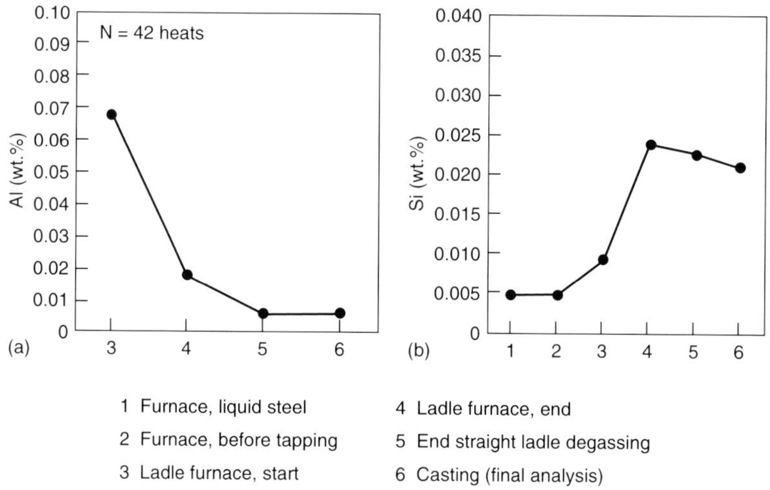

Fig. 1 Variation in the contents of aluminium and silicon during manufacture of superclean heats

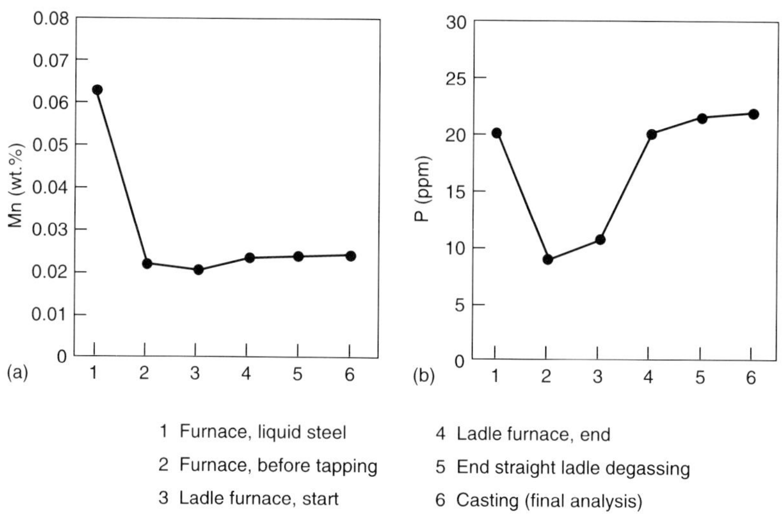

Fig. 2 Variation in the contents of manganese and phosphorus during manufacture of superclean heats

During treatment of the steel in the ladle furnace, there is only a slight increase in the Mn content (Fig. 2a) but a stronger increase in the P content (Fig. 2b). In spite of eccentric bottom tapping in combination with a residual pool in the electric arc furnace and the technically limited removal of the slag forming during tapping, some furnace slag and residual slag from the previous heat in the ladle is carried over into the new ladle slag. During the reduction of this slag, small quantities of Mn and P revert into the liquid steel. Moreover, small quantities of P contained in ferro-alloys are added into the heat during initial alloying.

Under oxidising conditions in the electric arc furnace, there is only an insufficient desulphurisation, as expected. The sulphur content drops from approx. 0.012 to 0.008% (Fig. 3). The average S content of the heat is further

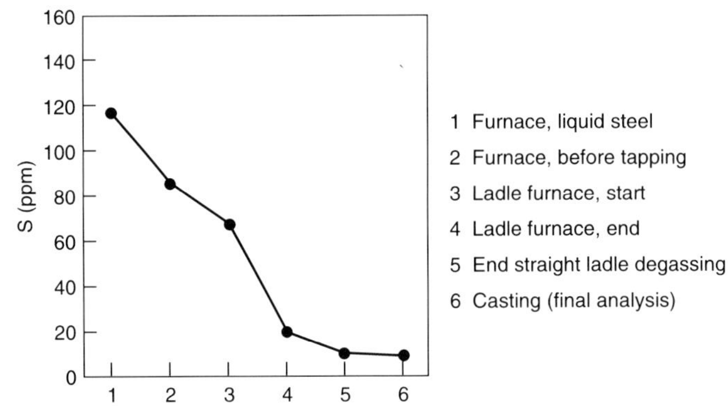

Fig. 3 Evolution of the sulphur content during manufacture of superclean heats

reduced to approx. 0.007% by tapping the heat almost slag-free from the furnace and adding 200 kg of CaO–CaF_2 mixture into the ladle in order to improve the fluidity of the slag resulting from the deoxidation reaction. During the slag work in the ladle furnace, the sulphur content in the liquid metal was further reduced to approx. 0.002%. Finally, the thorough mixing of steel and slag during the ladle degassing operation resulted in the very low average sulphur content of 0.0010%.

For the elements Mn, Si and P, which are taken into account in the calculation of the embrittlement indices, the contents in the cast samples are illustrated in Fig. 4. As can be seen, the average values are 0.024% for Mn, 0.019% for Si and 0.0022% for P. The scatter bands are shown in the respective diagrams.

As compared to this, the average contents in conventional heats are 0.225% for Mn, 0.055% for Si and 0.0045% for P, i.e. clearly higher (Fig. 5).

4.2 Trace elements Sn, As and Sb

The contents of these non-oxidisable elements are determined almost entirely by the choice of scrap (Fig. 6).* It is only at the initial alloying stage that small quantities of these elements can enter into the heat via ferro-alloys.

* The evolution of the Sb content throughout the melting operation was not established due to the high cost of the analytical processes.

Figure 7 summarises the contents of trace elements in the cast samples. The average contents are 26 ppm for Sn, 46 ppm for As and 4.5 ppm for Sb.

For conventional heats being subject to a less meticulous selection of scrap, the average contents are 50 ppm for Sn, 90 ppm for As and 10 ppm for Sb (Fig. 5), that means twice as high as for superclean steels.

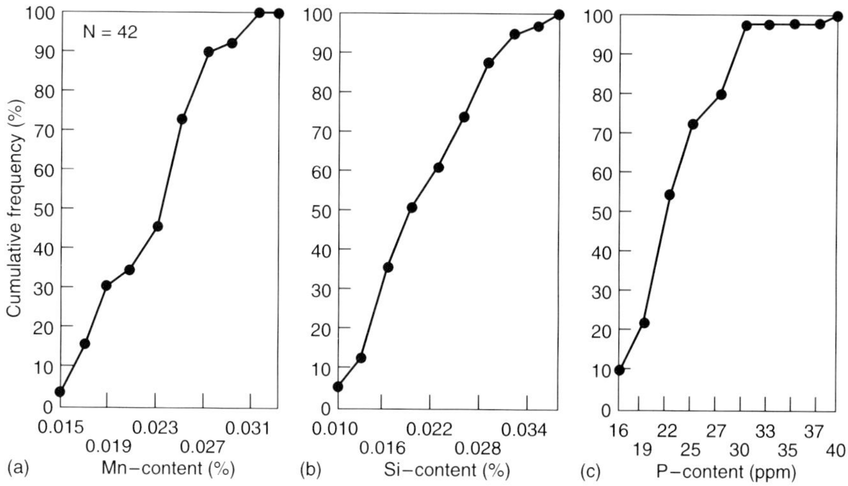

Fig. 4 Contents of manganese, silicon and phosphorus in the cast samples

	Analysis (wt%)	
	Conventional steel (Average of 100 heats)	Superclean steel (Average of 42 heats)
Mn	0.2250	0.0240
Si	0.0550	0.0190
Al	0.0065	0.0060
P	0.0045	0.0022
S	0.0025	0.0010
Cu	0.0660	0.0240
As	0.0090	0.0046
Sb	0.0010	0.00045
Sn	0.0050	0.0026

Fig. 5 Analysis of superclean and conventional 3–3.5% NiCrMoV steels

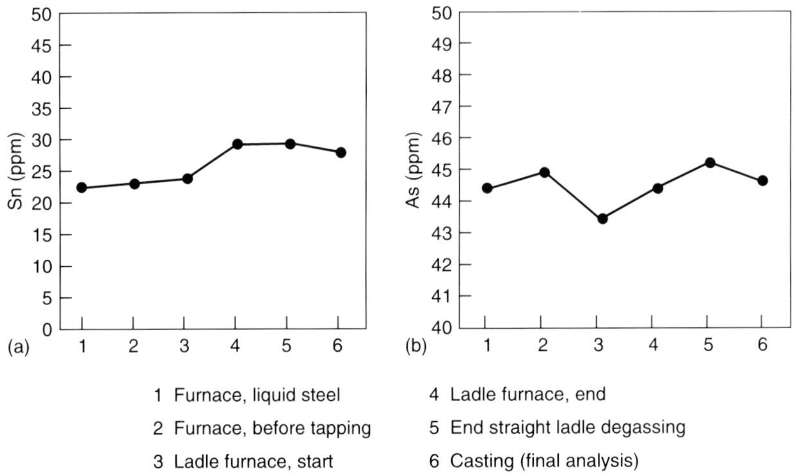

Fig. 6 Variation in the contents of tin and arsenic during manufacture of superclean heats

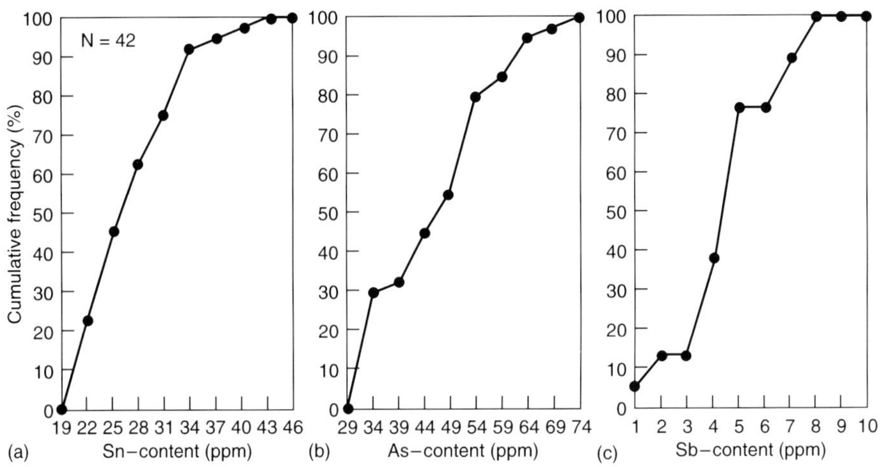

Fig. 7 Contents of tin, arsenic and antimony in the cast samples

4.3 Embrittlement factors

The present evaluation takes into account the following embrittlement factors:

J–factor

$$J = (\%Si + \%Mn)(\%P + \%Sn)$$

J'–factor

$$J' = (\%Si + \%Mn)(\%P + \%Sn + \%As + \%Sb)$$

$\bar{X}$–factor

$$\bar{X} = 10 \times \%P + 5 \times \%Sb + 4 \times \%Sn + \%As$$

K–factor

$$K = (\%Si + \%Mn)\,(10 \times \%P + 5 \times \%Sb + 4 \times \%Sn + \%As)$$

For the analysis values determined from the cast samples, the following average values are calculated:

2.2×10^{-4} $(1.3 - 3.1 \times 10^{-4})$	for the J–factor	(Fig. 8a)
4.3×10^{-4} $(2.3 - 6.9 \times 10^{-4})$	for the J'–factor	(Fig. 8b)
39×10^{-3} $(29 - 65 \times 10^{-3})$	for the $\bar{X}$–factor	(Fig. 9a)
1.7×10^{-3} $(1.0 - 2.6 \times 10^{-3})$	for the K–factor	(Fig. 9b)

5. EMBRITTLEMENT FACTORS AND TOUGHNESS PROPERTIES OF TURBINE STEELS AT ROOM TEMPERATURE

To clarify whether the decrease of the Si, Mn, P, Sn, As and Sb contents has effects on the toughness properties of 3% NiCrMoV steel at room temperature, the test results from forgings in superclean and conventional steel quality were compared. The comparison was conducted on similar gas turbine forgings, such as shaft ends, mid-length shaft sections, discs and barrels with the corresponding sample locations (Fig. 10). To guarantee a maximum comparability of the two melting processes the evaluation is subdivided in two parts.

The first part takes into account the average toughness and yield strength values of all forgings manufactured from the forging ingots of one heat.

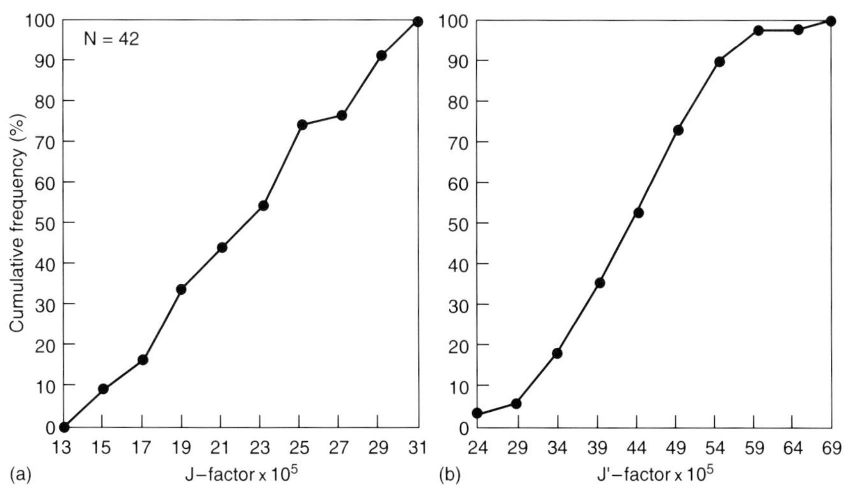

Fig. 8 Frequency distribution of embrittlement factors J and J' of superclean heats

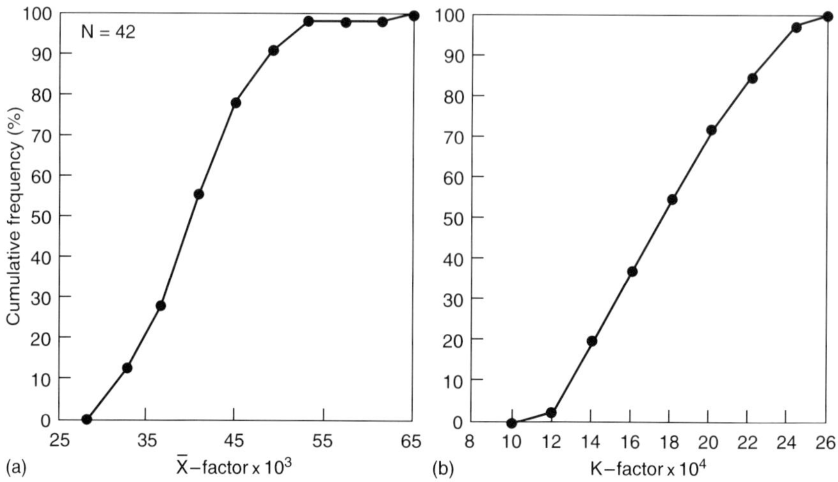

Fig. 9 Frequency distribution of embrittlement factors X̄ and K of superclean heats

Figure 11 shows the values for absorbed energy as a function of the yield strength. In comparison to conventional manufacture, the absorbed energy values for superclean steels are markedly higher and situated in a narrower scatter band. The embrittlement factors differ as shown in Fig. 12. For example, J–factor 2.2×10^{-4} for superclean steel and 27×10^{-4} for conventional steel. The reduction of area and elongation values are also influenced

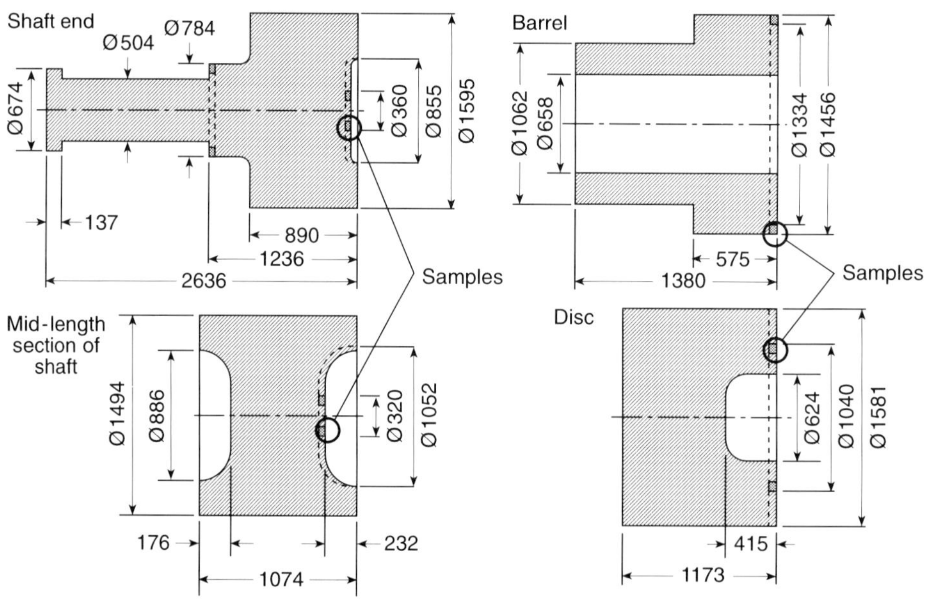

Fig. 10 Gas turbine parts showing sample location

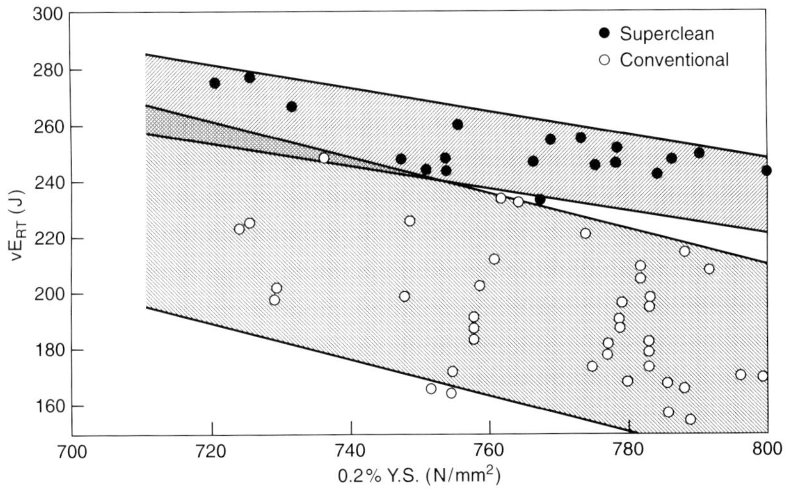

Fig. 11 Comparison of absorbed energy of turbine parts from superclean and conventional 3% NiCrMoV steels

	Number of heats	J-factor	J'-factor	$\bar{X}$-factor	K-factor
Superclean steel	42	2.2×10^{-4}	4.3×10^{-4}	39×10^{-3}	1.7×10^{-3}
Conventional steel	100	27×10^{-4}	53×10^{-4}	79×10^{-3}	22×10^{-3}

$J = (\%Si + \%Mn)(\%P + \%Sn)$

$J' = (\%Si + \%Mn)(\%P + \%Sn + \%As + \%Sb)$

$\bar{X} = 10 \cdot \%P + 5 \cdot \%Sb + 4 \cdot \%Sn + \%As$

$K = (\%Si + \%Mn) \times \bar{X}$

Fig. 12 Embrittlement factors of superclean and conventional 3–3.5% NiCrMoV steels

by reducing the embrittlement factors (Fig. 13). For superclean heats they are situated in the upper range of the scatter bands, which, however, are not shown.

To eliminate the influence of ingot segregation, forging geometry and size

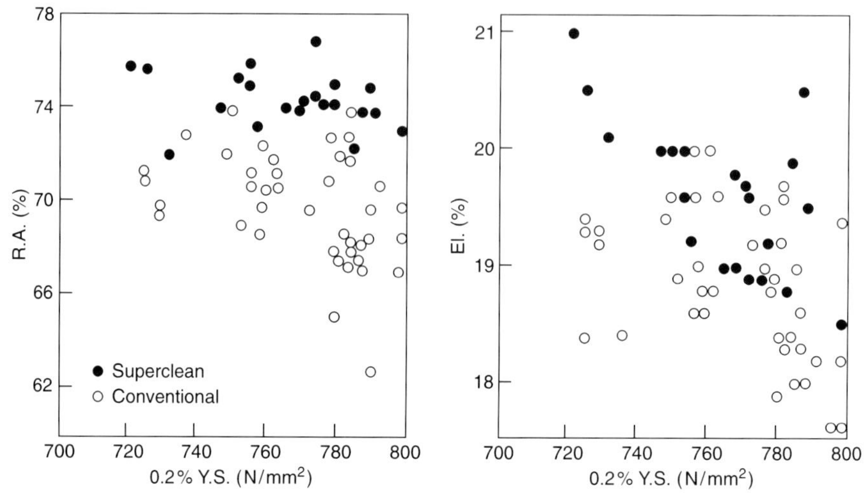

Fig. 13 Comparison of reduction of area and elongation of turbine parts from superclean and conventional 3% NiCrMoV steels

as well as sample location, the second part of the evaluation only covers the shaft end forged from the top portion of a 105–t ingot. The interrelationship between absorbed energy and yield strength is shown in Fig. 14. As expected,

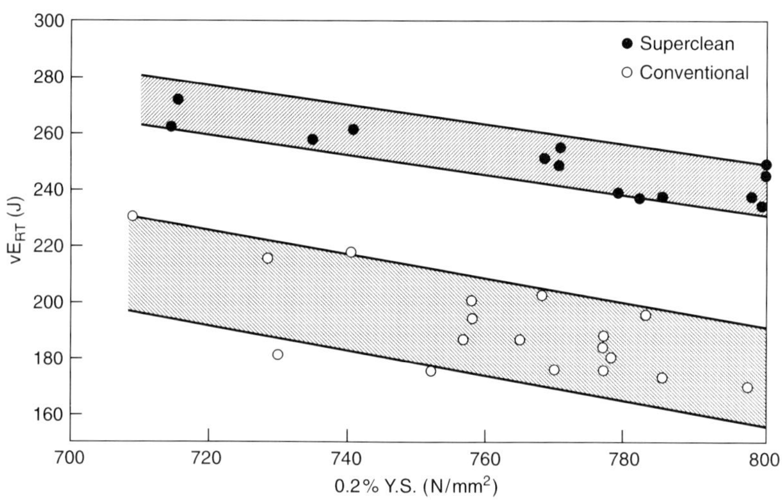

Fig. 14 Comparison of absorbed energy of shaft ends from superclean and conventional 3% NiCrMoV steels

there is no basically new finding. Only the scatter is somewhat less marked. The same holds true for the comparison of the reduction-of-area and elongation values (Fig. 15).

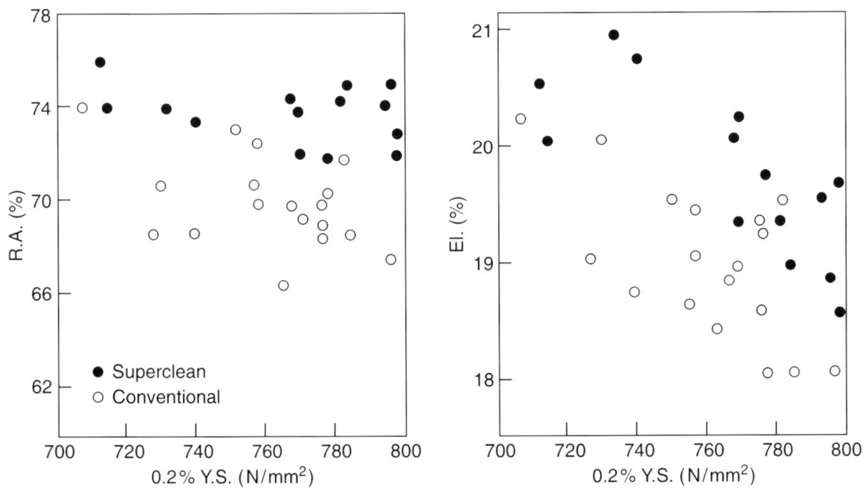

Fig. 15 Comparison of reduction of area and elongation of shaft ends from superclean and conventional 3% NiCrMoV steels

An additional evaluation was made to determine whether there was an influence of the embrittlement factors within the narrow scatter band for the absorbed energy values of the superclean steels. Such an interrelationship was not found.

6. SUMMARY

Saarschmiede as a major manufacturer of energy generation forgings worldwide has been supplying 3–3.5% NiCrMoV steels in superclean quality since 1989. Up to now, 42 heats were produced in a 125–t electric arc furnace and cast into forging ingots of 30 to 105 t weight.

This paper describes the production facilities and the metallurgical processes employed in the manufacture of superclean steels.

With the heats produced so far, average contents of 0.024% Mn, 0.019% Si, 10 ppm S, 22 ppm P, 26 ppm Sn, 46 ppm As and 5 ppm Sb were achieved. These low contents also resulted in low embrittlement factors, e.g. J–factor: 2.2 $\times 10^{-4}$, J'–factor: 4.3 $\times 10^{-4}$, $\bar{X}$–factor: 39 $\times 10^{-3}$ and K–factor: 1.7 $\times 10^{-3}$.

The ingots were forged into discs, shaft ends, mid-length shaft section, barrels and similar parts for gas turbines. The evaluation of the toughness properties of the forgings made from 3% NiCrMoV–steels shows, that the absorbed energy, elongation and reduction-of-area values are better than those obtained by conventional melting practices. Within the narrow scatter bands for the embrittlement factors of superclean steels, an influence of the factors on the toughness properties could not be shown.

Production and Properties of Superclean Monoblock LP Turbine Rotor Forgings

Y. IKEDA, H. YOSHIDA, Y. TANAKA and T. FUKUDA

The Japan Steel Work, Muroran Plant, 4 Chatsu-machi, Muroran, Japan

ABSTRACT

Since the first trial superclean rotor forging was manufactured by JSW in 1986, 28 superclean 3.5%NiCrMoV LP rotors have been manufactured and shipped by JSW. These are for the fossil fired power plants with output up to 700 MW, and the maximum ingot size so far to produce these forgings are 220 ton. Larger size forgings, however, have come to be needed for future application of superclean steel in 1000 MW class USC plant. JSW has initiated a study to produce large monoblock rotor for 1000 MW class plants by superclean version steel and recently two forgings have been produced. The evaluation tests of the forgings revealed that superclean chemistry has attained in all the part of the forging and good toughness and strength balance which is comparable to recent monoblock rotors produced from Mn bearing steel. Good resistance to temper embrittlement can be expected in the superclean monoblock LP rotor.

1. INTRODUCTION

In the fossil power generation plants, not only the size of steam turbine but also the steam temperature progressively increases in order to improve the thermal efficiency of power plants. This leads to the increase of service temperature of NiCrMoV LP rotor forging up to 400°C where the severe temper embrittlement may proceed depending on the contents of residual elements.

Around ten years ago, Electric Power Research Institute (EPRI) initiated a series of research projects on high purity steels with low Si, Mn, impurity contents and gas contents aiming at the elimination of temper embrittlement susceptibility of steels.[1] Since then, a considerable number of superclean, version 3.5NiCrMoV, steel forgings were manufactured and some of them have been in operation. Table 1 summarises the producing experiences of superclean rotor shaft forgings by JSW. The producible size of these superclean forgings has progressively increased and use of two ladle refining furnaces made it possible to produce a ingot over 200 tonnes. In this multiple pouring process, the refining schedule of each ladle refining furnace should be carefully adjusted in order to attain the low contents of impurity elements.

Table 1 Superclean NiCrMoV steel rotor shaft forgings produced by JSW

No.	Output (MWe)	Weight (ton)		Ingot wt. (ton)	Chemistry (ladle,wt.%)													J*
		Drum	Gashed		C	Si	Mn	P	S	Ni	Cr	Cu	Mo	V	As	Sn	Sb	
1	–	21.8	14.7	105	0.25	0.01	0.03	0.003	0.0012	3.66	1.77	0.03	0.41	0.12	0.006	0.003	0.0017	2.4
2	700	58.8	42.0	120	0.25	0.03	0.02	0.002	0.0008	3.64	1.75	0.02	0.43	0.13	0.003	0.003	0.0012	2.5
3	700	59.2	42.4	120	0.25	0.03	0.02	0.002	0.0012	3.65	1.75	0.03	0.42	0.13	0.003	0.003	0.0012	2.5
4	700	58.8	42.0	120	0.25	0.02	0.02	0.002	0.0010	3.61	1.75	0.03	0.42	0.13	0.003	0.003	0.0009	2.0
5	700	59.2	42.4	120	0.25	0.03	0.02	0.002	0.0007	3.64	1.76	0.03	0.42	0.13	0.003	0.003	0.0008	2.5
6	700	73.4	50.0	170	0.25	0.02	0.03	0.003	0.0010	3.86	1.75	0.03	0.40	0.13	0.004	0.003	0.0008	3.0
7	700	73.6	50.5	170	0.25	0.02	0.03	0.003	0.0020	3.92	1.78	0.02	0.41	0.13	0.004	0.002	0.0008	2.5
8	700	63.7	42.3	140	0.25	0.03	0.02	0.003	0.0020	3.63	1.74	0.03	0.41	0.13	0.003	0.003	0.0007	3.0
9	700	74.0	51.4	180	0.25	0.02	0.03	0.003	0.0015	3.88	1.74	0.03	0.42	0.12	0.003	0.003	0.0008	3.0
10	700	73.7	50.8	180	0.25	0.01	0.03	0.003	0.0015	3.86	1.74	0.04	0.41	0.12	0.004	0.003	0.0010	2.4
11	600	65.6	42.1	140	0.25	0.03	0.03	0.003	0.0014	3.65	1.76	0.04	0.42	0.13	0.003	0.003	0.0008	3.6
12	600	63.4	41.8	140	0.25	0.02	0.02	0.002	0.0017	3.65	1.75	0.04	0.43	0.13	0.004	0.003	0.0009	2.0
13	–	43.3	26.9	90	0.25	0.03	0.03	0.003	0.0014	3.71	1.76	0.03	0.42	0.13	0.003	0.003	0.0009	3.6
14	–	43.3	25.8	90	0.25	0.03	0.03	0.003	0.0013	3.64	1.72	0.04	0.42	0.13	0.005	0.003	0.0010	3.6
15	700	88.2	65.7	220	0.24	0.03	0.02	0.003	0.0013	3.86	1.74	0.02	0.39	0.10	0.003	0.003	0.0010	3.0

Table 1 *(continued)* Superclean NiCrMoV steel rotor shaft forgings produced by JSW

No.	Output (MWe)	Weight (ton)		Ingot wt. (ton)	Chemistry (ladle,wt.%)													J*
		Drum	Gashed		C	Si	Mn	P	S	Ni	Cr	Cu	Mo	V	As	Sn	Sb	
16	700	86.4	65.0	220	0.25	0.02	0.02	0.003	0.0014	3.84	1.73	0.02	0.39	0.10	0.003	0.003	0.0009	2.4
17	350	60.3	33.3	140	0.25	0.02	0.02	0.003	0.0015	3.60	1.72	0.03	0.40	0.10	0.003	0.003	0.0008	2.4
18	350	60.3	33.3	140	0.25	0.03	0.01	0.002	0.0015	3.62	1.74	0.02	0.41	0.10	0.003	0.003	0.0009	2.0
19	350	60.3	33.3	140	0.23	0.03	0.03	0.003	0.0007	3.58	1.71	0.03	0.40	0.10	0.003	0.003	0.0011	3.6
20	350	60.3	33.3	140	0.24	0.03	0.02	0.003	0.0008	3.64	1.74	0.03	0.41	0.10	0.003	0.003	0.0010	3.0
21	350	60.3	33.3	140	0.25	0.03	0.01	0.003	0.0007	3.74	1.72	0.02	0.41	0.10	0.003	0.004	0.0008	2.8
22	265	35.7	22.6	90	0.25	0.02	0.03	0.003	0.0005	3.60	1.70	0.03	0.41	0.10	0.004	0.003	0.0007	3.0
23	265	35.7	22.6	90	0.24	0.01	0.04	0.003	0.0006	3.65	1.79	0.02	0.42	0.11	0.004	0.003	0.0008	3.0
24	660	89.5	49.1	180	0.25	0.03	0.02	0.003	0.0009	3.86	1.75	0.03	0.41	0.10	0.003	0.003	0.0010	3.0
25	660	89.8	50.0	180	0.25	0.03	0.02	0.002	0.0008	3.85	1.76	0.03	0.42	0.11	0.003	0.003	0.0010	2.5
26	265	35.7	22.6	90	0.24	0.01	0.04	0.003	0.0008	3.71	1.74	0.04	0.42	0.10	0.004	0.003	0.0010	3.0
27	265	35.7	22.6	90	0.26	0.03	0.02	0.003	0.0019	3.64	1.77	0.04	0.44	0.10	0.004	0.004	0.0010	3.5
28	350	60.3	33.2	140	0.24	0.03	0.03	0.003	0.0006	3.64	1.74	0.03	0.42	0.10	0.003	0.003	0.0009	3.6
29	1,000	257.0	144.4	570	0.23	0.03	0.03	0.003	0.0014	3.57	1.66	0.03	0.42	0.10	0.003	0.003	0.0010	3.1
30	1,000	264.4	151.6	585	0.23	0.02	0.04	0.003	0.0017	3.56	1.64	0.04	0.41	0.11	0.004	0.003	0.0010	3.8

*J = (Si + Mn) × (P + Sn) × 10000

However, for the LP turbine of 1000 MW class fossil power plants, larger superclean forgings from 570 to 600 tonnes ingots will be required.

Production of these large ingots necessitates the multiple pouring of six heats of ladle refining furnace melt and attention should be paid to the refining schedule of each ladle furnace. Since some of the ladles must hold the molten steel for a longer time, than the time to produce a smaller superclean ingot, which may cause the increase of impurity elements in molten steel such as re-phosphorisation and re-manganisation. The lower yield of iron is another problem in such a long time holding.

In order to solve such questions and establish a manufacturing process for a large superclean monoblock LP rotor forging, an investigation on long time holding was performed and a superclean monoblock rotor shaft was produced. This paper summarises the production process and results of evaluation tests. These data are compared to the properties of smaller superclean forgings and conventional large monoblock LP rotor forgings.

2. PRODUCTION OF LARGE MONOBLOCK LP ROTOR FORGING

Based on the research conducted by JSW and Bethforge under a contract with EPRI, the excellent quality of high purity 3.5%NiCrMoV steel was confirmed and the chemistry of 'superclean' 3.5%NiCrMoV steel was specified. The 'superclean' is characterised by the chemistry free of Si, Mn, P, As, Sn, Sb, and gas contents. Table 2 shows the specifications of superclean 3.5%NiCrMoV steel by EPRI.[2]

In order to meet the required superclean chemistry for large monoblock rotor forging, the steel making process is the key technology. Therefore, extensive research on the steelmaking process was performed[3] and several preliminary studies were made prior to the production of a monoblock rotor shaft. Also, careful forging and heat treatment procedures are necessary for the production of a monoblock rotor, because of the difficult consolidation of porosity in a steel ingot and the easier growth of austenitic grains during heat treatment.

2.1 Ladle Refining Process

Figure 1 shows the outline of melting and ladle refining processes. The process is characterised by strong oxidising, reladling and ladle refining to reduce Mn, P and S.

In order to attain the high purity composition, adjustment of the timing of ladle refining is the important point. Excessively long heating in the ladle refining furnace may cause the increase the content of Mn. The longer holding may also promote the oxidation of iron and reduce the yield of iron which leads to the increase of the cost for superclean steel.

Table 2 Specifications of superclean 3.5%NiCrMoV steel by EPRI

	Chemistry/wt.%, (ppm for H, O and N)															
	C	Si	Mn	P	S	Ni	Cr	Mo	V	Al	As	Sn	Sb	H	O	N
EPRI Aim	0.25	0.02	0.02	0.002	0.001	3.50	1.65	0.10	0.10	0.002	0.002	0.002	0.001			
EPRI Spec.	0.30 max.	0.05 max.	0.05 max.	0.005 max.	0.002 max.	3.75 max.	2.00 max.	0.50 max.	0.15 max.	0.005 max.	0.005 max.	0.005 max.	0.002 max.	1.5 max.	35 max.	80 max.
ASTM A470 Class 7	0.28 max.	0.10 max.	0.20 ~0.60	0.015 max.	0.018 max.	3.25 ~4.00	1.25 ~2.00	0.25 ~0.60	0.05 ~0.15							

Figure 2 shows the results of preliminary investigation on the effect of long time holding. After the melting and oxidising in the basic electric furnace, reladling to ladle refining furnace was performed and the oxidising slag was removed completely to prevent the return of P and Mn. Then reducing refining and de-gassing were performed in the ladle refining furnace. Further reladling to another LRF was made and then the molten steels were held for another 8 hours. From the changes of P, S, Si and Mn content we can estimate that the superclean chemistries will be kept during the long time holding of molten steel in the LRF.

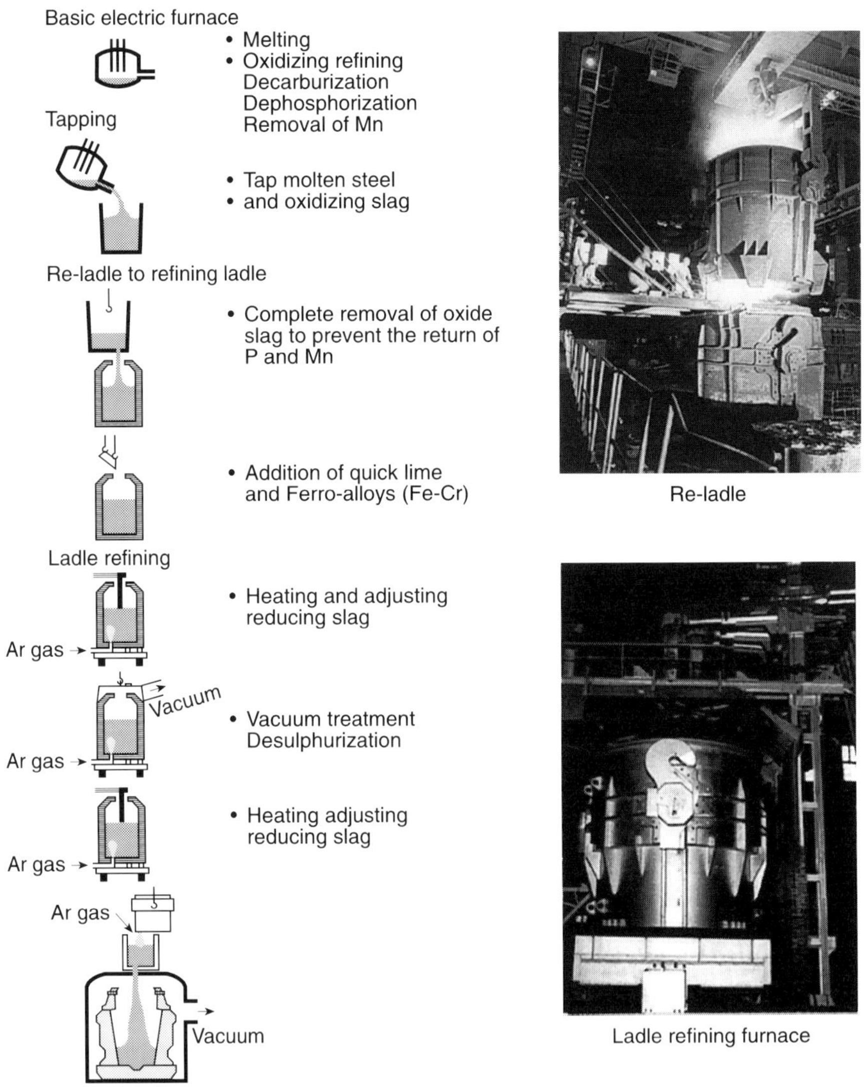

Fig. 1 Steelmaking process of superclean steel

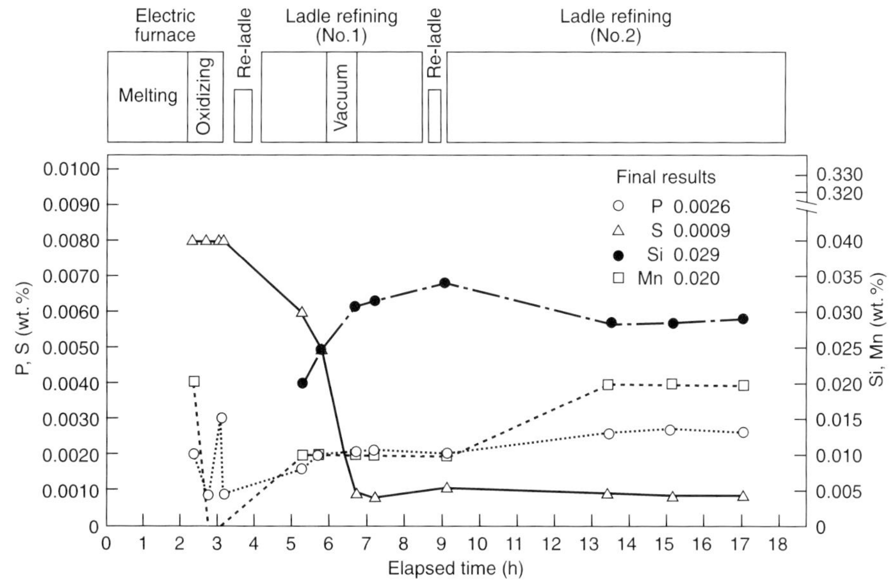

Fig. 2 Change of Si, Mn and impurity contents during refining of 600^T superclean steel

2.2 Production of 570 tonne ingot

Figure 3 shows the manufacturing sequence of the 570 tonne ingot. After the oxidising, refining in a basic electric furnace, the molten steel is reladled to LRF. Six (6) ladles were used to produce a 570 tonne ingot. The refining process in each LRF is the same as the process which is already reported.

2.3 Forging

Figure 4 shows the dimensions of a superclean monoblock LP rotor shaft forging and the forging sequence for the monoblock rotor shaft with a 2705 mm outer diameter at its body finish forging stage. The forging process was similar to that for a conventional large monoblock rotor forging. Numerical analysis by a non-linear finite element method makes it possible to predict the plastic deformation of the material during the forging operation. By applying this method to the problems of axiymmetric upsetting of cylinders with a void at various locations, the factors which dominates the porosity consolidation have been made clear, and the method has made it possible to evaluate the hot forging process in terms of void closure.[4] For the complete consolidation of voids, some new devices were added to the forging sequence during forging operation.

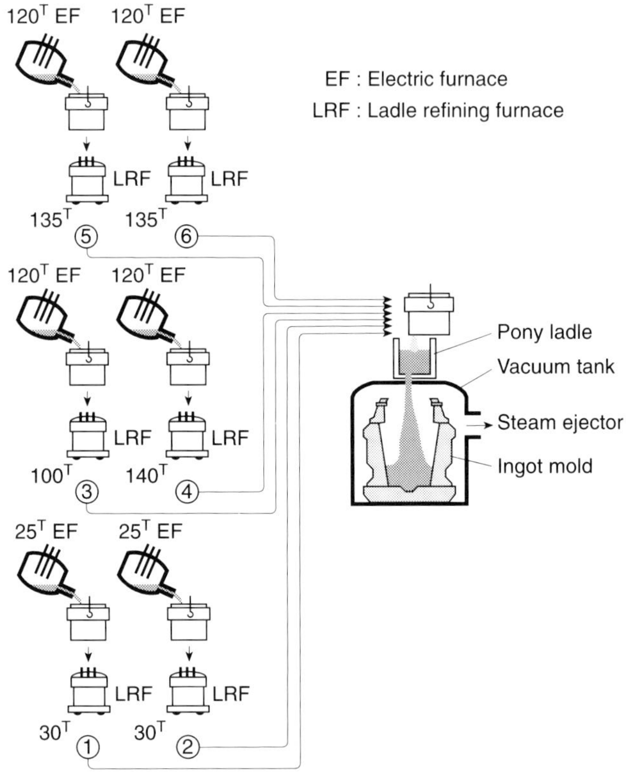

Fig. 3 Manufacturing sequence of superclean 570^T ingot for LP rotor shaft forging

2.4 Heat Treatment

In order to assure the reliability of large forgings, flaw detectability is one of the primary concerns. The heat treatment cycle for monoblock LP rotor shaft forgings was developed to obtain, by overcoming the mass effect of large diameters, the distribution of uniform fine grains. This was achieved by the repetition of normalising. In addition to this, optimisation of normalising temperatures is important to encourage the grain refining.[5] The superclean monoblock rotor shaft forging was subjected to three fold normalising before the quenching heat treatment. Figure 5 shows the minimum detectable flaw size by ultrasonic tests. The minimum detectable flaw size is equivalent to those of conventional ultralarge monoblock rotor shaft forgings. Figure 6 shows the appearance of a manufactured superclean monoblock LP rotor shaft forging.

3. EVALUATION OF TEST RESULTS

The superclean monoblock rotor shaft forging manufactured, showed a

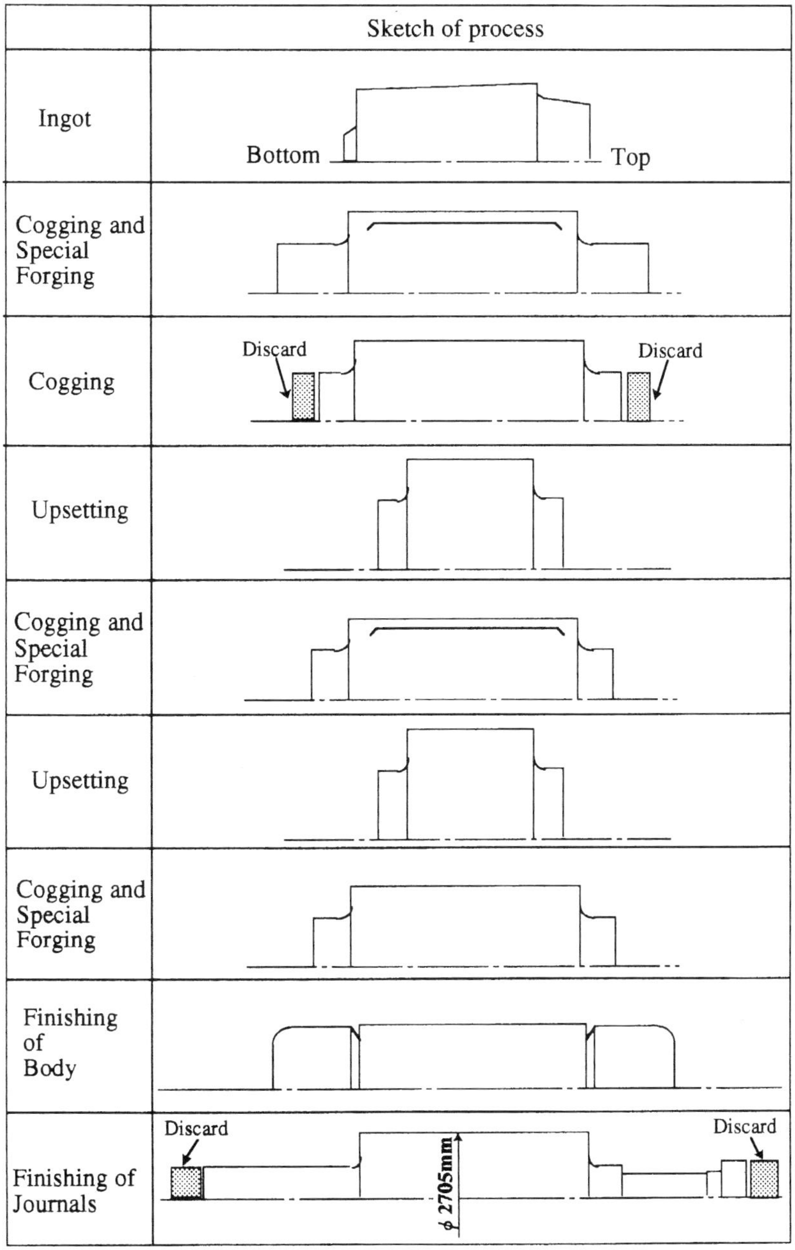

Fig. 4 Forging process of superclean monoblock LP rotor forging

sound internal quality by non-destructive examination and sufficient mechanical properties comparable to those of conventional monoblock rotor shaft forging.

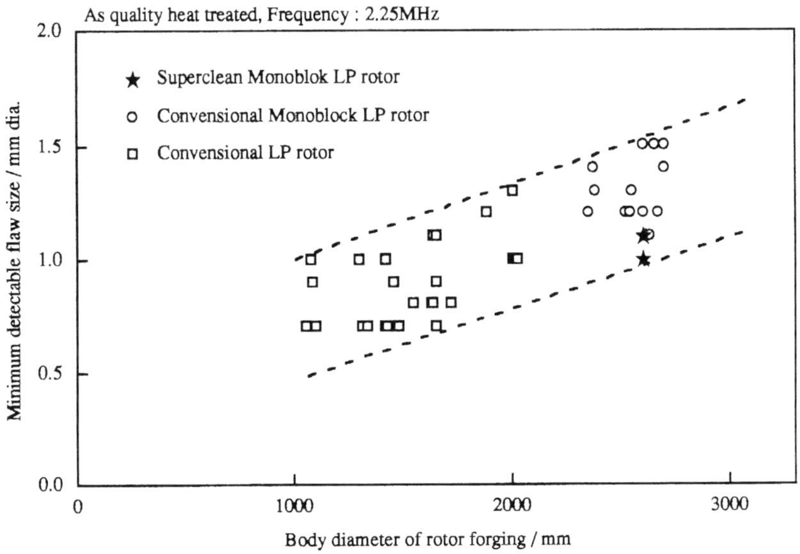

Fig. 5 Minimum detectable flaw size in ultrasonic test

Fig. 6 Appearance of superclean monoblock LP rotor shaft forging

3.1 Chemistry and Characteristics of Temper Embrittlement

Table 3 shows the chemistry of a superclean monoblock LP rotor shaft forging made from a 570 tonne ingot. All the chemistries of the six (6) ladle meet the required impurity content.

Figure 7 shows the distribution of elements at several locations of the forging. Uniform chemical distribution in the forging and low impurity content are confirmed in this figure. Figure 8 shows the history of impurity contents for each rotor forging. It should be noted that the impurity contents

Table 3 Chemistry of superclean monoblock LP rotor shaft forging

Ladle	Weight (ton)	wt.%												
		C	Si	Mn	P	S	Ni	Cr	Cu	Mo	V	As	Sn	Sb
①	30	0.36	0.014	0.030	0.0020	0.0008	3.55	1.60	0.03	0.50	0.10	0.004	0.003	0.0011
②	30	0.35	0.024	0.023	0.0025	0.0014	3.55	1.69	0.04	0.49	0.10	0.004	0.003	0.0010
③	100	0.32	0.014	0.033	0.0032	0.0012	3.59	1.64	0.03	0.49	0.10	0.004	0.003	0.0009
④	140	0.28	0.048	0.034	0.0024	0.0011	3.56	1.65	0.04	0.45	0.10	0.003	0.003	0.0008
⑤	135	0.20	0.018	0.022	0.0029	0.0012	3.58	1.71	0.03	0.40	0.10	0.003	0.003	0.0010
⑥	135	0.10	0.021	0.023	0.0029	0.0020	3.56	1.65	0.03	0.34	0.10	0.003	0.003	0.0011
Total	570 tonnes	0.23	0.025	0.028	0.0028	0.0014	3.57	1.66	0.03	0.42	0.10	0.003	0.003	0.0010
	Aim	0.23 0.25	Max. 0.03	Max. 0.03	Max. 0.003	Max. 0.0015	3.55 3.65	1.60 1.70	Max. 0.05	0.35 0.45	0.09 0.11	Max. 0.006	Max. 0.005	Max. 0.0015

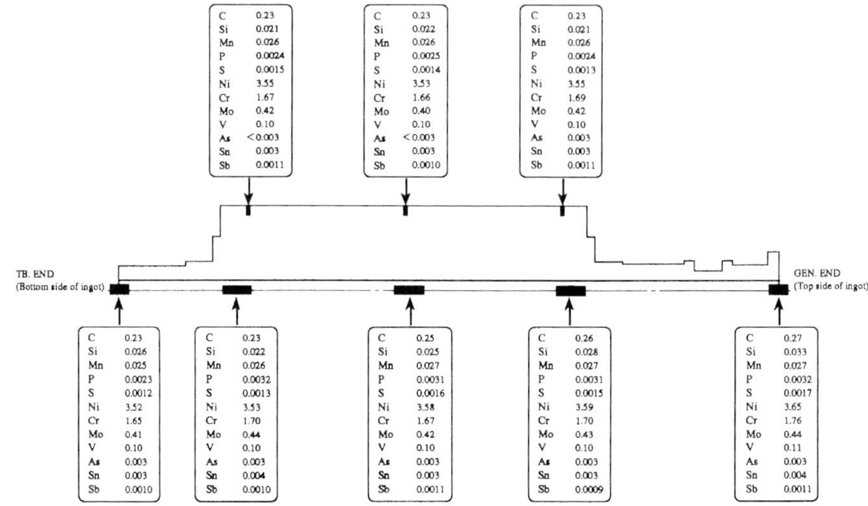

Fig. 7 Distribution of elements in the superclean monoblock LP rotor forging/ wt.%

are equivalent to the superclean rotor forgings produced from smaller ingots and these values meet the specification of superclean steels by EPRI.

Such low Si, Mn, and impurity contents are expected to lower the susceptibility to temper embrittlement of 3.5%NiCrMoV steel and to maintain the higher toughness level, while the LP rotor shaft is in operation for a long time in a fossil fired power plant. The J factor $\{J = (Si + Mn)(P + Sn) \times 10^4\}$ is widely used as an index of temper embrittlement susceptibility for low alloy steel.[6] This superclean monoblock rotor shaft forging demonstrates a very low J-factor of 3.1 to 3.8 as shown in Table 3. Figure 9 shows the relationship between J-factor and ΔFATT of conventional LP rotor shafts forging including those with superclean chemistry.[6] ΔFATT means the difference in fracture appearance transition temperature obtained by Charpy impact tests on test specimens taken from LP rotor shaft forging in between water quenched and step cooled conditions. Figure 9 indicates that ΔFATT decreases (fewer temper embrittlements), with the decrease of J-factor. According to this figure, it is estimated that the ΔFATT of the superclean monoblock rotor shaft forging will be very small, owing to its low J-factor of 3.1 to 3.8 and be immune to temper embrittlement.

3.2 *Strength and Toughness*

Figure 10 shows the mechanical and impact properties of the centre and rim positions of the monoblock rotor forging. Homogeneous distribution of the mechanical properties are achieved in the whole forging. Especially, excellent toughness is obtained even in the centre core bar of the body in spite of the

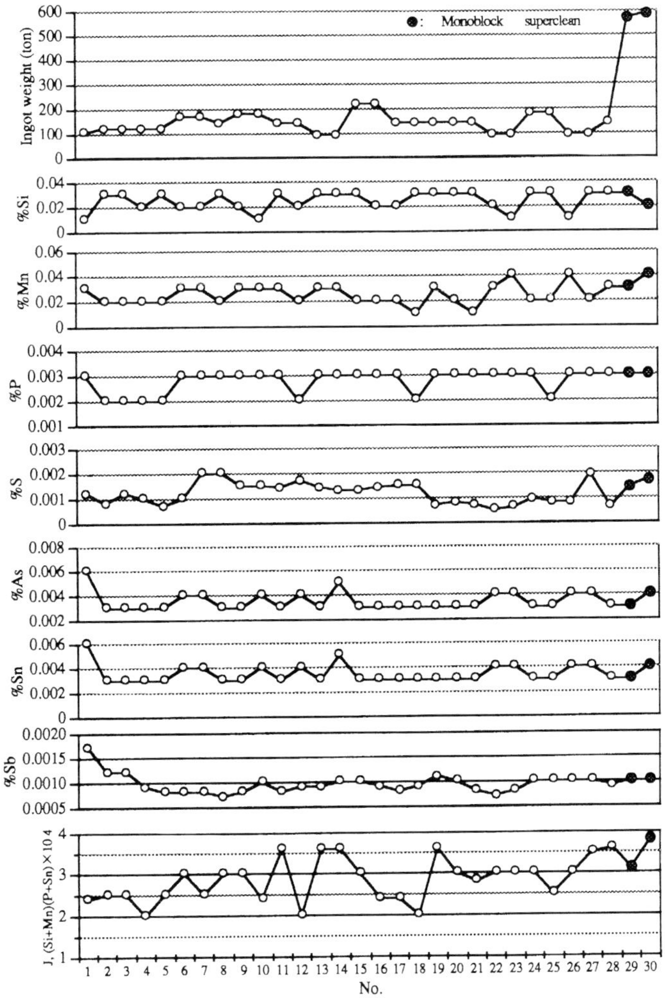

Fig. 8 Level of impurities and J-factor of superclean LP rotor forgings

large body diameter. The relationships between tensile strength and V-Charpy impact properties are also plotted. Figure 11 shows relationships between impact properties (FATT and absorbed energy) and the tensile strength at the body centre of the superclean monoblock LP rotor shaft forging in comparison with those of conventional monoblock shafts. The tensile strength versus impact properties is equivalent to those obtained in a monoblock LP rotor shaft forging. These qualities are the fruits of the careful selection of steel chemistries, advanced hot forging technology and well designed heat treatments.

The superclean LP rotor steels were reported to exhibit good creep rupture strengths which are equivalent or superior to those of conventional LP rotor

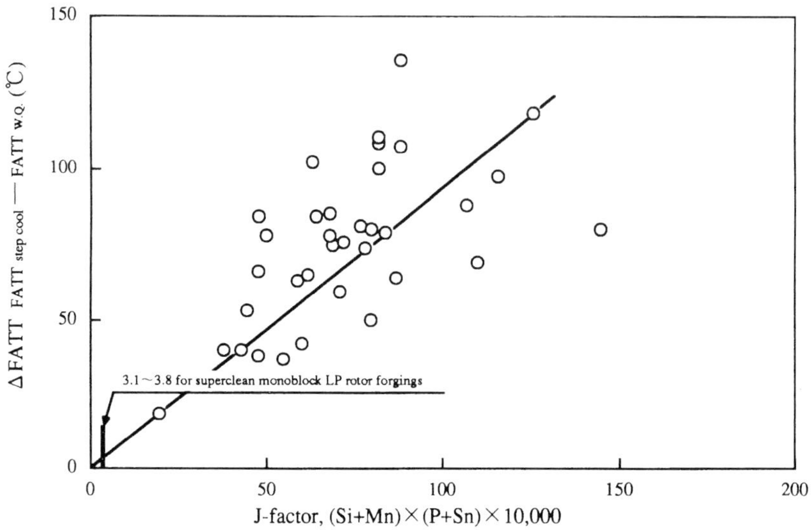

Fig. 9 Relationship between J-factor and ΔFATT of 3.5%NiCrMoV LP rotor steel forgings[(6)]
(Embrittlement test results by step cooling)

steels.[7] Such better creep rupture properties are thought to be caused by the effect of freedom from MnS. The superclean monoblock LP rotor shaft manufactured, is expected to exhibit high creep rupture strength at operation temperatures in addition to its higher resistance against temper embrittlement.

4. CONCLUSION

A monoblock LP rotor forging was manufactured successfully from 570 tonne ingot of superclean version 3.5%NiCrMoV steel. The mechanical and impact properties are equivalent to those of conventional ultra large monoblock LP rotor forgings. The evaluation tests conducted by using smaller size forgings revealed low sensitivity to temper embrittlement, superior creep rupture properties and some advantage in stress corrosion cracking and/or corrosion fatigue resistance of the superclean steels. These test results obtained so far suggests that the present superclean monoblock rotor shaft material has better reliability for long time operation. Such ultra large superclean monoblock LP rotor shaft forgings are applicable to 1000 MW class fossil power plants and will undoubtedly contribute to the efficiency of power plants.

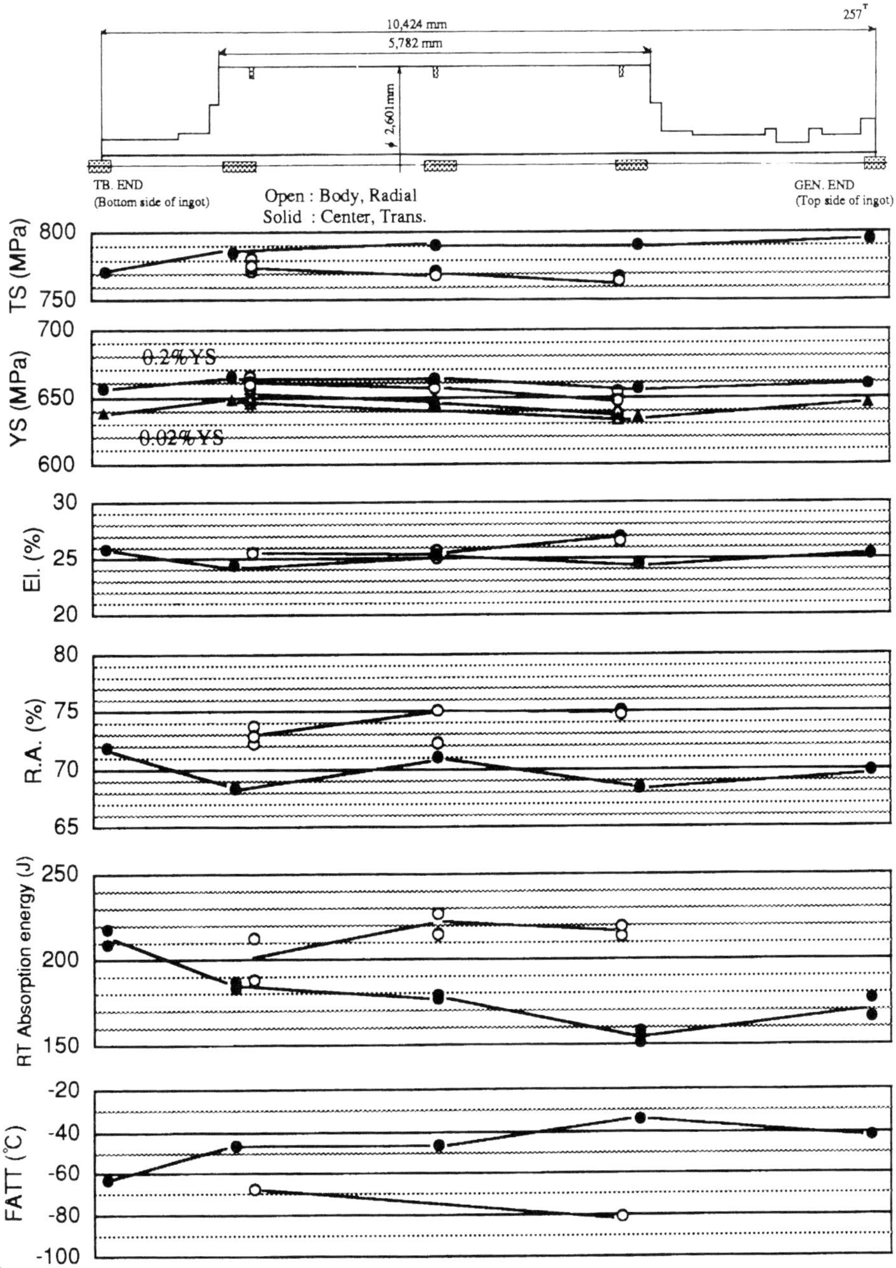

Fig. 10 Mechanical and impact properties in the superclean monoblock LP rotor shaft

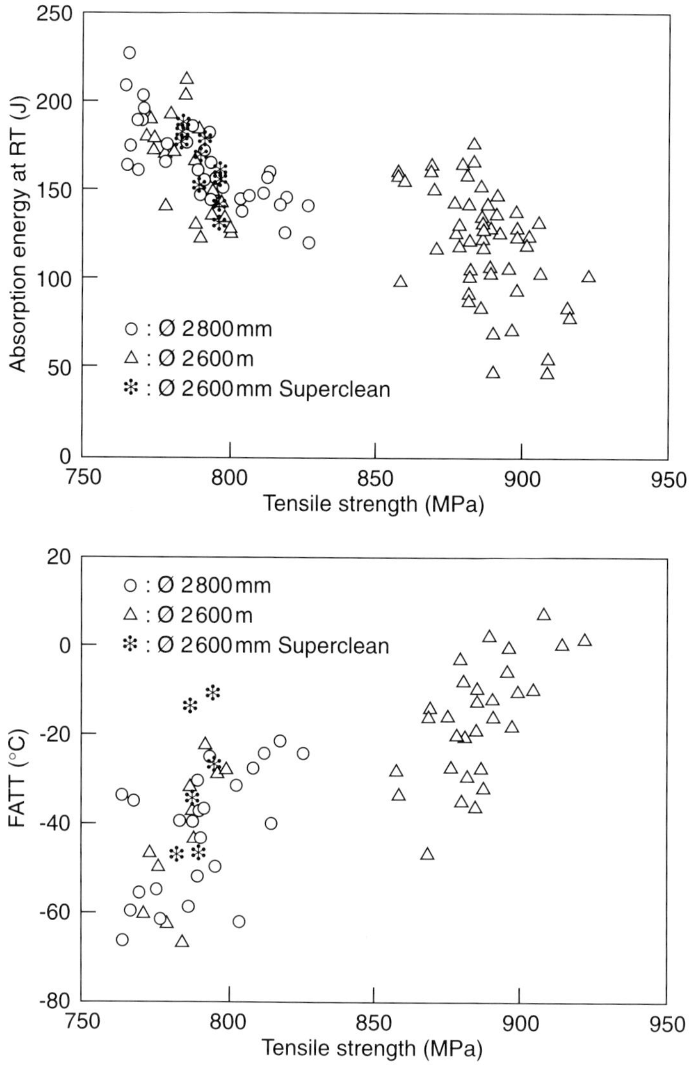

Fig. 11 Transverse toughness at body centre of monoblock LP rotor shaft forgings

REFERENCES

1. T. Ohhashi, R. L. Bodnar *et al: EPRI Report NP-5339*, September, 1987.
2. R. I. Jaffee *et al*: *Trans. ISS*, February, 1989, P. I & SM-45.
3. O. Watanabe *et al: The Thermal and Nuclear Power*, 1988, **39**, 607.
4. M. Tanaka *et al*: *Proc. North American Metal Working Research Conf.*, Hamilton, 1973.

5. Y. Tanaka *et al*: *Tetsu-to-Hagane*, 1990, **73**, 783.
6. J. Watanabe and Y. Murakami: *Proc. 46th API Midyear Meeting*, Chicago, 1981, 216.
7. R. I. Jaffee *et al*: *Proc. EPRI Superclean Review*, Sapporo, 1989.

Development of Production Technology and Manufacturing Experiences with Superclean 3.5 NiCrMoV Steels

W. MEYER, R. BAUER, G. ZEILER
Bohler Edelstahl GmbH, P.O. Box 96, A-8605 Kapfenberg, Austria

1. INTRODUCTION

Investigations carried out by several institutions under Projects RP559, RP2060, RP1403, RP2741 for the Electric Power Research Institute, Palo Alto, CA, resulted in a 3.5% NiCrMoV steel completely free from temper embrittlement as a material for low pressure rotors but also for combination property rotors, for example, integral high-pressure low-pressure (HILP) rotors for future steam plants. This steel contains very low percentages of residual elements including sulphur and phosphorus, and it's contents of aluminum, silicon, and manganese are also very low.

2. MELTING AND REFINING PROCESS

The challenge for the metallurgist melting such a steel is to deoxidise the melt without using deoxidisers such as aluminum, silicon or manganese, to maintain the required low contents in this element on the one hand and to obtain a high cleanliness with respect to non-metallics on the other hand. Good deoxidation is also a precondition for successful desulphurisation. Böhler solved these problems by applying a ladle metallurgical process based on diffusion deoxidation. For this process it is necessary to provide a slag with an extremely low iron oxide activity for the promotion of deoxidation. Furthermore the slag has also to maintain a low alumina activity to avoid aluminum pick up. After several experimental melts a lime alumina slag met the above described requirements.

Figure 1 shows the slag composition and the respective activities of alumina and iron oxide. Our experimental results are in accordance with results from thermodynamic calculations published by Gaye and Coulombet.[1]

This is demonstrated by Fig. 2. This graph also shows the necessity of providing a system with low alumina activity.[2]

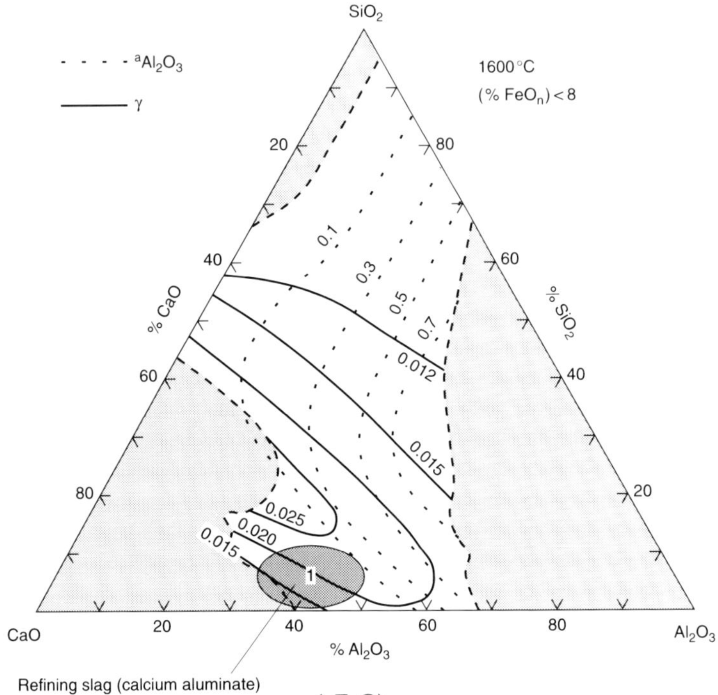

Fig. 1 The activity coefficient $\gamma = \dfrac{(aFeO)}{(\%FeO_n)}$ and the alumina activity $a_{Al_2O_3}$ in the system $CaO\text{-}Al_2O_3\text{-}SiO_2$

Two rotors were produced for EPRI by this method. An extensive test programme was carried out with the material from both forgings.[3,4] The results of step cooling and long-time activity test proved a complete resistance to temper embrittlement. This was the beginning of superclean production on industrial scale.

Meanwhile it has also become usual to make superclean steel by killing the melt with aluminum and removing the remaining aluminum content by controlled oxidation. We found that this method has disadvantages with respect to cleanliness and toughness of the produced forging.

3. PRODUCTION OF TURBINE DISCS USING THE BÖHLER ELECTRO-SLAG-TOPPING-PROCESS

Further development of this steel, sponsored by EPRI put Böhler into the position to apply the new material also for discs preferably for application up to 450°C. For the production of small discs Böhler developed a special

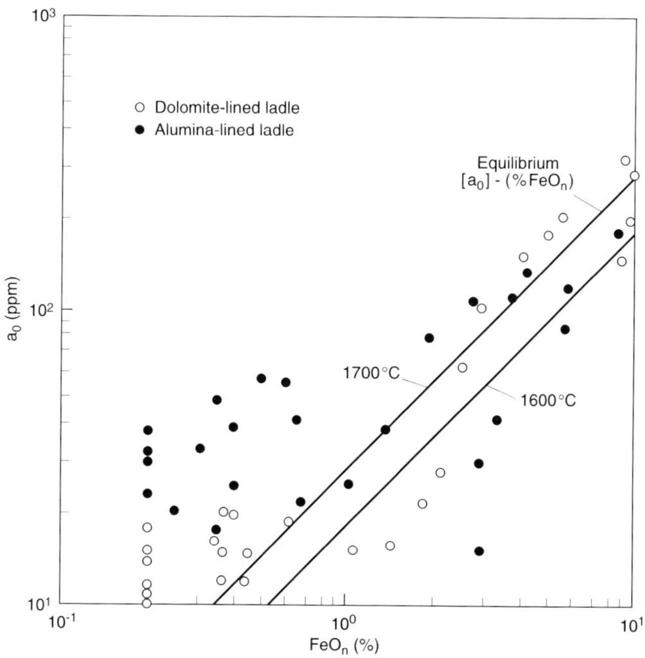

Fig. 2 Relationship between the FeO_n-content of the slag and the oxygen-activity of the melting bath

production route by applying the electro slag topping process on the solidification of superclean steel.

The principle of the Böhler Electro Slag Hot Topping process for heavy forging ingots is demonstrated in Fig. 3. A water cooled ring is placed at the top end of a forging ingot mould. The mould is filled with liquid steel by bottom pouring and the steel is covered with a metallurgically active slag similar to the ESR process. A consumable electrode is immersed in the slag, electrical energy is applied to the system, the slag is heated up and the electrode begins to melt. Droplets of the melting electrode are moving in the liquid pool of the solidifying ingot thus compensating the shrinkage of the solidifying ingot and influencing the solidification process.[5]

Ingots produced by this process are totally free from the v-segregations and shrinkage holes as can be seen from the autoradiograph of a 36 t ingot (Fig. 4). This autoradiograph demonstrates the solidification line after one, six, eight and eleven hours of solidification time. It can be seen that during the whole solidification time the liquid pool is open at the top so that non-metallics and impurities can be floated out even until the end of the process. Therefore ingots produced by this method have high cleanliness and especially no clustering of non metallic inclusions in the ingot centre.

Figure 5 shows the chemical composition at different test positions over the

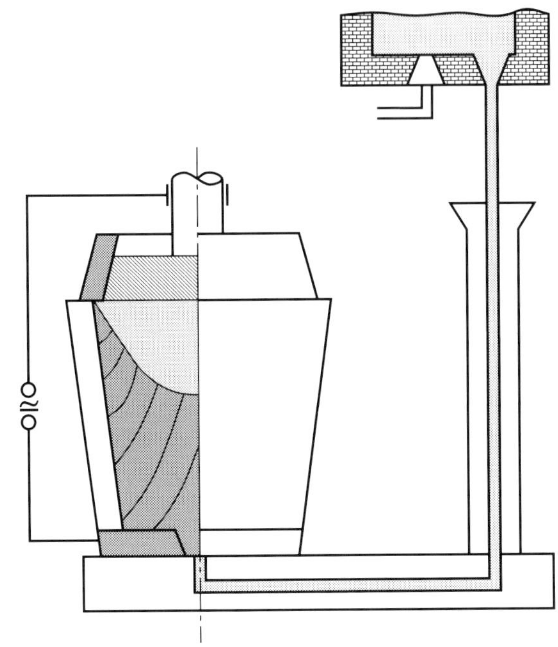

Fig. 3 Böhler electro slag topping (principle)

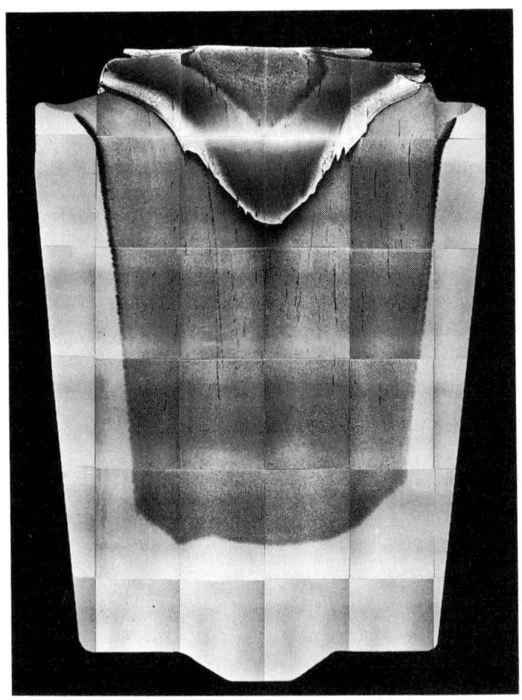

Fig. 4 Solidification of a 36t BEST Ingot demonstrated by Autoradiography

Fig. 5 Check analysis in different test positions
Heat R71184

	Test position and wt%										Heat Analysis
	T	TR	TM	TMR	M	MR	MB	MBR	B	BR	
C	0.31	0.31	0.31	0.31	0.31	0.31	0.30	0.31	0.30	0.31	0.30
Si	0.021	0.030	0.022	0.030	0.018	0.019	0.026	0.031	0.023	0.030	0.030
Mn	0.019	0.021	0.020	0.021	0.019	0.017	0.020	0.020	0.021	0.021	0.020
P	0.004	0.003	0.005	0.003	0.004	0.004	0.003	0.003	0.003	0.003	0.003
S	0.004	0.003	0.004	0.003	0.004	0.005	0.003	0.003	0.003	0.002	0.002
Cr	1.45	1.46	1.46	1.46	1.47	1.43	1.45	1.46	1.46	1.47	1.47
Mo	0.34	0.34	0.35	0.34	0.35	0.34	0.34	0.33	0.34	0.34	0.34
Ni	3.44	3.49	3.45	3.48	3.44	3.45	3.44	3.44	3.44	3.47	3.49
V	0.09	0.09	0.09	0.09	0.09	0.09	0.09	0.09	0.09	0.09	0.09
Al	<0.005	0.006	<0.005	0.006	<0.005	<0.005	0.006	0.006	0.006	0.006	0.005
As	<0.005	0.005	0.005	0.005	0.005	0.005	0.005	0.005	0.005	0.005	0.006
Sn	0.004	0.006	0.004	0.005	0.003	0.005	0.004	<0.003	0.004	0.005	0.005
Sb	<0.003	<0.003	<0.003	<0.003	<0.003	0.003	0.003	<0.003	<0.003	<0.003	<0.003

T = Top; M = Middle; B = Bottom; R = Rim

whole ingot. It is remarkable that neither the middle bottom position shows a negative segregation nor the middle top position a positive one. The chemical composition is very homogeneous.

The outstanding quality of the ingots manufactured according to this process offers the possibility to produce small discs as slices cut from forged bars. This offers also economic advantages. For our current production ingots of approximately 50 tonnes are used. The forging process is carried out on the 4500 tonne hydraulic press with repressing and cogging to rough forged bars of 960 mm in diameter. After normalising, premachining, spray-quenching and tempering to the required strength level the longitudinal bore is machined. The forged bars are sliced and the resulting discs are machined to the finished dimensions.

4. ACHIEVED MECHANICAL PROPERTIES AND MICROSEGREGATION BEHAVIOUR

To determine the mechanical properties and the chemical composition as well as the microsegregation behaviour over the whole bar, test samples from the drilling core in transversal direction and different positions according to Fig. 6 were taken. The microsegregation behaviour was evaluated by means of

TURBINE DISCS
3.5% NiCrMoV-superclean steel

Ingot weight: 37.700 kg (for 34 discs)

Disc–diameter: 750 to 930 mm

Quenched: Hardening temperature 850 °C (1560 °F)
Tempering temperature 630 °C (1166 °F)

Test positions related to the rough–forged ingot:

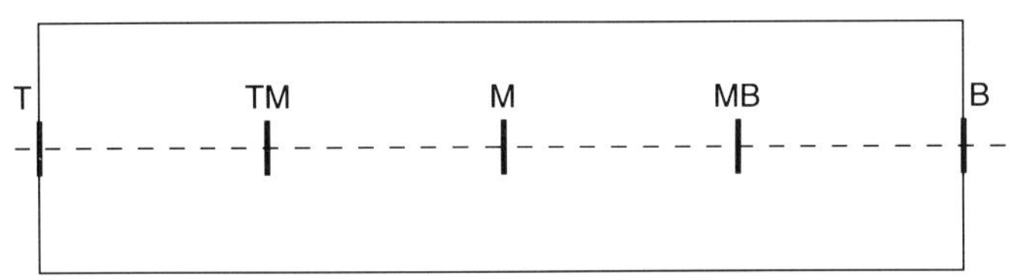

Fig. 6 Turbine discs cut from forged bars

microprobe. Figure 7 demonstrates a very low degree of segregation especially for the elements nickel and chromium. The insignificantly increased segregation rate of molybdenum corresponds to the higher dendrite arm spacing caused by a slower solidification towards the top area of the ingot.

This corresponds also to the increasing impact values from top to bottom as shown in Fig. 8. This graph gives the mean value and the scatter band of

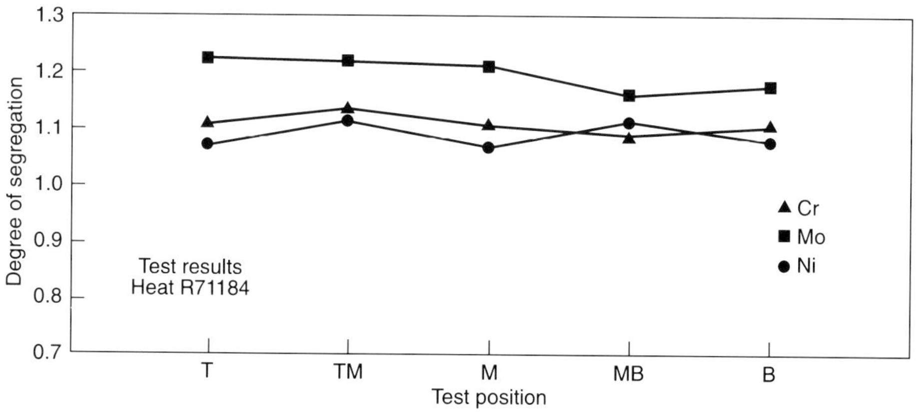

Fig. 7 Segregation behaviour (centre of bar/ingot)

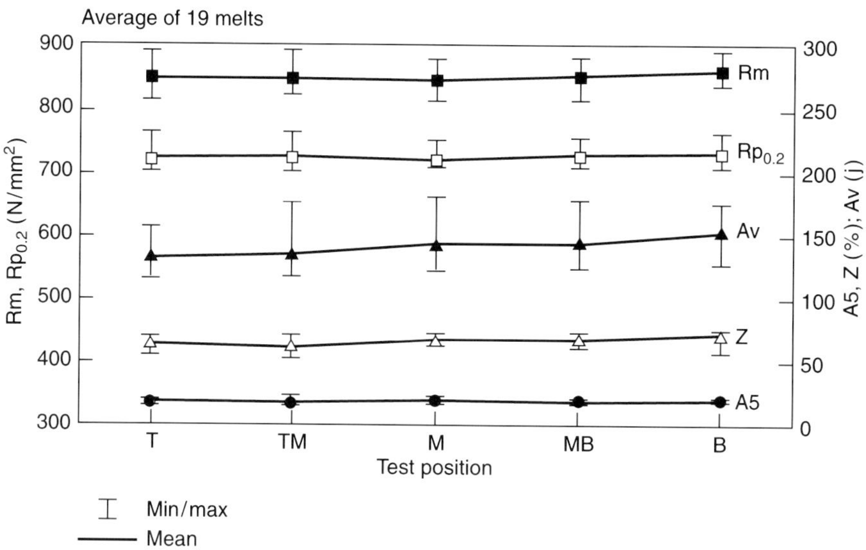

Fig. 8 Mechanical properties in the centre of the bars (transversal)

values obtained from tests of 19 heats. All other properties are very uniform over the whole lengths of the bar.

Figure 9 gives the scatter band of the FATT of the respectively 19 melts. It is apparent that FATT is lower at the bottom position. This may originate from the higher cooling rate and the resulting finer solidification structure.

Turbine discs with diameters larger than 1100 mm are forged in a conventional way. Figure 10 gives an overview over the production range, heat treatment and resulting mechanical properties for two strength levels. These reported properties are an average of values obtained from 20 discs of each strength level.

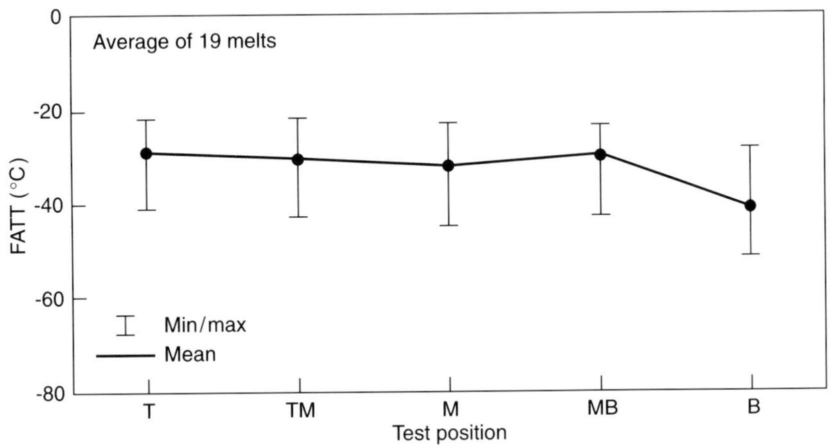

Fig. 9 Mechanical properties, FATT

- Ingot weight: up to 47.000 kg
  Disc dimentions: Ø 1500 x 470…1760 x 320 mm
  Hardening: 850 °C
  Tempering: 620/640 °C

- Test results: room temperature tensile test
  impact test
  FATT
  Specimen location: tangential, about mid radius

- Strength-level I:

	Rp0.2 (MPa)	Rm (MPa)	A5 (%)	Z (%)	Av (j)	FATT (°C) Center, rad
x	732	824	23	72	184	-70+5

- Strength-level II:

	Rp0.2	Rm	A5	Z	Av	FATT (°C)
x	859	946	19.5	67	156	-30+10

x…average of 20 discs (each strength level)

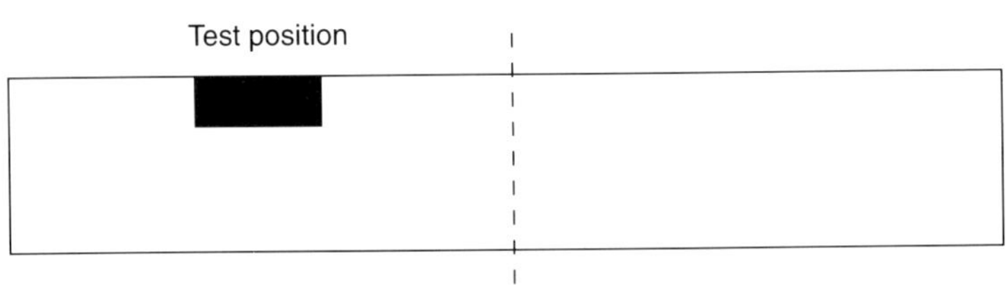

Fig. 10 Turbine discs, individually forged

Meanwhile more than 70 heats for discs and rotor production were performed. Figure 11 gives the scatter band of the important residual elements for the last 60 heats. The resulting J-factor for each of these heats can be seen in Fig. 12. This graph demonstrates that especially the production of the last two years maintains in a very uniform manner a J-factor of five. This also shows that the production of superclean 3.5 Ni steel has become proven

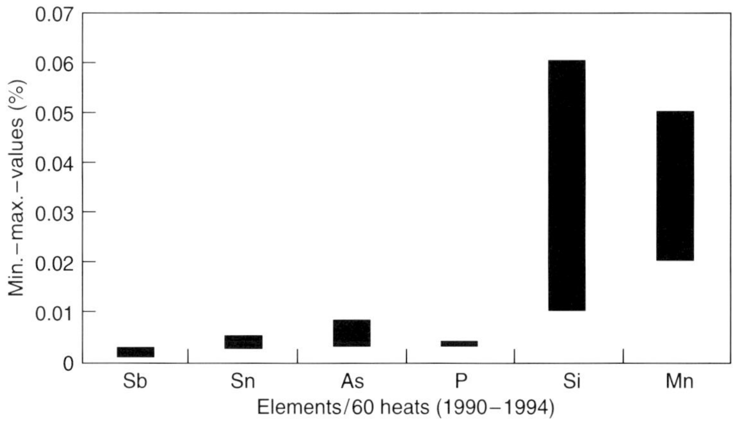

Fig. 11 Residual elements

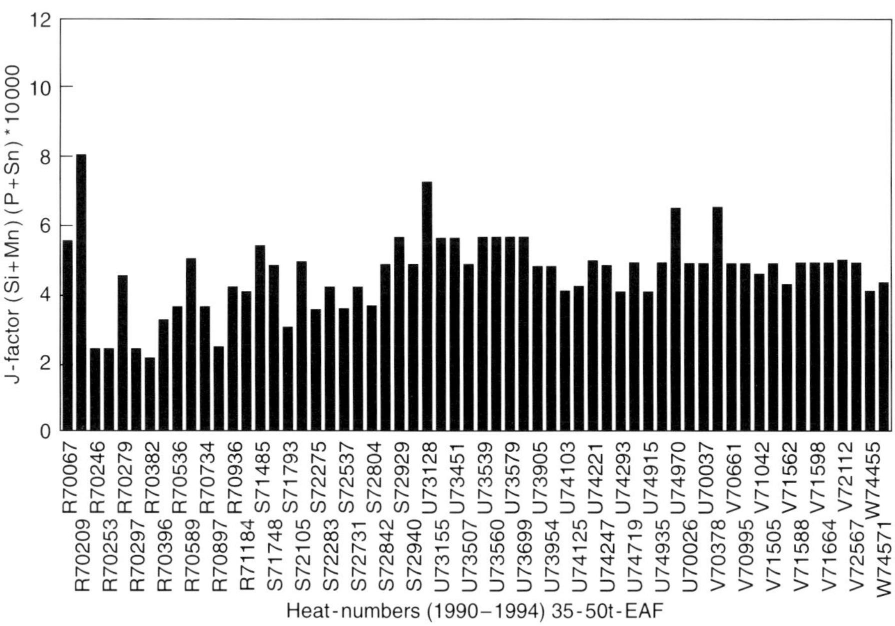

Fig. 12 J-Factor

industrial practice. This guarantees very uniform production conditions and material properties.

5. TRIAL TO PRODUCE 12% Cr STEEL IN SUPERCLEAN EXECUTION

The latest development of Böhler is aiming at a superclean 9–12% Cr steel. To obtain the desired low levels with respect to phosphorus, tin, antimony and arsenic contents it is necessary to apply the calcium carbide method.[6]

Figure 13 gives a scope of the production sequence of X20CrMoV12 1 1 steel

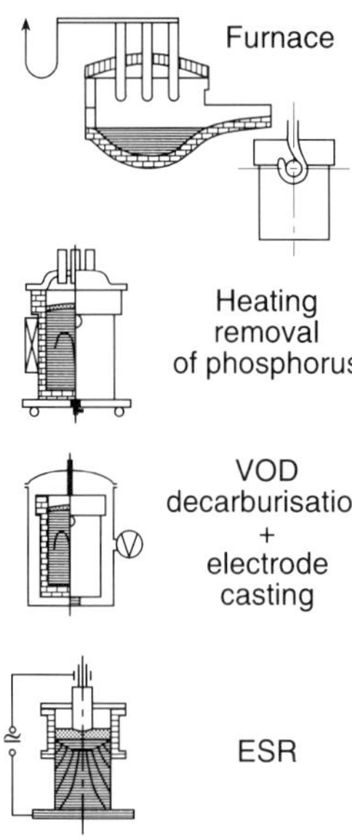

Chemical composition in % (after process completion)

	%C	%Si	%P	%S	%Al	%As	%Sn	%Sb
Furnace	0.47	0.10	0.013	0.024	n.d.	0.011	0.0150	0.006
Heating removal of phosphorus	1.47	0.16	0.003	0.001	n.d.	<0.001	0.0029	0.001
VOD decarburisation + electrode casting	0.22	0.14	0.005	0.002	0.005	<0.001	0.0029	0.001
ESR	0.22	0.05	0.005	0.001	0.005	<0.001	0.0040	0.0005

The contents of all other elements are identical to those of the BÖHLER T550 grade

Fig. 13 Phosphorous removal process in martensitic disc material

and the respective contents of residual elements. The effect of the method is evident.

The new method effects the FATT significantly as can be seen from Fig. 14.[7] This figure shows the comparison of the chemical composition and the resulting mechanical properties of a superclean heat and a heat produced by a conventional method. Long term ageing tests and creep rupture strength tests at 480 and 450°C respectively are in progress.

Because of the favourable segregation behaviour of superclean steels compared to conventional ones and the advantages of the BEST process it seems of interest to apply this process even on the solidification of superclean Cr steels which offers an additional economic advantage compared to the today applied ESR process.

Fig. 14 Trial of a superclean 12%Cr steel (Bö T550So)

Mechanical properties in comparison to conventional grade

			Results											
			Longitudinal						Transversal					
Heat No.	P content	Def. rate	TS (MPa)	0.2 YS (MPa)	Elong. (%)	RA (%)	Impact test (J)	FATT (°C)	TS (MPa)	0.2 YS (MPa)	Elong. (%)	RA (%)	Impact test (J)	FATT (°C)
V46710	0.005	6.3	920	756	19	54	54, 59, 67	11	921	765	18	51	42, 49, 50	12
V70370	0.021	4.4	941	779	16	55	35, 28, 28	28	934	771	15	49	24, 28, 24	31

Heat treatment: 1050°C/1h/OQ + 700°C/2h(M)/AC
(Laboratory)
Specimen location: 1/2 radius!

Chemical composition

	Elements [wt.%]														
Heat No.	C	Si	Mn	P	S	Cr	Mo	Ni	V	Al	Nb	N	As	Sb	Sn	Process
V46710	0.22	0.05	0.7	0.005	0.001	11.36	0.94	0.77	0.27	0.004	<0.01	0.029	<0.001	0.0005	0.004	ESR
V70370	0.21	0.31	0.72	0.021	0.001	11.32	0.94	0.68	0.26	0.025	0.014	0.038	0.008	0.005	0.01	EAF

6. CONCLUSION

It has been demonstrated that the development of the superclean technology in combination with the BEST process have led to the possibility to manufacture highly stressed gas turbine discs for temperatures up to 450°C in an economical way. By summarising the results it can be said that the achieved properties are very uniform over the whole length of the ingot. Applying this technology on the 12% Cr-steels it seems possible to manufacture turbine components in a more economical way than previously by at least the same level of properties.

REFERENCE LITERATURE

1. H. Gaye and D. Coulombet: *Rapport IRSID, RE 1064*, March 1984.
2. W. Meyer, A. Kucharz and G. Hochörtler: *1042 Residual and Unspecified Elements in Steel*, ASTM 8/P, 1986.
3. R. I. Jaffee, P. Machner, W. Meyer and J. E. Steiner: *Ironmaking Steelmaking*, 1986, **13**(6), 322–326.
4. P. Machner, E. Reisl, W. Meyer and R. I. Jaffee: *International Conference on Improved Coal Fired Power Plants*, Palo Alto, CA, 1986.
5. W. Meyer, P. Machner and R. Tarmann: *Proc. Conference on Clean Steel*, Balatonfüred, Hungary, 1986, 303.
6. K. Kitamura, *et al.*: *Transactions ISI, J*, 1984, **24**, 631–635.
7. F. W. Noble, B. A. Senior and B. L. Eyre: *Acta metall. mater*, 1990, **38**(5), 709–717.

Production of a HP/LP Rotor Forging of 2 CrMoV Steel

H. NAKASHIMA,* T. TSUCHIYAMA,*
M. MIYAKAWA*, M. OKAMURA,* P. MURPY† and
R. C. SCHWANT†

* *Kobe Steel Ltd., Takasago, Japan*

† *General Electric Company, Schnectady, New York*

ABSTRACT

The low Si–2Cr–1.7Mo–1.5Ni–0.2Mn–0.25V steel forging with high hardenability even for the rotor forging with over 1600 mm barrel diameter was developed and was applied to the combined high and low pressure rotor forging. An HP–LP rotor with a diameter of 1400 mm in LP section was made from ESR ingot.

The high homogenity in the rotor forging was observed due to ESR process.

Fracture appearance transition temperature (FATT) at the centre core in LP section was −10°C with a 0.02% proof strength of 63 kg mm^{-2}.

Creep rupture strength in high pressure section was comparable to that of a conventional 1Cr–1Mo–V high pressure rotor.

1. INTRODUCTION

Current trend indicates that a combined high pressure and low pressure turbine rotor shaft can be used for a large scale combined cycle turbine.

Because a 1Cr–1Mo–V steel forging has insufficient hardenability, the mechanical properties of the rotor forging with over 1600 mm barrel diameter are not adequate.

From the point of view, the 2Cr–1.7Mo–0.2Mn–V steel was developed[1] and the actual rotor forging was manufactured from conventional ingot.[2] Recently an actual HP–LP rotor forging was manufactured from ESR ingot.

This paper describes the manufacturing results of the actual HP–LP rotor forging made from 72 tons ESR ingot.

2. MANUFACTURING PROCESS OF THE HLP ROTOR FORGING

Figure 1 shows the outline of the manufacturing process of a combined HP–LP rotor forging. The ingot was vacuum carbon deoxidated and cast into the ingot. The ingot was electro-slag-remelted (ESR). After ESR, the ESR ingot was forged by using the 13 000 ton hydraulic press and the preliminary heat

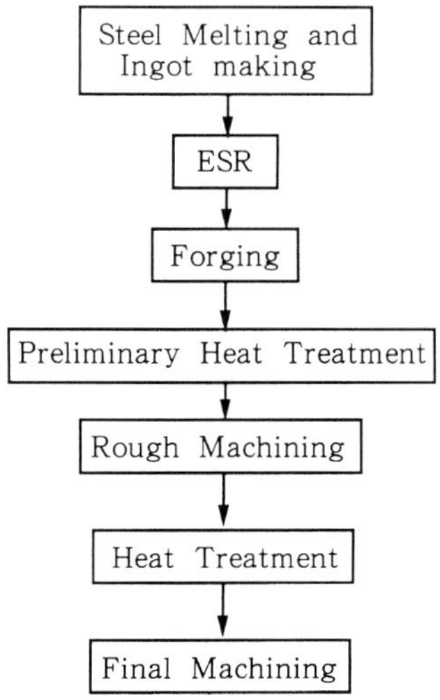

Fig. 1 Manufacturing Process

treatment was performed. Prior to quality heat treatment, the rotor forging was rough machined. As for the quality heat treatment, the differential heat treatment was applied in a vertical electric furnace as shown in Fig. 2. The LP section was austenitised at 910°C and water-quenched to get higher toughness, while the HP section was austenitised at 940°C and then forced air cooled to get higher creep rupture strength. The tempering temperatures

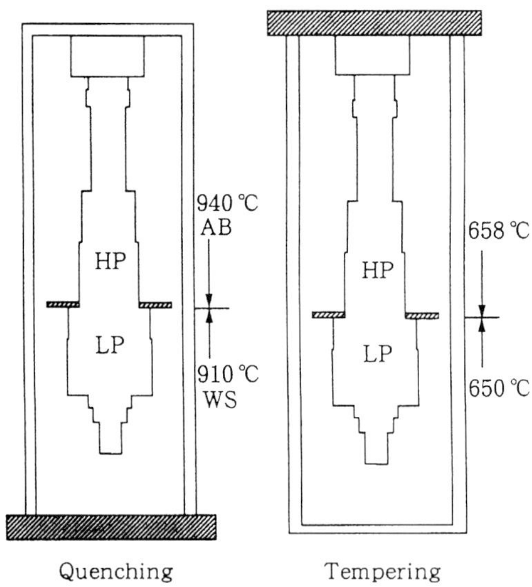

Fig. 2 Differential Heat Treatment

were 650°C for the LP section and 658°C for the HP section to get the 0.02% PS of 63 kg mm^{-2}.

Figure 3 shows the configuration of the rotor forging at the quality heat treatment operation. The diameter of the forging were 1356 mm at LP section and 968 mm at HP section.

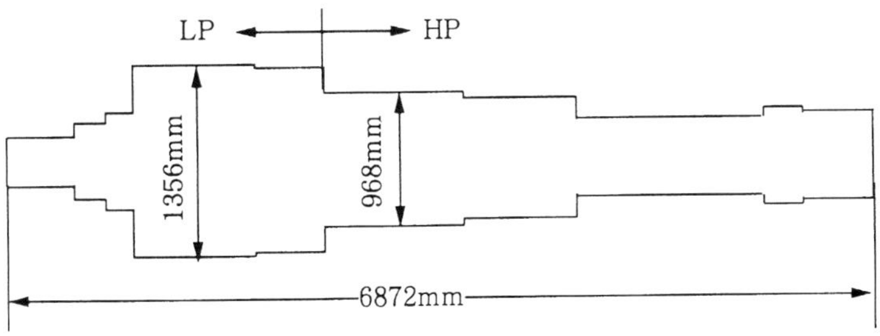

Fig. 3 Configuration of the Rotor

3. PROPERTIES OF THE ROTOR FORGING

3.1 *Chemistry and Microstructures*

Table 1 shows the ladle analysis of the rotor forging. Low impurity contents were achieved. Also, high homogenity was observed in the rotor forging as shown in Fig. 4.

Table 1 Ladle Analysis of the Rotor Forging (wt%)

C	Si	Mn	P	S	Ni	Cr	Mo	V	Al
0.23	0.01	0.19	0.002	0.001	1.46	2.04	1.65	0.25	0.002

Figure 5 shows the macrostructures for the rotor forging. Sound macrostructures were observed in the centre core. No ghosts were recognised.

Figure 6 shows the microstructures for the rotor forging. Fully bainitie microstructures were obtained in the centre core. The grain size were ASTM 6.5–7 in the LP section and ASTM 5.5–6 in the HP section. The fine grains were obtained at the forging.

3.2 *Mechanical Properties*

Figure 7 shows the mechanical properties of the rotor forging. The 0.02% proof stress (PS) and the tensile strength were approximately 63 and

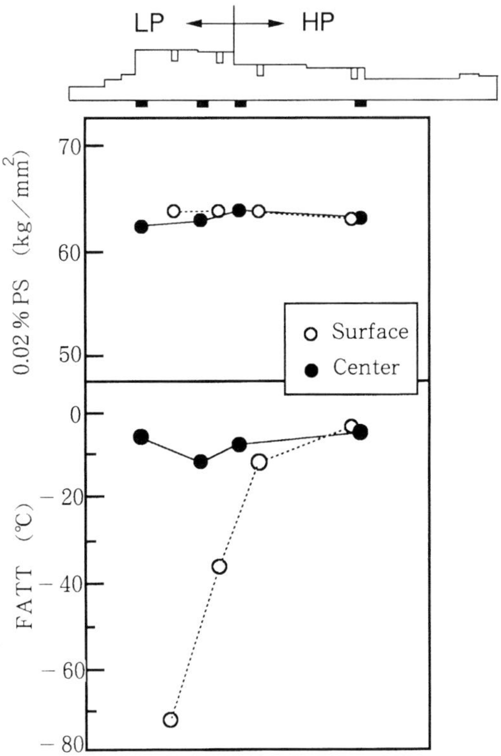

Fig. 4 Mechanical Properties of the HP/LP Rotor

80 kg mm^{-2}, respectively. No difference of the 0.02%PS and tensile strength with the positions were observed. The Fracture Appearance Transient Temperature (FATT) at the centre core of the forging were from −4 to −10°C. The FATT were −30 to −73°C at the surface of LP section. Good values were obtained in comparison with those of the conventional CrMoV rotor.

The fracture toughness test at the centre core of the rotor forging was conducted at room temperature. The J_{IC} test was carried out with 0.8 RCT test specimen. The J_{IC} values obtained were converted to the K_{IC}. The K_{IC} values was 800 kg mm$^{-3/2}$. It is thought that the fracture toughness is comparable with those of a high pressure turbine rotor.

3.3 Creep Rupture Properties

Creep rupture tests were carried out at temperatures of 550, 600 and 650°C. The specimens were taken from the surface of the HP section.

Figure 8 shows the creep rupture strength as a function of Larson-Miller parameter. It is concluded that creep rupture strength of the forging is

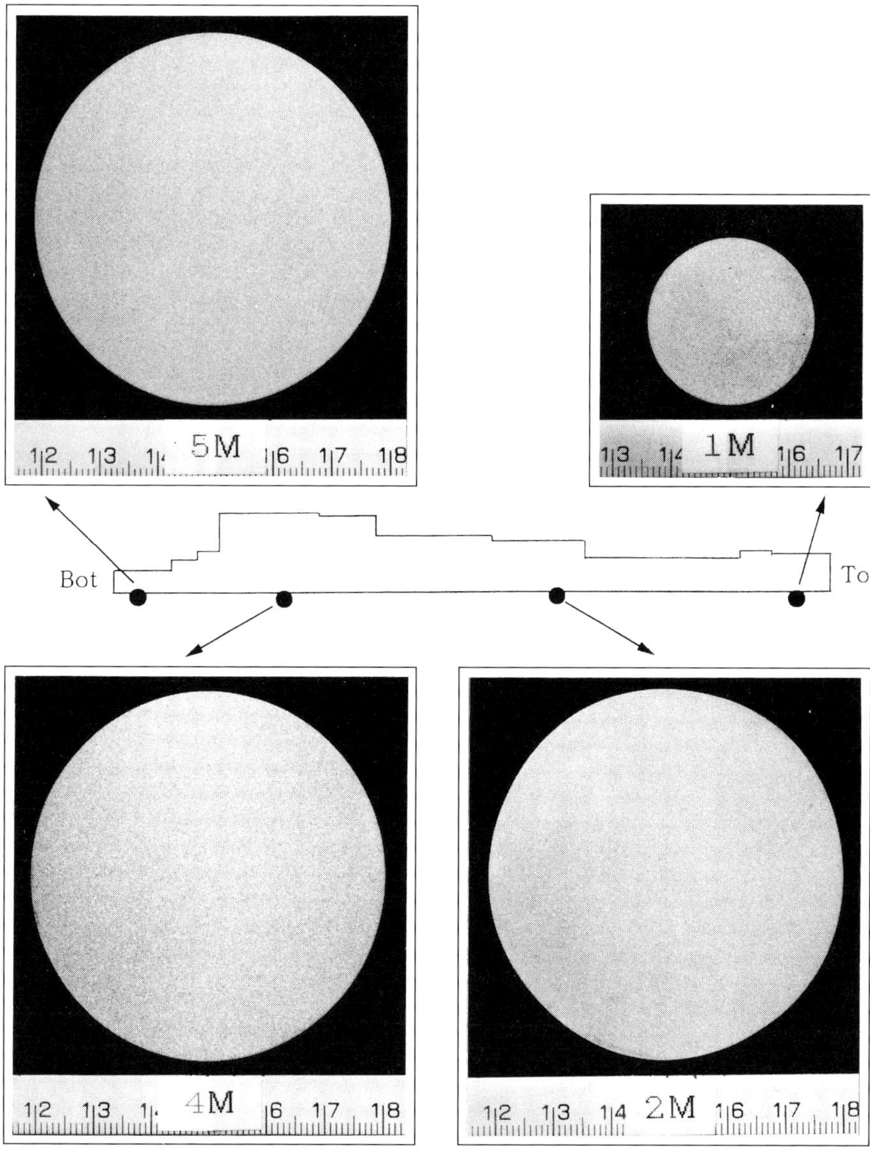

Fig. 5 Macro Structures of the Centre Core

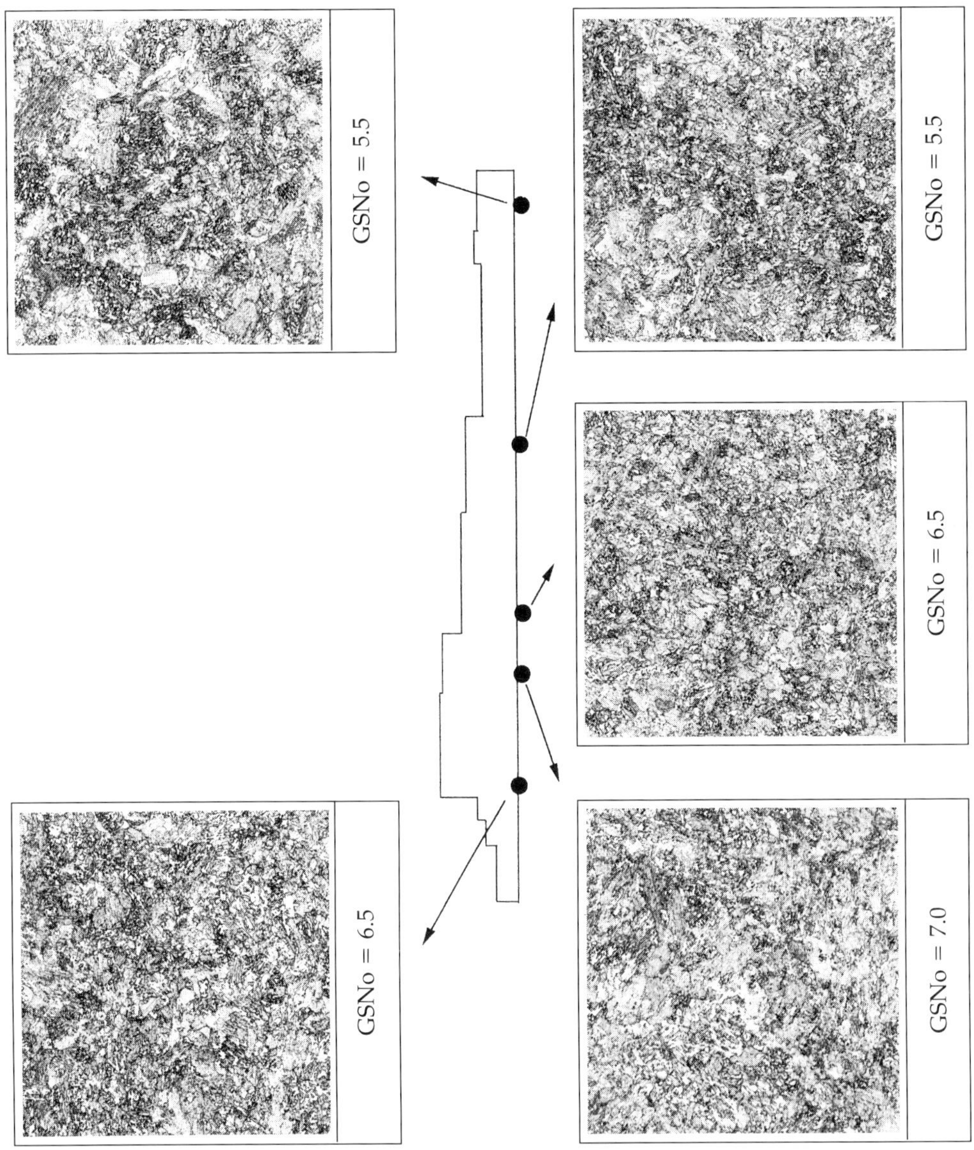

Fig. 6 Microstructures and Grain Size of the Centre Core

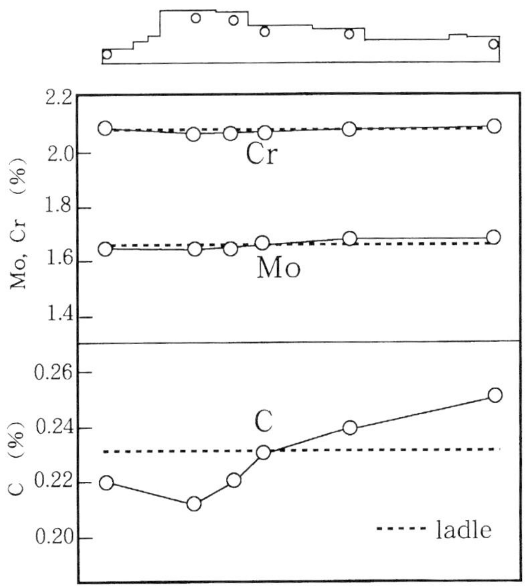

Fig. 7 Distribution of Chemical Composition

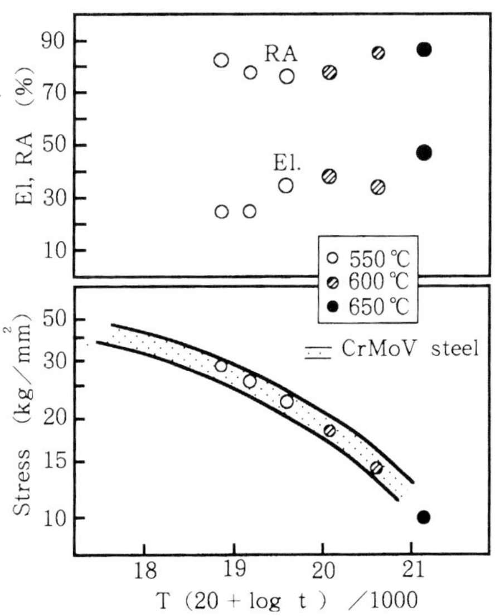

Fig. 8 Creep Rupture Properties

comparable to that of the CrMoV rotor. For the other hand, creep rupture ductility are also good.

4. CONCLUSION

1. The high and low pressure combined rotor forging with a diameter of 1356 mm was made from 72 tons ESR ingot of 2Cr–1.7Mo–1.5Ni–0.2Mn–V steel.
2. The sound macrostructures were observed in the centre core due to the ESR ingot.
3. The FATT in the centre core of the forging were -5 to -10°C with a 0.02% proof strength of 64 kg mm^{-2}.
4. Creep rupture strength of HP section is comparable to those of the conventional CrMoV rotor forging.

REFERENCES

1. T. Tsuchiyama, M. Kohno, M. Okamura and H. Ohizumi: *Materials for Combined Cycle Power Plant*, Sheffield, 1991.
2. T. Tsuchiyama, M. Miyakawa, M. Okamura, M. Morita, T. Yamashita and S. Nishida: *Clean Steel Technology*, Chicago, 1992.

A New Superclean High Pressure–Low Pressure Rotor Steel for Advanced Combined Cycle Power Plants

M. YAMADA,* Y. TSUDA,* M. MIYAZAKI,*
Y. TANAKA,† T. AZUMA,† and Y.IKEDA†
** Toshiba Corporation, Yokohama, Japan*
†The Japan Steel Works, Ltd., Muroran, Japan

ABSTRACT

A new steel, 2.25Cr–1.7Ni–MoVNbW steel has been developed for the high pressure (HP)–low pressure (LP) single cylinder steam turbine rotor and the producibility and the mechanical properties have been verified by manufacturing the trial rotor forging (70 ton ingot) using the vacuum carbon deoxidation (VCD) process. The rotor forging revealed material properties suitable for the advanced HP–LP single cylinder steam turbines. However, a slight chemical segregation was observed in the centre of the trial rotor forging. Then, as the next step, the increasing of Nb content and the application of a superclean process using the electroslag remelting (ESR) process were conducted in order to reduce the chemical segregation and improve the toughness. A trial ESR ingot (43 tonnes) was produced and the internal quality was evaluated. The chemical segregation was considerably reduced and the proof stress and the toughness were estimated to be improved by conducting the simulated heat treatment tests in the centre of LP portion. This newly developed steel forging will be produced in the near future for HP–LP single cylinder steam turbine rotors of the advanced combined cycle power plants.

1. INTRODUCTION

Steam turbines generally consist of 2 or 3 turbine cylinders (Fig. 1). However, when a rotor forging with a good combination of creep rupture strength, proof stress and toughness is available, the steam turbine can be designed as a single cylinder type, which makes the unit compact and simple. 1CrMoV steel rotor forgings currently used at high temperatures are excellent in creep rupture strength but poor in toughness, while 3.5NiCrMoV steel rotor forgings used at low temperature are excellent in toughness but poor in creep rupture strength. It is required for HP–LP single cylinder turbine rotor forgings to have both high creep rupture strength and good toughness which are mutually contradictory properties, as a rule. One of this type of turbine rotor steels is 1CrMoVNiNb steel.[1] However, it would be applicable only for the turbine rotors of 100 MW or a smaller unit with its last stage blades up to

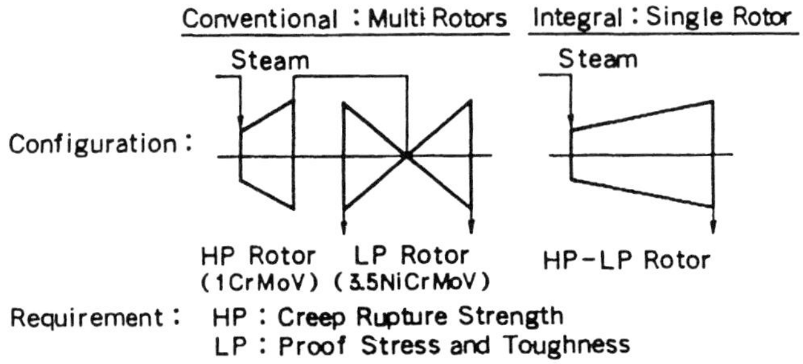

Fig. 1 Comparison of Conventional and Integral HP–LP Turbines

660 mm (26 inches) in length because of the limit of the proof stress and toughness of this steel.

When a turbine of 100 to 200 MW class is constructed as HP–LP single cylinder type, the length of the last stage blade might be of the 850 mm (33.5 inches) or 1070 mm (42 inches) class. Therefore, new HP–LP single cylinder rotor forgings are required to have a higher proof stress and toughness than conventional single cylinder rotor forgings to withstand the increased centrifugal stress. Research and development (R&D) have been carried out on a new steel which has mechanical properties suitable for the advanced HP–LP single cylinder turbine rotors. The R&D on the optimisation of chemical composition and heat treatment condition with many small size test blocks was performed and subsequently the production and verification study of the trial product-size rotor forgings were carried out. A part of the above test results has already been reported.[2,3,4] This paper covers the summary of the verification test results on the VCD trial rotor forging and the examination on the reduction of chemical segregation of the ingot and the further improvement of toughness in the centre of the LP portion.

2. PRODUCTION AND EVALUATION OF TRIAL ROTOR FORGING

2.1 Production of Trial Rotor Forging

A product-size trial rotor forging was produced on the basis of knowledge obtained so far from the evaluation of small size test materials. The drum diameter of the trial rotor forging was 1750 mm in LP portion and 1000 mm in HP portion, as shown in Fig. 2. The diameter of this rotor is large enough for covering HP–LP single cylinder rotors for turbines of 200 MW plants. As shown in Table 1, the optimum chemical composition for the trial rotor forging was chosen to be 0.24C–low Si–0.5Mn–2.25Cr–1.7Ni–1.1Mo–0.2V–0.015Nb–0.2W–Fe (Bal.) from the viewpoint of material properties and chemical segregation at the centre region of the rotor. Nb content was suppressed to the 0.015% level because eutectic Nb (C, N) might have formed

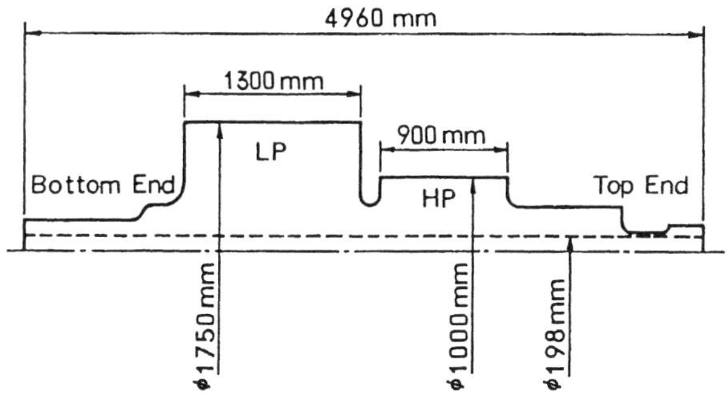

Fig. 2 Diagram of HP–LP Trial Rotor Forging

at the centre of the rotor in the preliminary solidification test. Moreover, the Mn content was determined to be at the 0.5% level in reflection of the preliminary solidification test result that the superclean process (ultra-low level of Mn content) promoted the chemical segregation.[5] Impurities such as P, S, As, Sn and Sb were determined to be suppressed to the lowest industrial levels.

The outline of the manufacturing process of the trial rotor forging is shown in Fig. 3. Ladle refining was carried out after basic electric arc furnace melting and VCD was performed during vacuum pouring. The ingot was 70 tons in weight and the height-to-diameter ratio (H/D) was enlarged to quicken solidification for decreasing the chemical segregation. High temperature annealing was performed at 1100°C before quality heat treatment so that the pearlite structure might be fully formed. The quality heat treatment consists of quenching and double tempering. A differential quenching was carried out, as being confirmed in the small size test materials. The HP portion was austenitised at 970°C and afterwards forced air cooled. The LP portion was austenitised at 900°C and afterwards water spray cooled. Thereafter, two stage temper treatments were employed to improve the proof stress due to the removal of residual austenite. The first tempering was carried out at 580°C and the differential second tempering at 640°C and 625°C were applied for HP portion and LP portion respectively. Thereafter, stress relieving was

Table 1 Target and Ladle Analysis of Chemical Composition for HP–LP Trial Rotor Forging

	C	Si	Mn	P	S	Ni	Cr	Mo	V	Nb	W	Cu	Al	As	Sn	Sb
Target	0.24	max 0.03	0.50	max 0.003	max 0.0015	1.70	2.25	1.10	0.20	0.015	0.20	max 0.05	max 0.005	max 0.004	max 0.004	max 0.0015
Ladle Analysis of Trial Rotor	0.24	0.02	0.45	0.004	0.0009	1.69	2.22	1.08	0.19	0.015	0.19	0.04	0.005	0.004	0.004	0.0010

(wt%)

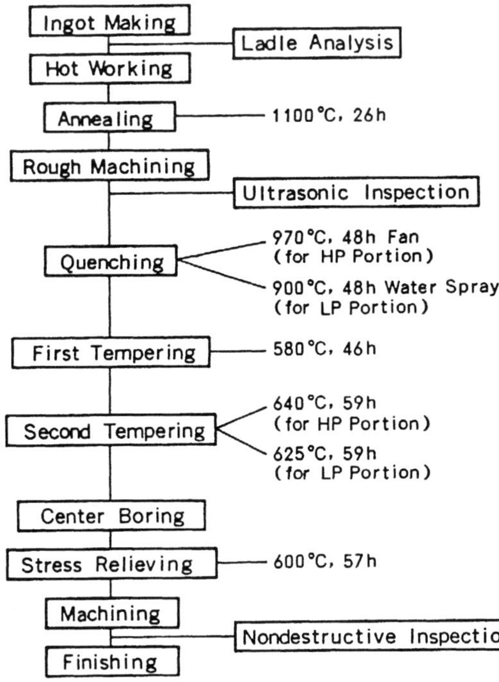

Fig. 3 Production Sequence of HP–LP Trial Rotor Forging

performed at 600°C over the whole rotor, and the production of the trial rotor forging was completed.

2.2 Evaluation of Trial Rotor Forging

Ultrasonic inspection and magnetic particle inspection were performed from the outer surface and the centre bore of the trial rotor forging. No indications were detected and it was confirmed that the rotor forging had excellent quality.

The result of ladle analysis was approximately consistent with the target composition shown in Table 1. Meanwhile, the distribution of the principal alloying elements in the centre portion of the rotor forging is shown in Fig. 4. The carbon content tends to be gradually more concentrated approaching the top end from the bottom end of the ingot, also, the carbon segregation is observed in the centre portion of the ingot. Figure 5 shows the macrostructure in the HP centre portion of the rotor forging. A lot of spots which correspond to the chemical concentrated area are observed in this macrostructure. The microstructure was tempered bainite which was typical in the low alloy rotor steels and no apparent eutectic Nb(C, N) was found.

Tensile and Charpy impact properties in various regions of the fully heat-

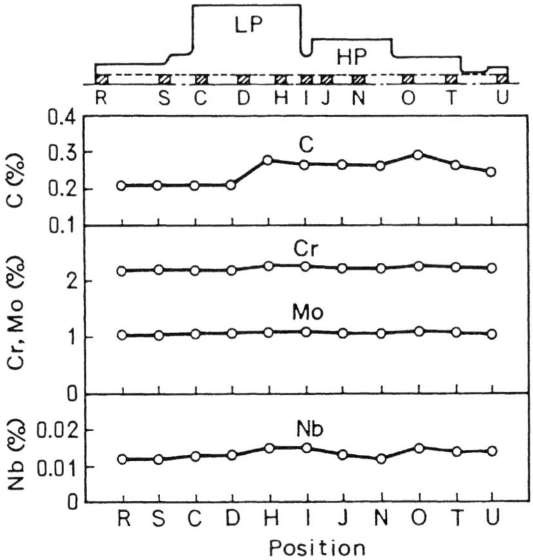

Fig. 4 Distribution of Principal Alloying Elements in Centre Portion of HP–LP Trial Rotor Forging

treated (after stress relief) trial rotor forging are shown in Fig. 6. The FATT in the centre core of HP portion was 40°C, and its toughness was considerably superior to that of 1CrMoV steel rotor forgings. In the centre core of the LP portion (position D), FATT of 23°C and 0.02% proof stress of 655 MPa were obtained. Figure 7 shows the creep rupture strength in the centre core and the outer surface of the HP portion. Good creep rupture strength equivalent to that of conventional 1CrMoV steel rotor forgings was obtained.

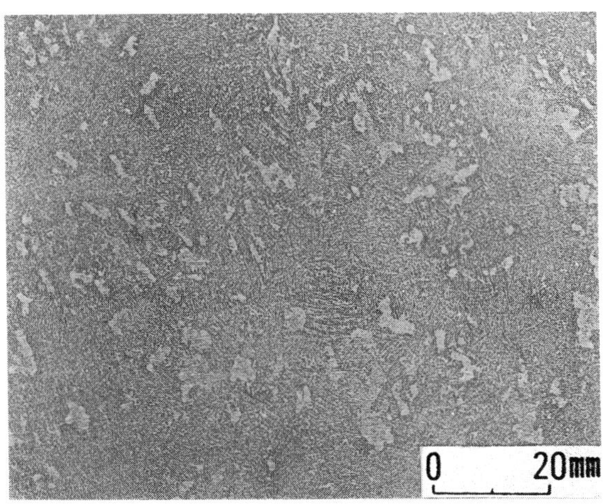

Fig. 5 Macrostructure in HP Centre Portion of HP–LP Trial Rotor Forging

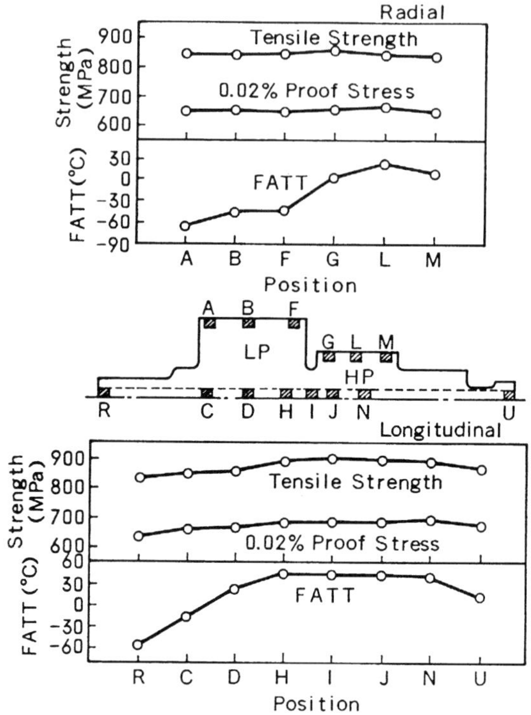

Fig. 6 Distribution of Tensile and Charpy Impact Properties in HP–LP Trial Rotor Forging

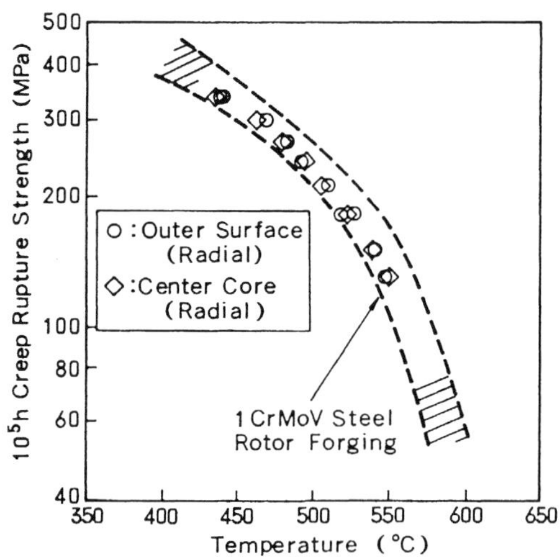

Fig. 7 Creep Rupture Strength of HP–LP Trial Rotor Forging

3. REDUCTION OF CHEMICAL SEGREGATION AND IMPROVEMENT OF TOUGHNESS

The estimated degree of chemical segregation was observed in the centre portion of the above trial rotor forging that was manufactured using VCD process. The FATTs in the centre core of LP portion were −16°C (position C) and 23°C (position D) respectively. The toughness in position D was less than the estimation by the preliminary simulation test. In order to reduce the chemical segregation in the centre of the ingot and to improve the toughness of the LP centre portion, the application of the ESR process, the modification of chemical composition and the accompanied optimisation of heat treatment condition (quenching condition) were considered as shown in Fig. 8.

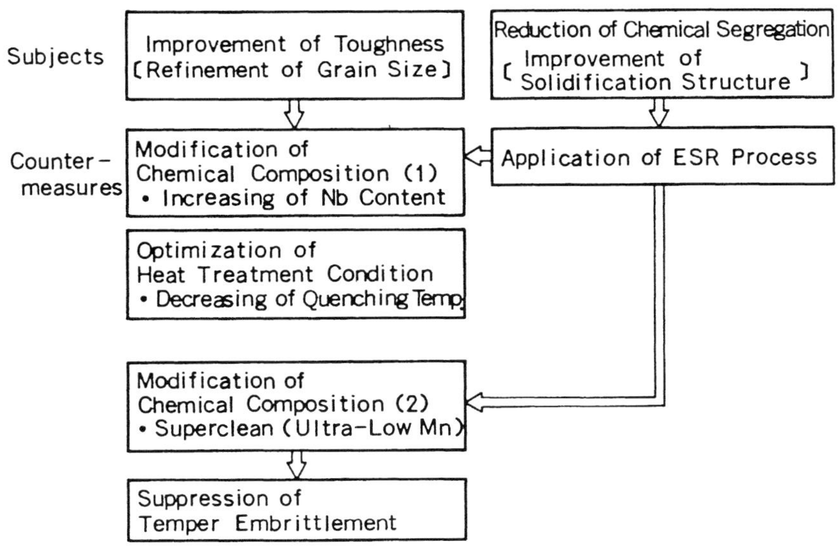

Fig. 8 Countermeasures for Improving the Properties in LP Portion of HP–LP Rotor Forging

3.1 Chemical Composition

The modified chemical composition for the next step was shown in Table 2. Nb content was increased to 0.030% from 0.015% in order to refine the grain size. Mn content was reduced to the ultra-low level (superclean) from the viewpoint of the suppression of temper embrittlement during long time operation.

Table 2 Modified Chemical Composition for Trial ESR Ingot

	C	Si	Mn	P	S	Ni	Cr	Mo	V	Nb	W	Cu	Al	As	Sn	Sb
Target	0.24	max 0.03	max 0.05	max 0.003	max 0.0015	1.70	2.25	1.10	0.20	0.030	0.20	max 0.05	max 0.005	max 0.004	max 0.004	max 0.0015
Trial ESR Ingot	0.24	0.09	0.03	0.003	0.0005	1.70	2.25	1.12	0.20	0.030	0.20	0.04	0.005	0.004	0.004	0.0011

(wt%)

3.2 Heat Treatment Condition

It was thought that the toughness of VCD trial rotor forging was less than the estimation obtained by the preliminary simulation test due to the difference of holding time when austenitising. The holding time was 3 hours in the simulation test and 40 hours in manufacturing of the trial rotor forging respectively. In order to clarify the differences of toughness, the following investigation was performed. First, the grain growth behaviour during the austenitising was investigated using the material sampled from the VCD trial rotor forging. The test result is shown in Fig. 9. The grain was coarsened in accordance with

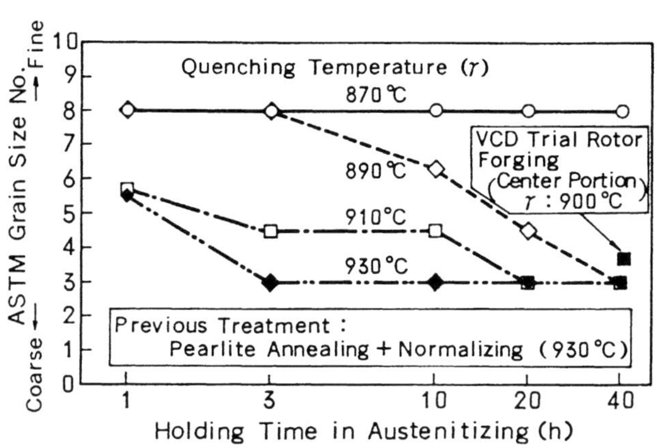

Fig. 9 Grain Growth Behaviour during Austenitising of VCD Trial Rotor Forging Material

increasing of the holding time of austenitising over 890°C. Secondly, the grain growth behaviour was similarly investigated using the material sampled from the trial ESR ingot as shown in the next paragraph. The test result is shown in Fig. 10. It was clarified that the grain size was more refined compared with the results of Fig. 9. By the increasing Nb content and the application of the superclean process; that the grain size could be maintained in the refined condition even with an austenitising temperature of 890°C.

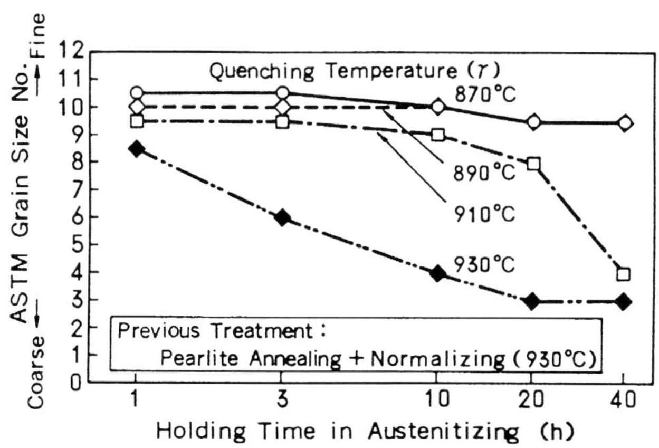

Fig. 10 Grain Growth Behaviour during Austenitising of ESR Trail Ingot Material

3.3 Application of ESR Process

The application of the ESR process was considered to reduce the chemical segregation in the centre of the ingot. Table 3 shows the electrode size, the

Table 3 Specification for Production of ESR Ingot

Electrode Size	ϕ 1275 mm × 4850 mml (46.3 ton)
Mold Size	ϕ 1850 mm × 5500 mmH
Ingot Size	ϕ 1800 mm × 2290 mml (43 ton)

mold size and the ingot size of the ESR ingot. The appearance of the ESR ingot is shown in Fig. 11. Several metallurgical investigations were performed in the longitudinal and the transverse cross sections of the ESR ingot after being cut at the position shown in Fig. 12.

There was no significant difference of cleanliness of non-metallic inclusions at each position. The ESR ingot had excellent cleanliness (the percentage of area occupied by the non-metallic inclusions such as sulphides, silicates, alumina, etc.) of 0.004% to 0.016%. The apparent eutectic Nb(C,N) was not observed. Figure 13 shows the macrostructure at the centre of the ESR ingot. The macrostructure of the ESR ingot was fairly fine and the degree of macro-segregation was considerably reduced from that of the VCD ingot, as shown in Fig. 5. The check analysis was carried out at many positions in the longitudinal and the transverse cross sections. Figure 14 shows the distribu-

Fig. 11 Appearance of ϕ 1800 mm ESR Ingot

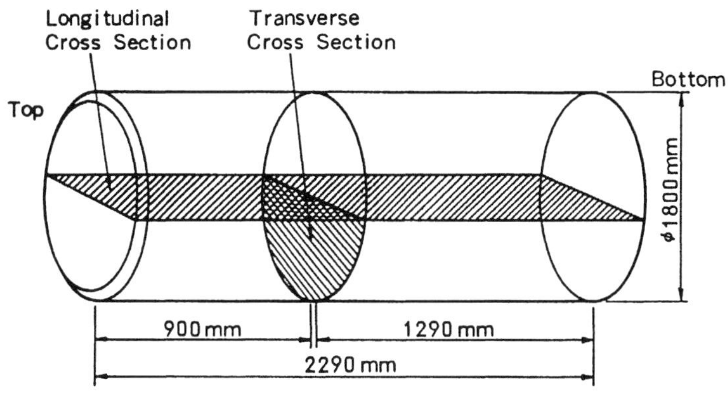

Fig. 12 Investigated Cross Sections of ESR Ingot

tion of C content along the longitudinal axis direction at the centre of the ESR ingot compared with that of the VCD trial rotor forging. A heavy C segregation was not observed in the ESR ingot. As for the other alloying elements, the degree of chemical segregation was similarly low and a relatively homogeneous ingot was obtained.

3.4 Simulated Mechanical Properties

The test block sampled from the centre region of the ESR ingot was forged (forging ratio 1:4) and heat-treated nearly in accordance with the LP portion

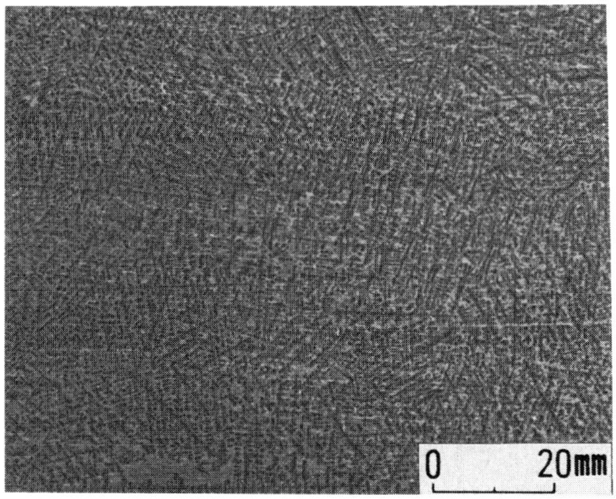

Fig. 13 Macrostructure in Centre of ESR Ingot

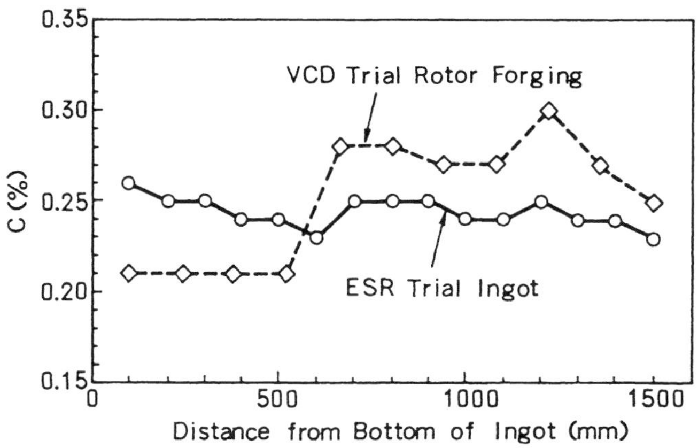

Fig. 14 Distribution of C Content at Centre of ESR Ingot Compared with that of VCD Trial Rotor Forging

of turbine rotor forgings as shown in Fig. 15. The pearlite treatment and the normalising were performed before the quality heat treatment. The austenitising temperature before quenching was determined to be 890°C at which the grain size was not changed during a long holding time. The quenching rate was simulated to be the water spray cooling rate in the centre region of an LP portion with a drum diameter of 1600 mm. Two kinds of tempering conditions were selected in order to obtain the balance of strength (tensile strength, proof stress) and toughness.

The grain size number of these test specimens was 8.2 (average of ASTM

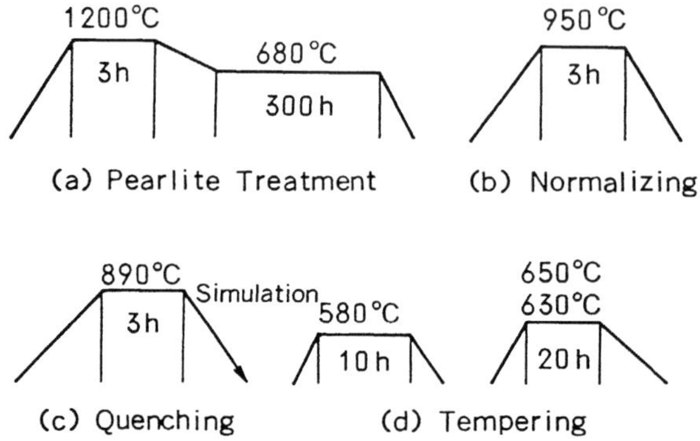

Fig. 15 Simulated Heat Treatment Diagram

No.) that is to say fairly fine. The correlation between FATT and 0.02% proof stress of the ESR steel forging is shown in Fig. 16 compared with the results of VCD steel forging. The ESR steel forging shows a good combination of FATT and proof stress superior to those of VCD steel forging, that is, the ESR steel forging has better toughness than the VCD steel forging when compared at the same proof stress. It is estimated that the improvement of toughness in the ESR steel forging is due to the refinement of the grain size (ASTM No.: 6.8 → ASTM No.: 8.2) as a result of increasing the Nb content (0.015% → 0.030%) and the reduction of quenching temperature (900°C → 890°C).

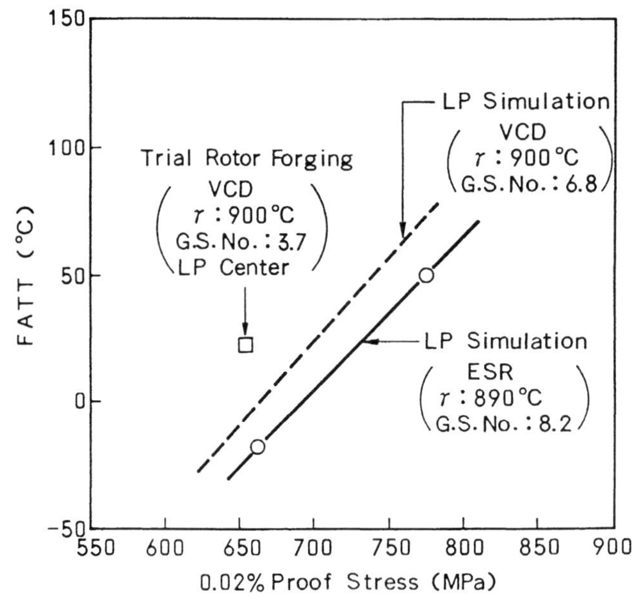

Fig. 16 The Correlation between FATT and 0.02% Proof Stress

From the above results, the improvement in toughness would be obtained by the reduction of chemical segregation using the application of the ESR process and the refinement of grain size with the increasing of Nb content and the accompanied optimisation of quenching temperature. Moreover, it would be anticipated that the reduction of chemical segregation due to the application of the ESR process made it possible for the steel to be superclean and to suppress the temper embrittlement during long time operation.

4. CONCLUSIONS

A new steel, 2.25Cr–1.7Ni–MoVNbW steel has been developed for HP–LP single cylinder steam turbine rotors by improving the chemical composition and the heat treatment conditions. This steel shows a good creep rupture strength equivalent to the conventional 1CrMoV rotor steel and proof stress and toughness considerably superior to the conventional HP–LP single cylinder rotor steels. The reproducibility/producibility and the mechanical properties have been verified by manufacturing a product-size trial rotor forging (70 tonne ingot) using the VCD process. The mechanical properties satisfied the target. However, a slight chemical segregation was observed in the centre of the trial rotor forging.

The application of the ESR process and the modification of the chemical composition and heat treatment conditions were considered in order to reduce the chemical segregation and to improve the toughness. A considerable reduction of chemical segregation was confirmed by manufacturing and evaluating a trial ESR ingot 43 tonnes of weight. The proof stress and toughness were estimated to be improved by conducting the simulated heat treatment tests in the centre of the LP portion. Moreover, the suppression of temper embrittlement during long time operation would be anticipated by conforming to the superclean composition.

This newly developed steel forging will be produced in the near future for HP–LP single cylinder steam turbine rotors of the advanced combined cycle power plants with a gas turbine inlet temperature of 1300°C.

REFERENCES

1. M. Yamada, M. Miyazaki, O. Watanabe and M. Kawai: *ASTM STP903 Steel Forgings*, 1986, 59.
2. O. Watanabe, M. Yamada, Y. Tsuda and M. Miyazaki: *Symposium on Materials for Combined Cycle Power Plant*, 1991, 60.
3. M. Yamada, Y. Tsuda, O. Watanabe, M. Miyazaki, Y. Tanaka, T. Takenouchi and Y. Ikeda: 'R. I. Jaffee Memorial Symposium on Clean Materials Technology', *ASM/TMS Materials Week*, 1992, 161.

4. Y. Tsuda, M. Yamada, O. Watanabe, M. Miyazaki, Y. Tanaka, T. Takenouchi and Y. Ikeda: *JSME-ASME International Conference on Power Engineering-93*, 1993, 91.
5. H. Yamada, T. Takenouchi, M. Funazaki, T. Takahashi, T. Iwadate and S. Nakada: *TETSU-TO-HAGANE*, 1995, **81**, 83.

Development of Superclean 12Cr Steel Disc for Heavy Duty Gas Turbines

Y. FUKUI, M. SHIGA, S. NAKAMURA and K. HIDAKA

Hitachi Research Laboratory, Hitachi, Ltd., Japan

1. INTRODUCTION

High efficiency is a major concern for gas turbine power plants because of the solution of global warming problems including CO_2 and reduction of fossil fuel usage. Hitachi, Ltd. manufacture heavy duty gas turbines (H25) whose inlet gas firing temperature is 1260°C. In order to achieve higher efficiency, the inlet gas firing temperature should be higher. However, higher temperature operation causes a severe environment for the materials of construction.

Although CrMoV steel is used for conventional gas turbine disc material, Ni base alloys are used in the state of art gas turbines because of higher heat resistant characteristics. The disc temperature of the H25 gas turbine reaches 450°C thus the conventional CrMoV steel cannot be used. Ni base alloys are used reluctantly for gas turbine discs at disc temperatures above 450°C.

The authors previously developed a heat resistant 12Cr steel as a disc material for H25 gas turbines, which ran over 3 years without replacement in a commercially based power plant. The new heat resistant 12Cr steel has been developed for higher disc temperature applications. This new 12Cr steel is designed by the alloying of high purity elements under clean processing conditions.

2. RESULTS AND DISCUSSION

2.1 Optimisation of Alloying Elements for Developing the 12Cr Steel Disc Material

Table 1 shows chemical composition of the developed 12Cr steel disc (HGTD1) which is manufactured from ingot by electroslag remelting. Nb is added to the conventional 12Cr steel for the improvement of grain refining and toughness. The amount of Mo is higher than that of conventional 12Cr steel in order to increase creep rupture strength. Figure 1 shows grain size refining as a function of Nb content. The Nb addition causes a remarkable grain refining effect. Figure 2 shows an increase of both creep rupture strength and toughness. It can be considered that a dispersion of fine NbC

Table 1 Analytical Result of Chemical Composition for Trial Disc Forgings

(wt%)	C	Si	Mn	Ni	Cr	Mo	V	Nb	N
Developed Material (HGTD1)	0.12	0.09	0.28	2.59	11.4	2.11	0.24	0.09	0.04
Conventional Material (12Cr Steel)	0.12	0.17	0.76	2.81	11.7	1.82	0.38	–	0.04

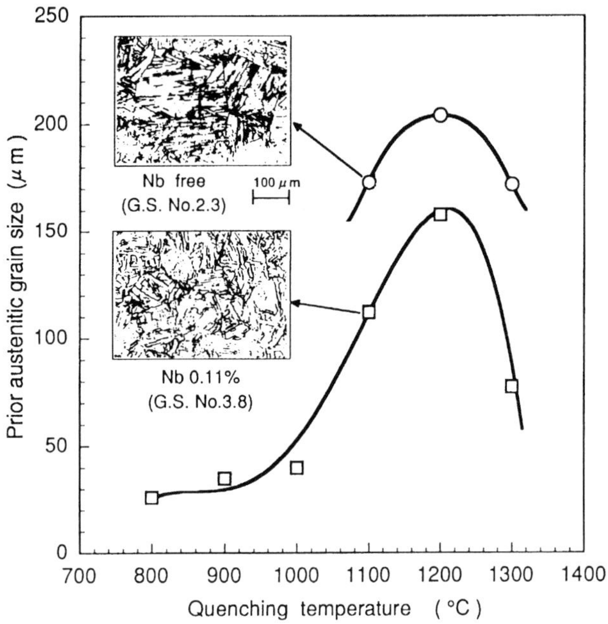

Fig. 1 Effect of Quenching Temperature and Nb Addition on Grain Size

particles in the matrix before quenching introduces grain refining which causes improvement of the mechanical properties. Figure 3 shows a comparison of creep rupture strength between HGTD1 and conventional 12Cr steel. The HGTD1 shows remarkably high creep rupture strength such as 568 MPa. One of the authors[1] previously reported that large numbers of Mo carbide precipitates are observed by transmission electron microscopy of extraction carbon replica in higher Mo content steels. It can be concluded that Mo carbide precipitation improves creep rupture strength. Because the Mo addition causes high creep strength, it was decided to add a higher amount of Mo to the developed 12Cr steel than is normal in conventional 12Cr steel. The specification for disc materials used in H25 gas turbines requires 500 MPa of creep rupture strength at 450°C. HGTD1 satisfies the specification of creep rupture strength.

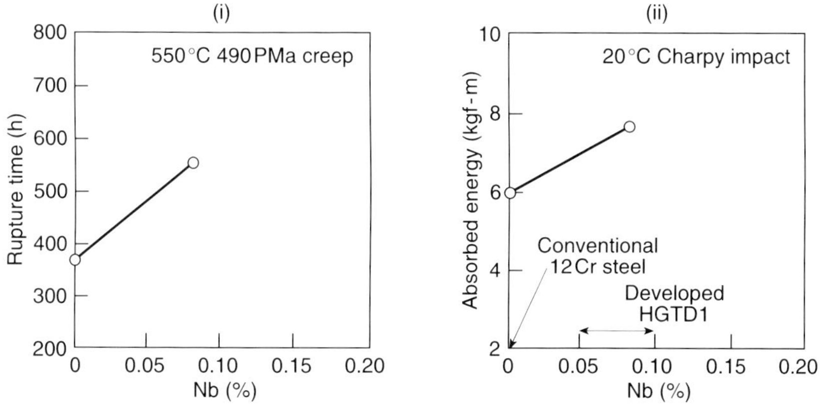

Fig. 2 Effect of Nb on (i) Creep Rupture Time and (ii) Charpy Impact Absorbed Energy

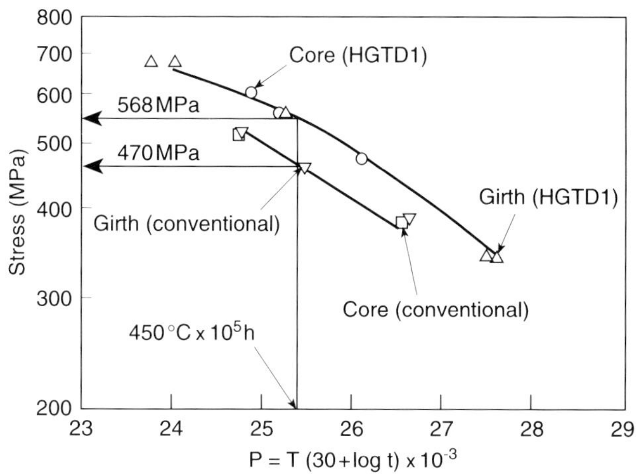

Fig. 3 Creep Rupture Strength of HGTD1 and Conventional 12Cr Steel Disc Forging

Figure 4 shows the fracture appearance temperature transition (FATT) of HGTD1 and conventional 12Cr steel with fractographs as a function of ageing time. The conventional 12Cr steel shows an intergranular fracture surface after ageing. However, the HGTD1 shows a transgranular fracture surface even after ageing.

2.2 Improvement of Ageing Embrittlement of the HGTD1 with Superclean Materials

2.2.1 Reduction of Si content

Figure 5 shows FATT of 16 pieces of forged discs as a function of Si content for HGTD1. The FATT is improved as the Si content decreases in the range from

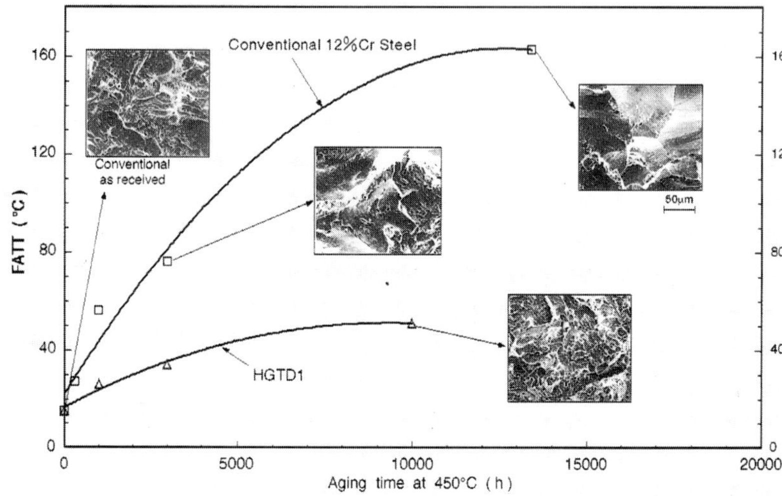

Fig. 4 Embrittlement of Conventional 12%Cr Steel and Developed HGTD1 Steel

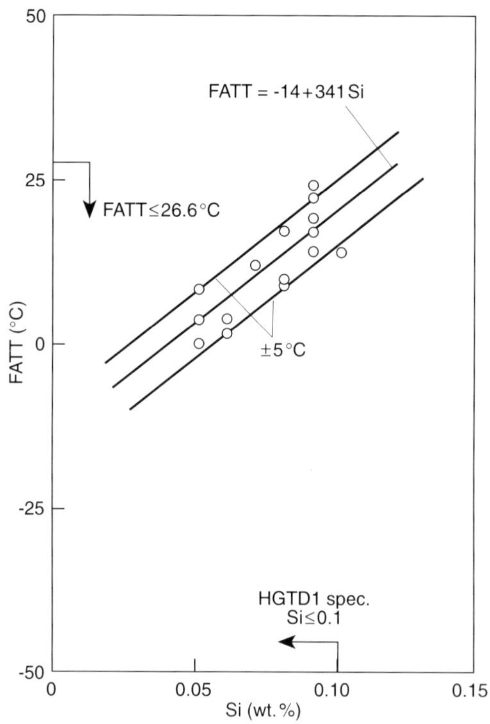

Fig. 5 Effect of Si on Toughness (FATT)

0.05 to 0.1 wt%. It is expected that the lower Si content in HGTD1 shows a better FATT. Chemical composition, tensile strength, and FATT of the 16 pieces of forged disc (HGTD1) are listed in Table 2.

Table 2 Range of Chemical Composition, Tensile Strength, and Toughness of 16 Disc Forgings

	Range of Chemical Composition (wt%)								Tensile	
	C	Si	Mn	Ni	Cr	Mo	V	Nb	Strength (MPa)	FATT (°C)
16 Disc Forgings	0.11	0.05 ~ 0.10	0.20	2.6	11.4	2.0	0.20	0.08	1090 ~ 1154	0 ~ 24

2.2.2 HGTD1 with superclean materials (HGTD1S)

Chemical composition of HGTD1 with superclean materials (HGTD1S) and with very pure superclean materials are listed in Table 3. The superclean

Table 3 Chemical Composition of the Studied Superclean Steels (wt%)

	C	Si	Mn	P	S	Ni	Cr	Mo	V	Nb	N
No. 1 (Ultra-superclean)	0.11	0.023	0.06	0.002	0.002	2.65	11.36	1.77	0.20	0.088	0.033
No. 2 (Superclean)	0.11	0.044	0.20	0.006	0.007	2.64	11.32	1.74	0.20	0.089	0.036

(34 kg vacuum melting)

samples (34 kg) were prepared with high purity materials by high frequency induction furnace and forged into square bar, 30 × 30 × 900 mm. The ultra-superclean sample was labelled as No.1, and the commercial based superclean sample as No. 2. Figure 6 shows the heat treatment cycle of these samples. The results of mechanical tests on these samples are listed in Table 4. FATT of Nos 1

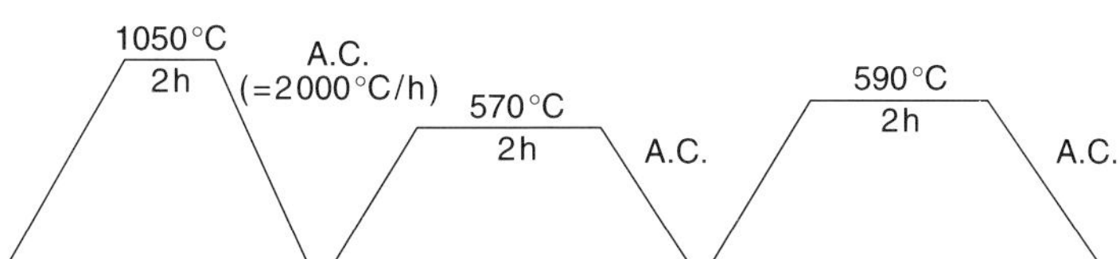

Fig. 6 Heat Treatment of Studied Superclean Steels

Table 4 Tensile, Impact and Creep Rupture Test Results of the Studied Superclean Steels

	T.S. (MPa)	0.2%Y.S. (MPa)	El. (%)	R.A. (%)	Absorbed energy (J)	FATT (°C)	450°C, 10^5 h rupture st. (MPa)
No. 1	1069	921	19.6	68.1	140	−48	522
(Ultra-superclean)	1070	920	19.8	67.4			
No. 2	1082	917	19.3	66.2	113	−29	531
(Superclean)	1106	915	19.3	66.3			

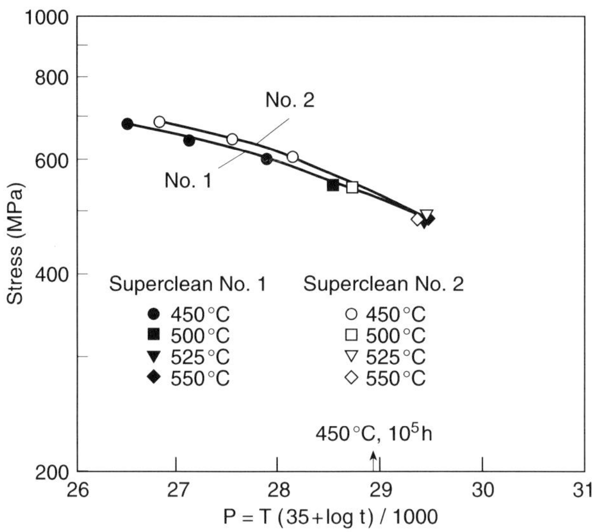

Fig. 7 Creep Rupture Test Results of Superclean Steels. (Master Rupture Curve on Larson-Miller Plot)

and 2 samples are −48°C and −29°C, respectively. Both samples show much better FATT than HGTD1. Figure 7 shows the creep rupture strength of samples 1 and 2. Both samples satisfy the specification for H25 gas turbine disc materials. Figure 8 shows FATT as a function of ageing time in terms of samples 1 and 2. Both samples show very little temperature rise in FATT during ageing. The temperature rise saturates after 10 000 hours ageing.

2.2.3 Trial manufacturing of gas turbine disc with HGTD1S

Table 5 shows the chemical composition of superclean 12Cr steel disc forging. The superclean material was prepared by vacuum carbon deoxidation (VCD) in ladle refining (LF) process in order to reduce Si and Mn contents. Figure 9 shows the manufacturing process, heat treatment, and appearance of

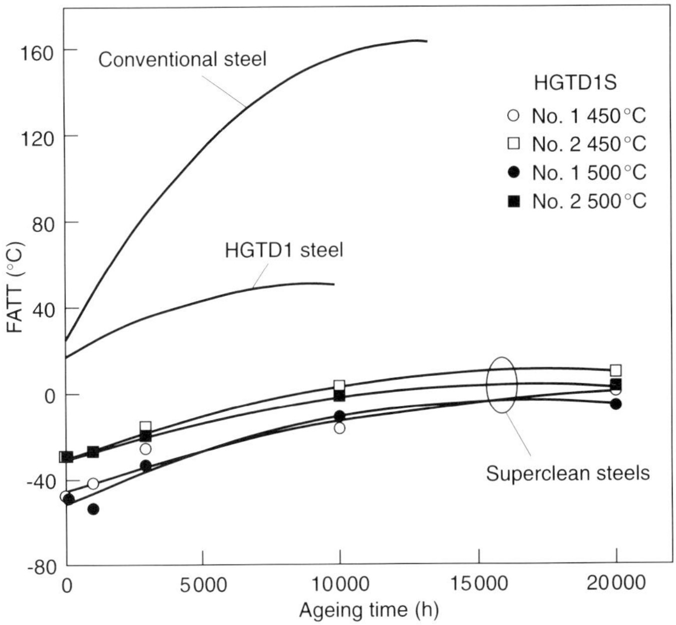

Fig. 8 Change in FATT after Ageing for Superclean Steels

Table 5 Chemical Composition of Conventional and Superclean HGTD1S, HGTD2S Trial Disc Forging (wt%)

Steel	C	Si	Mn	P	S	Ni	Cr	Mo	V	Nb	N
Conventional	0.11	0.20	0.74	0.017	0.006	2.78	11.54	1.84	0.37	–	0.044
HGTD1S	0.12	0.01	1.13	0.010	0.004	2.68	11.36	1.85	0.19	0.06	0.036
HGTD2S	0.12	0.01	0.15	0.010	0.004	2.67	11.32	2.32	0.22	0.11	0.036

HGTD2S trial disc forging. Another superclean material is alloyed with a higher Mo content (2.32 wt%) in order to increase the creep rupture strength (HGTD2S). Iseda *et al.*[2] reported that a higher Mo addition requires a reduction of the Si content to prevent Laves phase formation during ageing. This is because a reduction of Si increases the solubility limit of the matrix for Mo. They give an empirical equation in terms of Mo and Si contents for 8–12Cr steel as

$$(\text{Mo})\text{solute} = 0.84 - 0.89\text{Si}(\text{wt\%}) \quad (\text{at } 550°\text{C}).$$

Table 6 shows the results of tensile and Charpy impact tests on HGTD1S, HGTD2S, and for a previous conventional 12Cr steel. The FATT for superclean discs show satisfactory results, such as −9°C for HGTD1S, and −2°C for HGTD2s. The superclean disc forging shows a much better FATT than that of

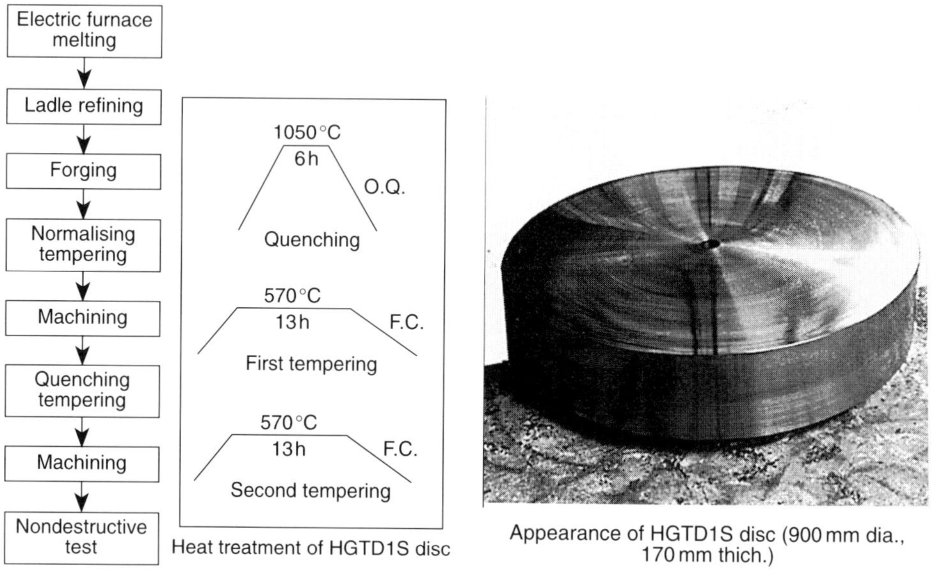

Fig. 9 Manufacturing Procedure and View of HGTD1S Trial Disc Forging

Table 6 Tensile Properties and Toughness of Conventional and Superclean Trial Disc Forgings

		T.S. (MPa)	0.2%Y.S. (MPa)	El. (%)	R.A. (%)	Absorbed Energy (J)	FATT (°C)
Superclean Trial Disc	HGTD1S	1083	955	19.4	58.6	89	−9
	HGTD2S	1090	994	18.6	60.0	61	−2
Conventional Disc		1060	957	17.2	48.2	54	22

the HGTD1 disc forging. Figure 10 shows the creep rupture strength of these materials given in Table 6. The same creep rupture strength of HGTD1S is found some 25°C higher in steel HGTD2S. It can be concluded that the higher Mo addition for superclean 12Cr steel improves the creep rupture strength without a decrease in the FATT. The HGTD2S is expected to find application in gas turbine discs where the inlet gas firing temperature is higher than 1260°C.

3. CONCLUSION

Development of 12Cr steel is preferred for application for gas turbine discs instead of Ni base superalloys. The following results were obtained.

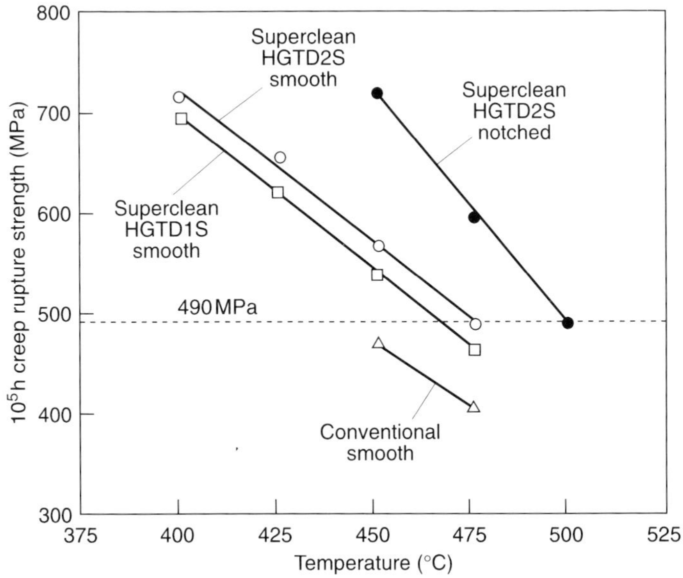

Fig. 10 Creep Rupture Streng.. ~f Conventional and Superclean Disc Forgings

1. Nb addition caused a remarkable increase of toughness and resistance to ageing embrittlement.
2. Superclean 12Cr steel discs were applicable to 1300°C class gas turbine.
3. Superclean 12Cr steel with higher Mo addition was applicable to gas turbine discs where the inlet gas firing temperature is higher than 1260°C.

REFERENCES

1. K. Hidaka, M. Shiga, S. Nakamura and Y. Fukui: 'Development of 12Cr Steel Disc for Heavy Duty Gas Turbines', *AMSE 94-GT-199*, The Hague, 1994.
2. A. Iseda, K. Yoshikawa and H. Teranishi: 'Effect of Si and Mo on Mechanical Properties of 9%Cr Heat Resistant Steels After Long-Term Heat Exposure', Report of the 123rd Committee on Heat-Resisting Metals and Alloys, *JSPS*, 1986, **27**, 39–54.

Stress Corrosion in 3.5 NiCrMoV Welded Joints

S. R. HOLDSWORTH,* M. NOUGARET† and B. VITTEMANT‡

** GEC Alsthom LST, Rugby, UK*

†GEC Alsthom LST, La Courneuve, France

‡GEC Alsthom Cerm, Belfort, France

1. INTRODUCTION

The mechanism of stress corrosion crack initiation in low to medium strength 3.5 NiCrMoV LP turbine rotor steels involves the nucleation and development of microcracking from pits formed at the sites of surface breaking manganese sulphide bearing inclusions.[1] The probability of microcrack formation is influenced by:

- material condition (eg. inclusion characteristics, transformation product, strength)
- applied stress (eg. $\sigma_{app}/Rp^{0,02}$) and
- environment.

There is evidence to show that the risk of SCC can be reduced in medium strength 3.5 NiCrMoV steels produced using superclean technology,[2] and this is just one reason why turbine builders are pursuing the feasibility of employing higher strength superclean steels to manufacture LP rotor forgings. This interest extends to the use of superclean 3.5 NiCrMoV to fabricate welded LP rotor assemblies.

Regions of HAZ, even in stress relieved welded joints, can exhibit higher hardness and therefore be potentially more prone to SCC than adjacent parent and weld metal microstructures. However, while the density and types of inclusion in the HAZ are identical to those of the parent steel, the number of surface breaking non-metallic particles associated with this region could be relatively low (since the surface area of the HAZ exposed to the environment is low). In addition, the stress acting on the HAZ of post weld heat treated weldments is dictated by the properties of the lower strength parent and weld materials, and is consequently at a lower fraction of $Rp_{0.2}$ than in those adjacent microstructures. The risk of stress corrosion microcrack nucleation and growth in 3–3.5 NiCrMoV course grain heat affected zone (CGHAZ) microstructure is therefore likely to be lower than might be first anticipated.

The project set out to compare the SCC initiation properties of a welded joint manufactured from superclean 3.5 NiCrMoV LP rotor steel with those displayed by a conventional 3 NiCrMoV forging. This work is a part of the European research programme COST 501 WP8.

2. MATERIAL DETAILS

2.1 Parent Materials

The test programme was based on two LP turbine rotor steels. The first was a conventional $Rp_{0.2}$ 700 MPa 3 NiCrMoV trial disk manufactured by Creusot[3] while the second originated from a superclean 875 MPa 3.5 NiCrMoV trial forging manufactured by Böhler.[4] The chemical compositions, heat treatment details and mechanical properties of the two parent materials are given in Table 1a.

The inclusion population in the superclean steel is low and those non-metallic inclusions (NMIs) which are present are mainly calcium and aluminium bearing. While the inclusion content of the conventional steel is not high, there are significantly more NMIs evident than in the superclean alloy. These are mainly MnS bearing, although there is also evidence of Ca, S, Al bearing particles.

2.2 Weldment

A weldment was prepared using the superclean steel and the details are also summarised in Table 1a. The weldment was examined in two PWHT conditions, the first with a 560°C and the second with a 620°C soaking temperature. The purpose of the higher temperature treatment was to obtain a section of superclean weldment with comparable strength characteristics to conventional 700 MPa 3 NiCrMoV weldments. Peak HAZ hardnesses for the superclean weldment in both PWHT conditions are given in Table 1a. These compare with a value of around 330 VPN for conventional 700 MPa 3 NiCr MoV weldments. The weldment[3] microstructures adjacent to the fusion boundary are shown in Fig. 1.

The number of NMIs in the weld metal is significantly higher than in the parent steels. While the density of non-metallic particles is high, they are very fine and predominantly Mn, Si bearing with traces of Mg.

2.3 Simulated CGHAZ Materials

In addition to the material conditions described, simulated CGHAZ structures for the two steels were prepared. A 0.25h/1000°C/AC simulation treatment was adopted following trials to reproduce the transformation product, grain size and hardness of the CGHAZ microstructure observed

Table 1a Material Details

	Conventional 3NiCrMoV		Superclean 3.5NiCrMoV				
	Parent	Simul CGHAZ	Parent	Weldment 1	Simul CGHAZ	Weldment 2	Weld Metal
Composition							
C	0.24	0.24	0.27	0.27	0.27	0.27	0.07
Si	0.11	0.11	0.03	0.03	0.03	0.03	0.36
Mn	0.49	0.49	0.03	0.03	0.03	0.03	1.07
S	0.006	0.006	0.0002	0.002	0.002	0.002	0.008
P	0.006	0.006	0.004	0.004	0.004	0.004	0.011
Cr	1.83	1.83	1.71	1.71	1.71	1.71	0.66
Mo	0.46	0.46	0.42	0.42	0.42	0.42	0.44
Ni	2.89	2.89	3.80	3.80	3.80	3.80	2.23
V	0.11	0.11	0.09	0.09	0.09	0.09	0. 02
Cu	0.07	0.07	0.08	0.08	0.08	0.08	0.19
As	0.018	0.018	0.008	0.008	0.008	0.008	0.002
Sn	0.005	0.005	0.006	0.006	0.006	0.006	0.005
Heat Treatment							
	18h 850C. WQ 24h 620C. AC	0.25h 1000C. AC 20h 620C. AC	18h 850C. WSQ 30h 585C. AC	18h 850C. WSQ 30h 585C. AC 20h 560C. AC	0.25h 1000C. AC 20h 560C. AC	18h 850C. WSQ 30h 585C. AC 20h 620C. AC	
Properties							
$Rp_{0.2}$ (MPa)	698	830	874	854	971	711	
UTS (MPa)	820	919	991	994	1108	829	
Elong (%)	20	18	18	17	17	17	
E (GPa)	205	214	210	215	205	199	
PM VPN	255	310	310	305	375	278	W1: 255; W2: 220
CGHAZ VPN				383		308	
Weldment Details							
SAW filler				NiCrMoV, low basic solid flux 3.2 mm diam wire @ 14kJ/cm for first 10–12 mm 4.0 mm diam wire @ 18.7kJ/cm for remainder			

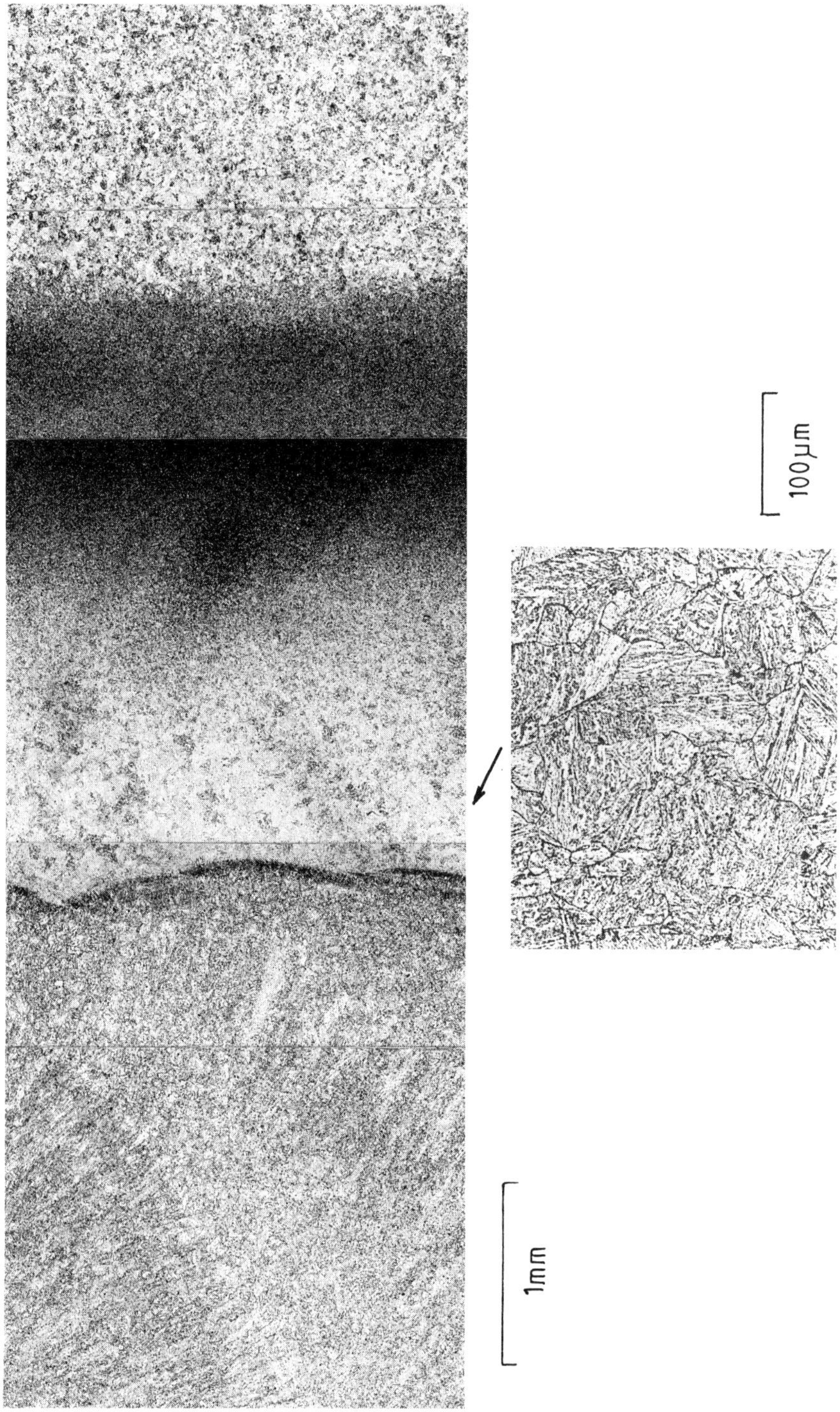

Fig. 1 Coarse grain heat affected zone microstructure in superclean weldment

adjacent to the fusion boundary of the superclean weldment. The target was a lower bainitic to martensitic matrix structure with a prior austenite grain size of ~100 μm and a hardness of 420–450 VPN immediately after the simulation treatment. The uniformity in microstructure at the edge and centre of 25 mm thick blocks of the two simulated CGHAZ materials is shown in Fig. 2.

The grain size of the CGHAZ simulated superclean material is consistent with the target of ~100 μm. The grain size of the CGHAZ simulated conventional steel is closer to 40 μm, the lower grain size being anticipated in view of

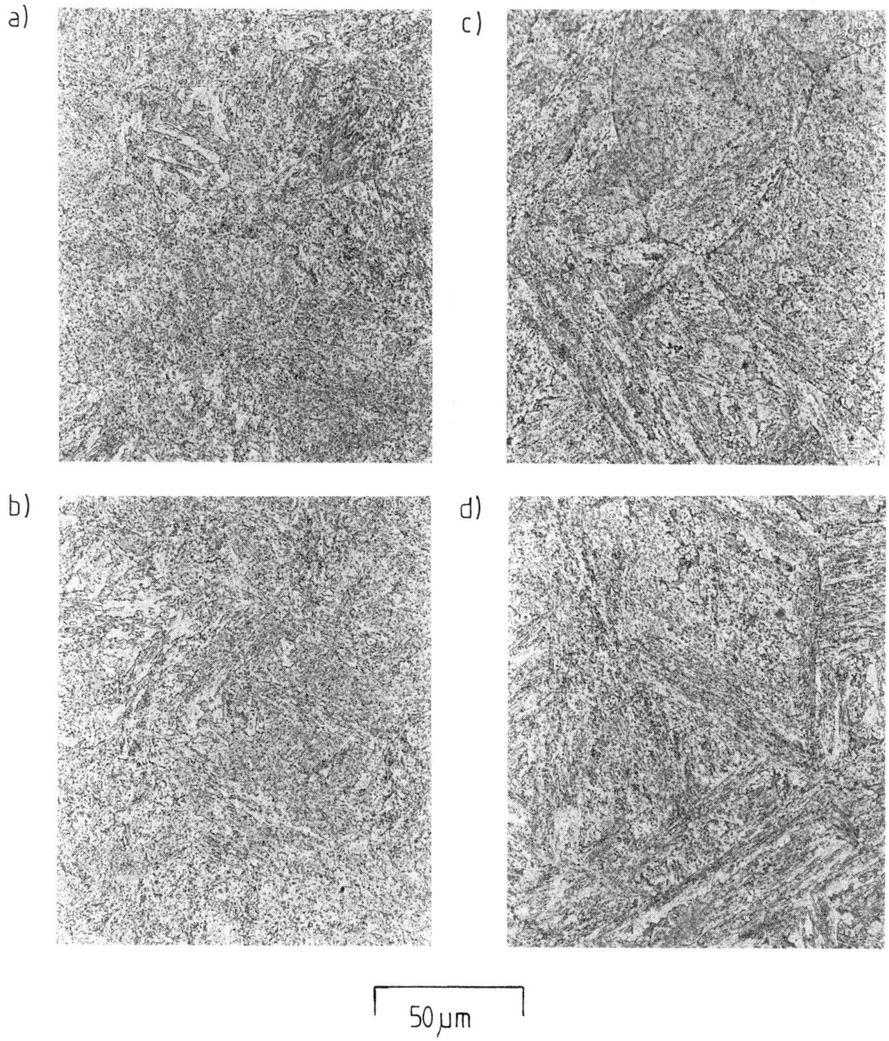

Fig. 2 Typical microstructures at (a) the edge & (b) the mid section of the conventional 3 NiCrMoV CGHAZ simulated blocks and (c) the edge & (d) the mid section of the superclean 3.5 NiCrMoV CGHAZ simulated blocks.

Table 1b Material Details for WP8 Superclean Steel

Composition	
C	0.27
Si	0.04
Mn	0.03
S	0.003
P	0.004
Cr	1.51
Mo	0.32
Ni	3.61
V	0.11
Cu	0.06
Al	0.001
Heat Treatment	
	2h 850°C, OQ
	20h 600°C, FC
	20h 610°C, FC
RT Properties	
$Rp_{0.2}$ (MPa)	752
UTS (MPa)	867
Elong (%)	19
E (GPa)	205

the fact that superclean alloys are notoriously less resistant to grain growth than conventional material.

Following the austenitising treatment, the conventional and superclean steels were respectively post weld heat treated for 20h/560°C/AC and 20h/620°C/AC, and hence the final hardness values of 375 and 310 VPN (Table 1a).

2.4 WP8 Superclean Steel

A small number of check tests were performed in pure steam on material from the Böhler superclean rotor evaluated extensively by the remaining COST 501 WP8 partners. The chemical composition, heat treatment details and mechanical properties of this steel are summarised in Table 1b. The material is fully characterised elsewhere.[6]

3. SCC TEST PROGRAMME

The test matrix was devised to examine the effect of (a) material condition and

HAZ surface area, (b) applied stress and (c) environment on the SCC initiation properties of 3.5 NiCrMoV (Table 2).

3.1 *Material condition*

Details of the alloy chemistry, inclusion characteristics and strength level of the conventional purity and superclean 3–3.5 NiCrMoV parent materials and the post weld heat treated superclean 3.5 NiCrMoV weldments have already been introduced in the previous section. Initially all tests were performed using T6 tensile specimens with a 50 mm long × 6 mm diameter gauge length.

Table 2 Stress Corrosion Test Matrix

	Applied Stress (MPa)			
Pure Steam				
Conventional 3NiCrMoV				
Parent Material ($Rp_{0.2}$=700MPa)	650 (×5)	550 (×4)	400 (×4)	200 (×4)
Simulated CGHAZ	550 (×4)	400 (×4)	200 (×4)	
Superclean 3.5NiCrMoV				
Parent Material ($Rp_{0.2}$=875MPa)	850 (×5)	650 (×5)	400 (×4)	200 (×4)
Simulated CGHAZ (380VPN)	550 (×5)	400 (×4)	200 (×4)	
Cross Weld (PWHT 560C)	550 (×5)	400 (×4)	200 (×4)	
Cross Weld (PWHT 620C)	550 (×5)	400 (×4)	200 (×4)	
Pure Water				
Conventional 3NiCrMoV				
Parent Material ($Rp_{0.2}$=700MPa)	650 (×4)			
Simulated CGHAZ	550 (×4)			
Superclean 3.5NiCrMoV				
Parent Material ($Rp_{0.2}$=875MPa)	850 (×4)			
Simulated CGHAZ (380VPN)	550 (×4)			
Cross Weld (PWHT 560C)	550 (×4)			
Cross Weld (PWHT 620C)	550 (×4)			
Simulated Early Condensate				
Conventional 3NiCrMoV				
Parent Material ($Rp_{0.2}$=700MPa)	650 (×4)			
Simulated CGHAZ	550 (×4)			
Superclean 3.5NiCrMoV				
Parent Material ($Rp_{0.2}$=875MPa)	850 (×4)			
Simulated CGHAZ (380VPN)	550 (×4)			
Cross Weld (PWHT 560C)	550 (×4)			
Cross Weld (PWHT 620C)	550 (×4)			

Notes: Steam test metallographic inspections at 1 000, 3 000, 6 000, 10 000 & 20 000h for (×5) and 1 000, 3 000, 6 000 & 10 000h for (×4). Water test metallographic inspections at 100, 300, 1 000, & 3 000h.

However, a number of check stream tests were conducted using both the 4 mm diameter T4 tensile and 0.8 mm thick bent beam testpieces used in earlier studies.[1,2]

The HAZ surface area effect was examined by comparing the results from cross weld specimens and simulated CGHAZ specimen tests. All these tests were performed using the larger T6 tensile testpiece geometry. The width of the weld was approximately 20 mm and the cross weld specimens were sectioned with the meld metal located in the middle of the gauge length. As a consequence, the exposed HAZ on cross weld specimens was in the form of two rings ~6–7 mm wide × 6 mm diameter (ie. ~250 mm^2 surface area). This compared with the full gauge length surface area of the simulated CGHAZ specimens (ie. ~950 mm^2).

3.2 *Applied Stress*

A range of applied stress levels were adopted and these are listed in Table 2. In the single microstructure specimens, the $\sigma_{app}/Rp_{0.2}$ ratios ranged between 0.29 and 0.97.

In the cross weld specimens, the $\sigma_{app}/Rp_{0.2}{}^{PM}$ ratios changed between 0.28 and 0.64. For a given cross weld specimen, the $\sigma_{app}/Rp_{0.2}{}^{HAZ}$ ratio was typically around 10% lower and the $\sigma_{app}/Rp_{0.2}{}^{WM}$ approximately 20% higher than the $\sigma_{app}/Rp_{0.2}{}^{PM}$ ratio.

3.3 *Environment*

SCC tests were conducted in three test environments. The first was the same 95°C pure low oxygen condensing steam environment adopted by GECA in previous studies, the second was flowing pure low oxygen water at 90°C, and the third was flowing low oxygen water dosed with 100 ppb Na_2SO_4 to simulate early condensate in a low pressure steam turbine (also at 90°C). The conductivity and oxygen concentrations of the pure low oxygen steam feedwater/condensate and water environments were less than 0.1 μS/cm 10 ppb respectively. The conductivity of the simulated early condensate was >2.5 μS/cm.

The steam test specimens were pre-loaded in individual stainless steel tubular fixtures,[1,2] and the assemblies stacked on shelves in a single environmental chamber. In the water tests, specimens with local environmental cells attached to shoulders either side of the gauge length were loaded in dead weight loading frames.

3.4 *Inspection*

In order to gather the metallographic evidence necessary for comparison with the SCC crack development diagrams, a multi-specimen test procedure was adopted. This approach was employed to enable the destructive metallographic inspection of exposed specimens at various fractions of time to macrocrack formation (Table 2) and the process of crack initiation to be determined.

In low to medium strength 3.5 NiCrMoV LP turbine rotor steels, pit formation is an integral part of the process of SCC crack initiation.[1,2] Hence, the inspection procedure included an evaluation of pit size distribution in addition to the examination for cracking.

After removal from the environment, specimen gauge length surfaces were examined by inspecting three metallographic planes of section, ie. at the mid-plane and 0.5 mm to either side of the mid-plane.

All specimens were routinely inspected visually every 1 000 h for evidence of macrocracking. Cracks were referred to as macrocracks (or engineering size cracks) when their surface length exceeded 1 mm. Typically, such cracks were >250–500 μm deep. Cracks were referred to as microcracks if their depth as-measured on a metallographic section was less than 250 μm.

4. PITTING

During the course of metallographic inspection, observed trends relating to pit depth and pit density were recorded.

4.1 Pit growth Kinetics

In pure condensing steam, maximum pit depths increase with time, with the deepest pits being observed in the conventional purity steel (Fig. 3). In pure flowing water, the depth of pitting is notably less than that due to steam exposure and there is no significant difference in the pit growth kinetics of the conventional and superclean steels (Fig. 3). The pit growth kinetics of the two steels in flowing water with 100 ppb Na_2SO_4 appear to be more similar to the characteristics observed in pure steam (Fig. 3).

There is no consistent influence of applied stress on pit depth in the three environments.

The pit growth kinetics of simulated and actual HAZ microstructure are similar to those of the respective parent steels in the same environment. In cross weld tests, the depth of pitting observed in the weld metal is marginally higher than that observed in the other constituent microstructures of the joints (Fig. 4).

4.2 Pitting Density

The density of pitting is also observed to increase with time, although the

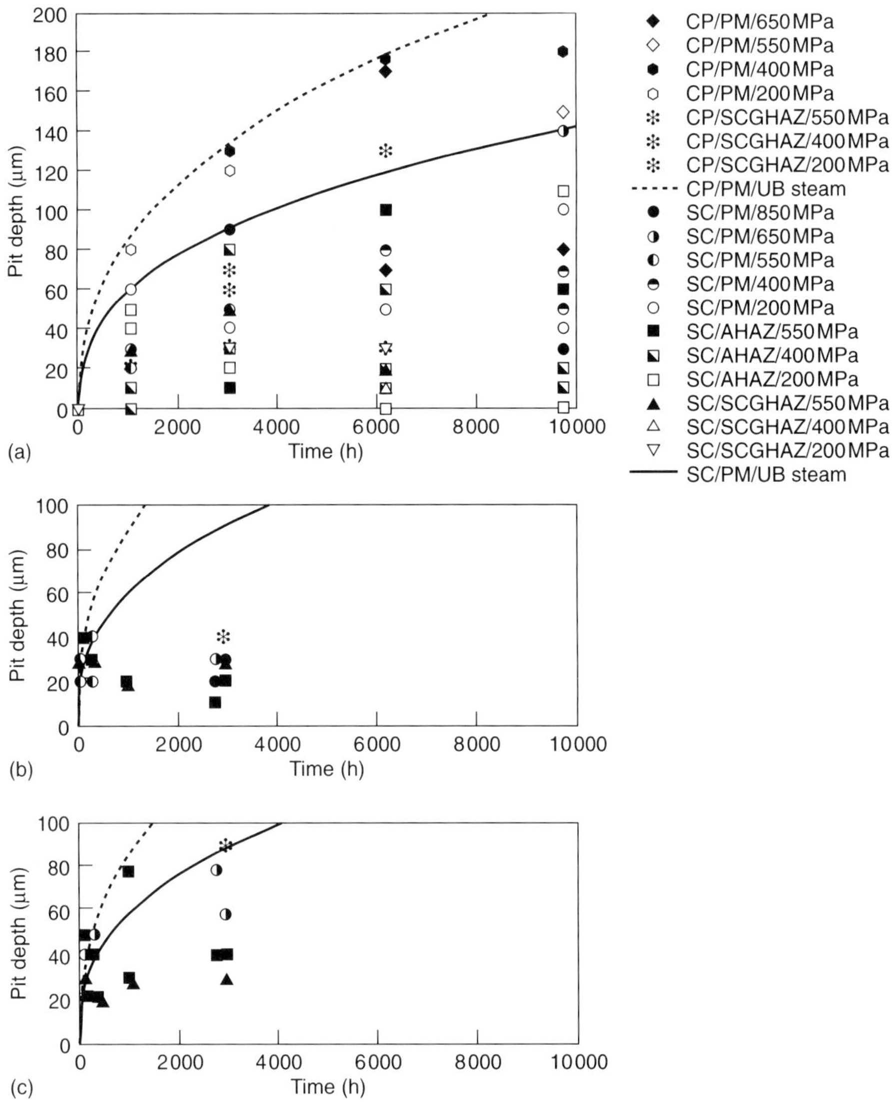

Fig. 3 Comparison of pit growth kinetics in (a) pure condensing steam, (b) pure flowing water and (c) flowing water with Na_2SO_4

tendency diminishes increasingly for t >1 000 h (Figs 5, 6). Pit densities in the superclean steel are lower than in the conventional steel. However, differences in pit density are not nearly as marked as those observed in terms of pit depth. Similarly, any differences in pit densities resulting from exposure to the three environments are less notable than the equivalent pit depth differences (Fig. 5).

In cross weld tests, the density of pitting in the parent material and HAZ is

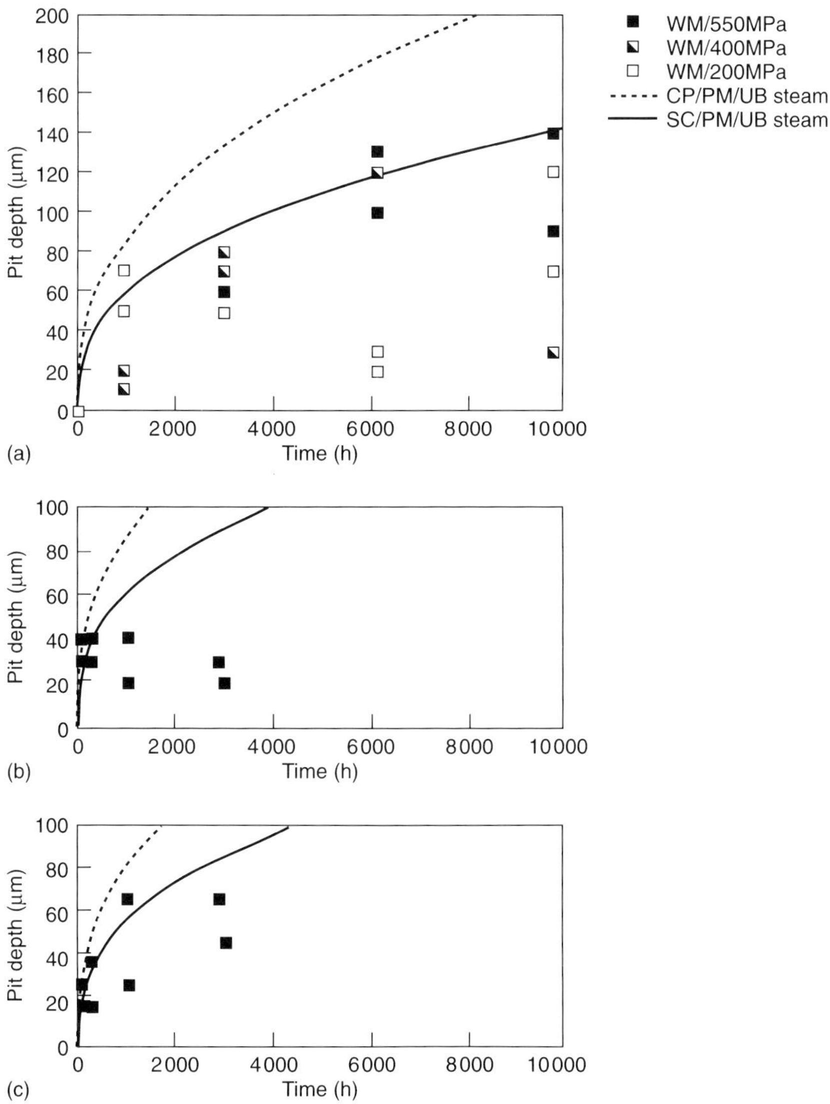

Fig. 4 Comparison of pit growth kinetics in (a) pure condensing steam, (b) pure flowing water and (c) flowing water with Na_2SO_4 for weld metal

similar, but significantly higher in the weld metal (Fig. 6). The density of pitting is particularly high on weld metal surfaces exposed to pure condensing steam (Fig. 6a).

5. CRACK INITIATION

The results of the SCC initiation tests are represented in figures 8 to 15. These

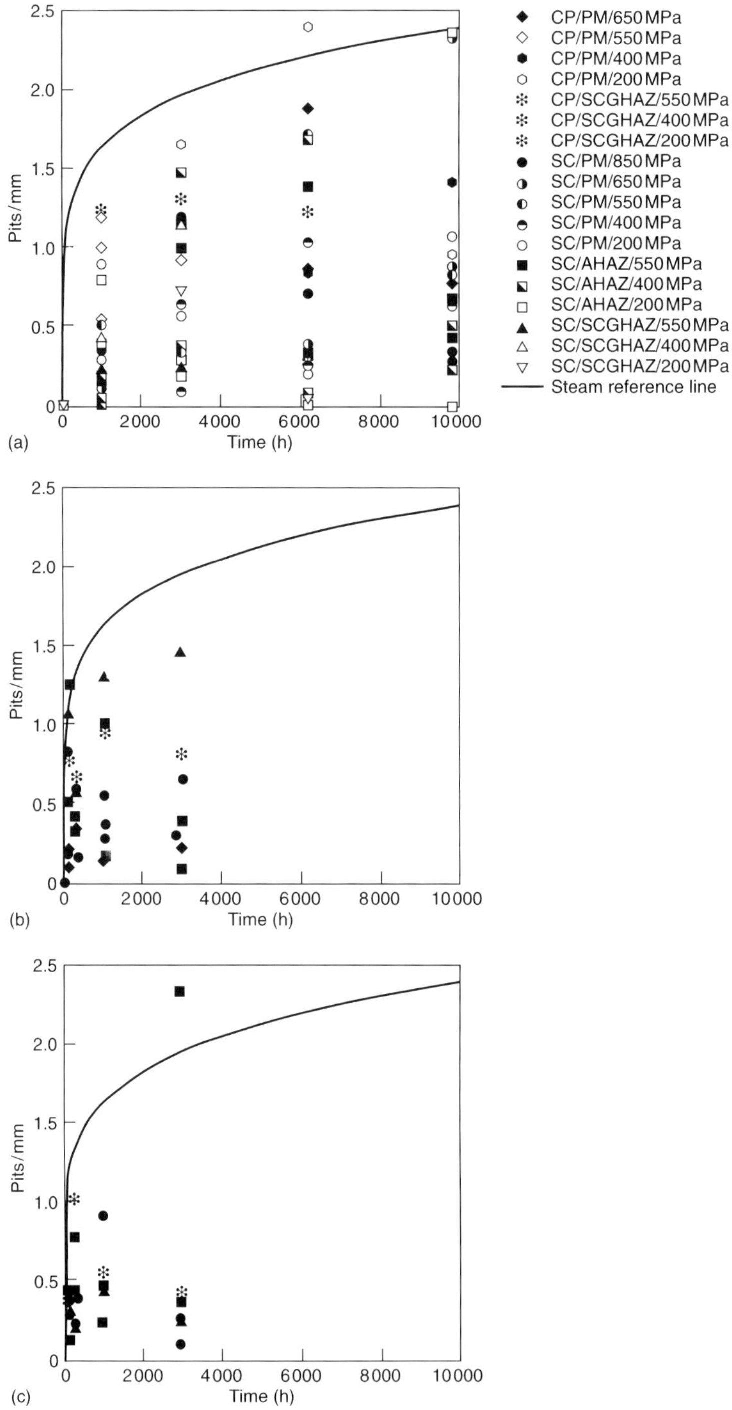

Fig. 5 Comparison of pit densities in (a) pure condensing steam, (b) pure flowing water and (c) flowing water with Na_2SO_4

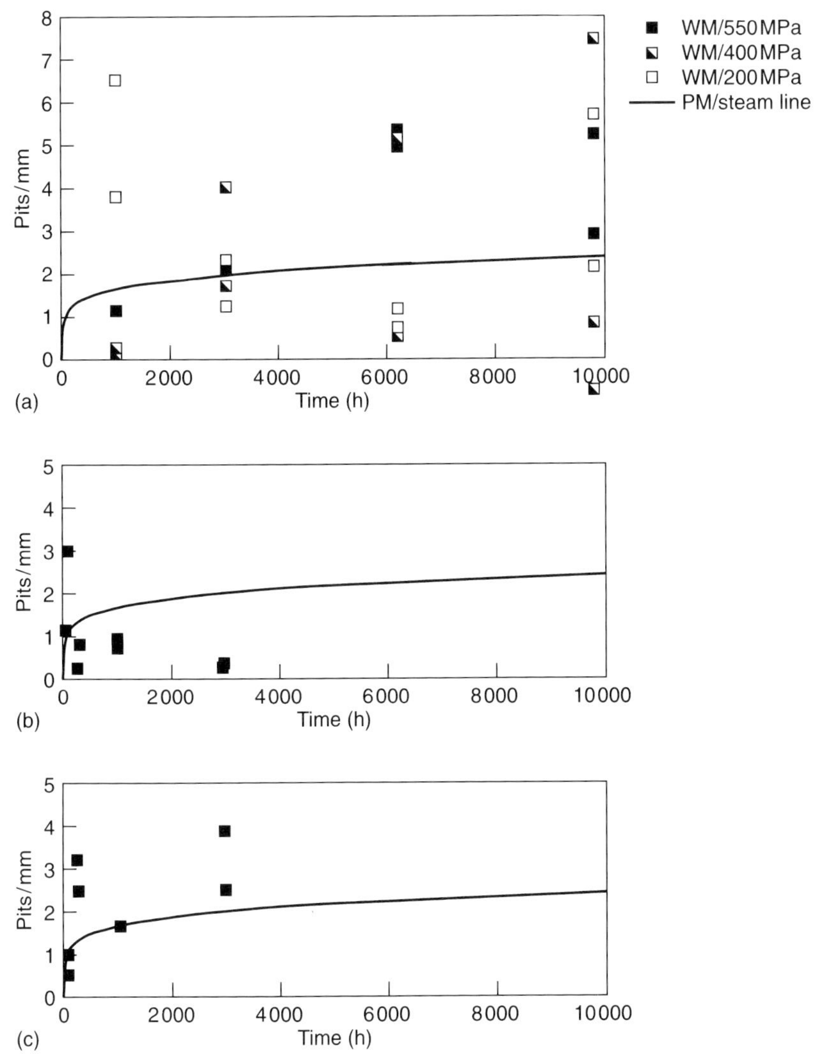

Fig. 6 Comparison of pit densities in (a) pure condensing steam, (b) pure flowing water and (c) flowing water with Na_2SO_4 for weld metal

are compared with previous experience for conventional and superclean 3–3.5 NiCrMoV steels exposed to pure low oxygen condensing steam at 95°C, this being conveniently summarised in terms of the SCC crack development diagram (Fig. 7).[1,2]

The SCC crack development diagram, which continues to be validated by evidence from a variety of sources,[5,7], acknowledges two regimes of behaviour. In the case of medium strength steels with $Rp_{0.2} < \sim 850$ MPa, SCC microcracks almost invariably nucleate form pits formed at the sites of surface

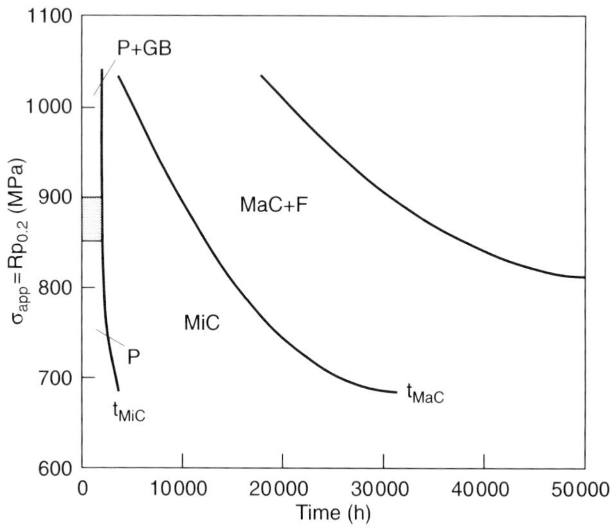

Fig. 7 Pure steam SCC crack development diagram for 3–3.5 NiCrMoV steels at 95°C[1,2]

breaking NMIs. Depending on strength level, these may initiate after only 1 000 to 3 000 h but take at least another 10 000 to 30 000 h to develop into engineering size cracks (>1 mm). At higher strength levels (ie. $Rp_{0.1}$<~ 900 MPa), microcracks increasingly initiate from other metallurgical discontinuities (eg. surface breaking grain boundaries). The greater number of potential initiation sites and the inherently lower resistance to crack propagation displayed by the higher strength steels are largely responsible for their shorter and notably less scattered microcrack initiation and crack development times.

In referring to the crack development diagram, it is important to remember that (a) microcracks do not nucleate from every pit and (b) every microcrack does not develop into a macrocrack.[1,2]

5.1 Parent Steels

The results of the metallographic inspections performed are summarised in diagrams of applied stress versus time and compared with the lowerbound times to microcrack (t_{MiC}) and macrocrack (t_{MaC}) formation in pure low oxygen condensing steam, as predicted by the crack development diagram (eg. Fig. 8).

One aim of this work was to compare the SCC initiation resistance of a conventional 700 MPa 3 NiCrMoV and a superclean 875 MPa 3.5 NiCrMoV steel. Previous experience had suggested that the probability of SCC initiation in medium strength superclean material is significantly lower than that in conventional steel. In pure steam the results indicate that, while the conven-

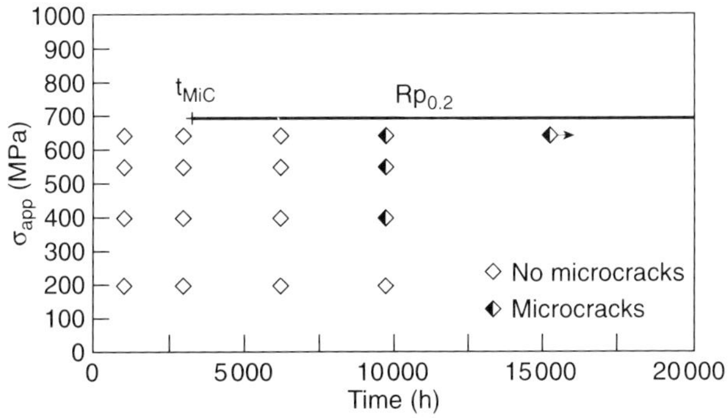

(a) Conventional 3NiCrMoV steel

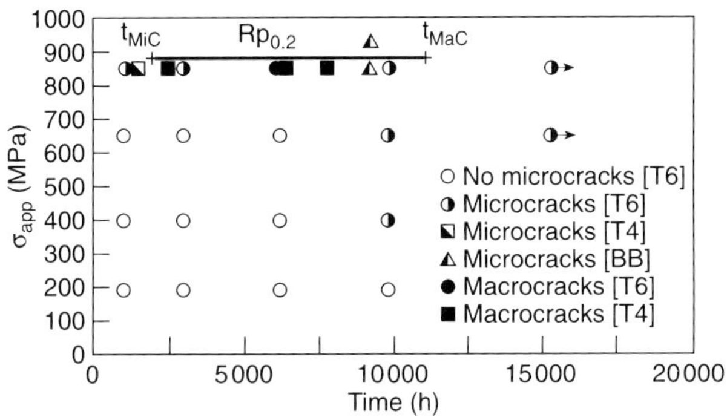

(b) Superclean 3.5NiCrMoV steel

Fig. 8 SCC tests in pure low oxygen condensing steam at 95°C

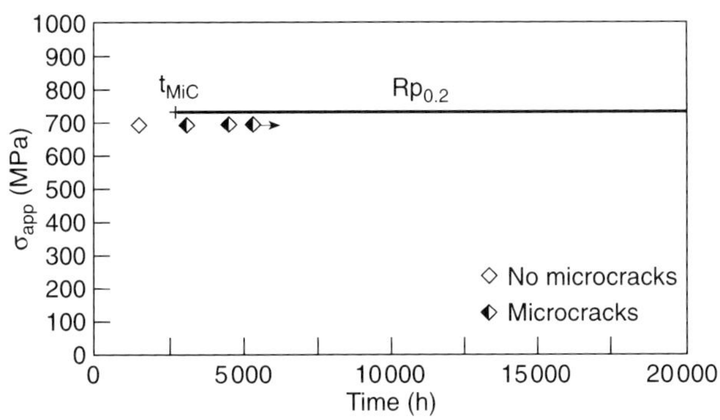

Fig. 9 SCC tests on WP8 superclean steel in pure low oxygen condensing steam at 95°C

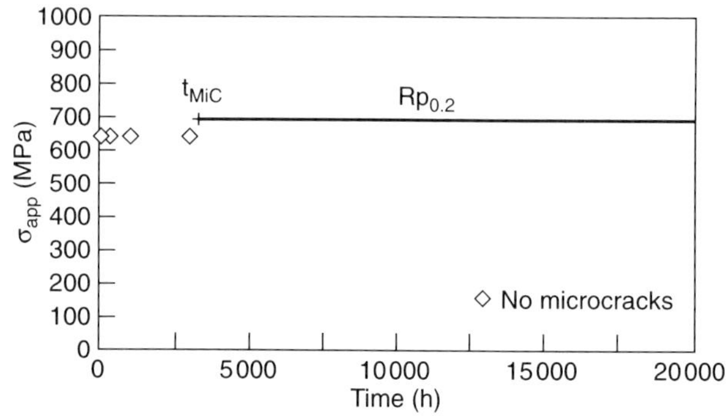

(a) Conventional 3NiCrMoV steel

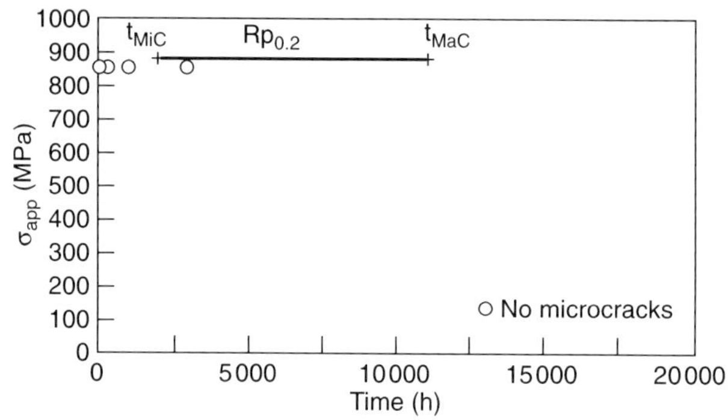

(b) Superclean 3.5NiCrMoV steel

Fig. 10 SCC tests in pure flowing water at 90°C

tional steel behaviour is consistent with previous experience (Fig. 8a), the higher strength superclean steel is not only susceptible to early microcracking in the SCC microcrack initiation regime (Fig. 8b) in a way untypical of previously investigated superclean steels.[2] The most highly stressed $Rp_{0.2}$ 875 MPa superclean specimens also reveal evidence of extensive macrocracking in the microcracking regime (Fig. 8b).

These observations were the result of tests using T6 Tensile testpieces, whereas the crack development diagram had been established on the basis of results from several hundred T4 tensile and 0.8 mm thick bent beam specimens. The σ_{app} = 850 MPa superclean specimen tests were therefore repeated using the traditional testpiece designs. The T6 specimen results for the 875 MPa superclean steel were reproduced in the T4 specimen tests (Fig. 8b).

It has always been acknowledged that superclean offers no advantage over conventional 3–3.5 NiCrMoV for $Rp_{0.2}$ >~900 MPa and the strength of the 875 MPa superclean material is very close to this value. Nevertheless, these facts do not account for the remarkably low SCC resistance of this steel.

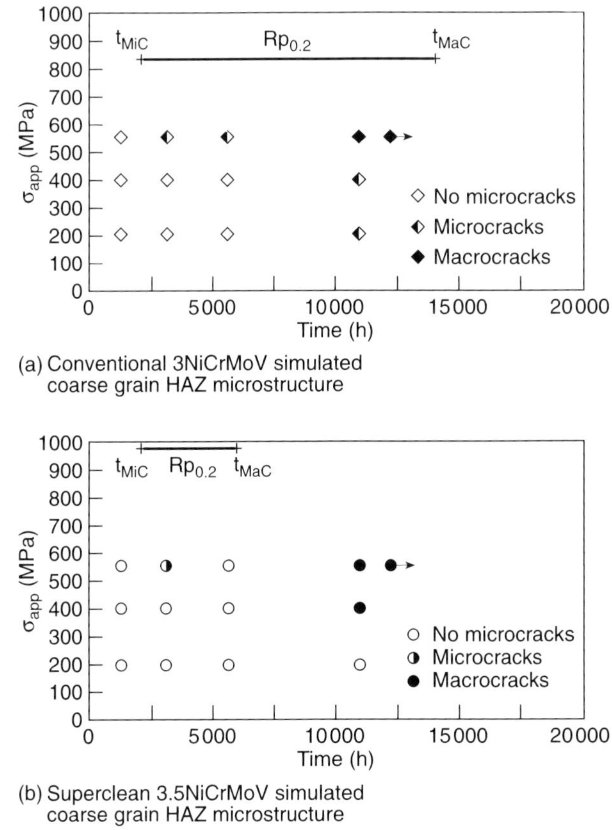

Fig. 11 SCC tests in pure low oxygen condensing steam at 95°C

At lower applied stresses, only microcracking is observed and the SCC susceptibility of the higher strength superclean steel is apparently no worse than that of the 700 MPa conventional steel. No microcracking is reported at an applied stress of 200 MPa (Fig. 8).

As a consequence of the experience described, high applied stress tests were performed on the 740 MPa superclean steel. ABB SCC tests on this material had indicated it to be superior to conventional steel of a similar strength level,[6] consistent with the findings of reference 2. The GECA pure steam tests on the WP8 steel are not as far advanced as those for the steels, but the results to date show that the properties of the 750 MPa superclean steel are not yet inconsistent with previous findings (Fig. 9).

The results of tests performed on the conventional and superclean steels in pure flowing water (Fig. 10) are consistent with the behaviour predicted by the pure steam crack development diagram in that only microcracking is observed to exposure durations of 3 000 h.

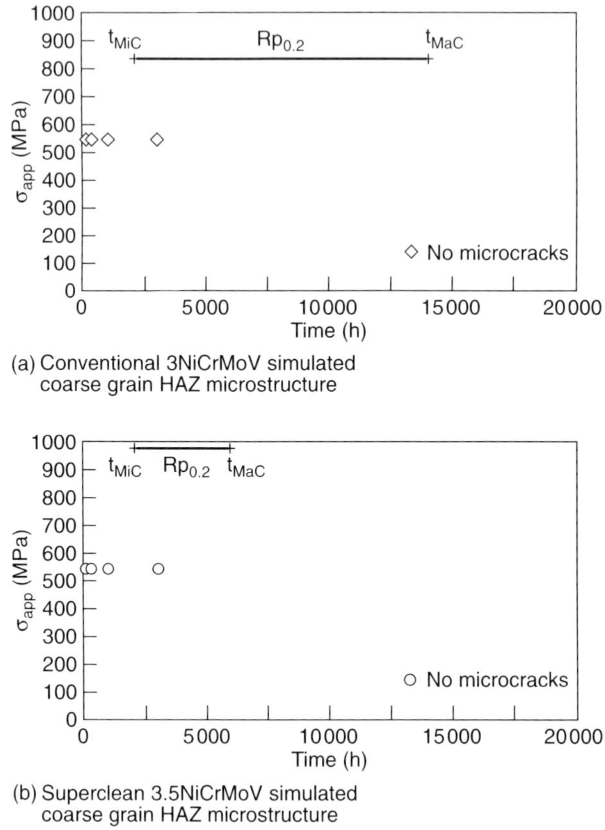

Fig. 12 SCC tests in pure flowing water at 90°C

In summary, the superclean 875 MPa 3.5 NiCrMoV parent steel is untypically prone to SCC at high applied stress levels close to $Rp_{0.2}$, while at lower applied stress levels its resistance to environmental cracking is no worse than that of conventional 700 MPa NiCrMoV (Fig. 8). There is strong experimental evidence to indicate that this superclean steel is particularly susceptible to low temperature creep when loaded to yield point stress levels. It is postulated that the material's low SCC resistance in these circumstances is linked to the repeated applications of strain to overcome stress relaxation during initial specimen loading in a regime where hydrogen embrittlement is likely to be an influential component of the cracking mechanism.

5.2 Welded Joints

5.2.1 Simulated coarse grain HAZ structures

The SCC initiation resistance of welded joints has been evaluated in two ways. In the first, it is assumed that the most vulnerable region of a NiCrMoV

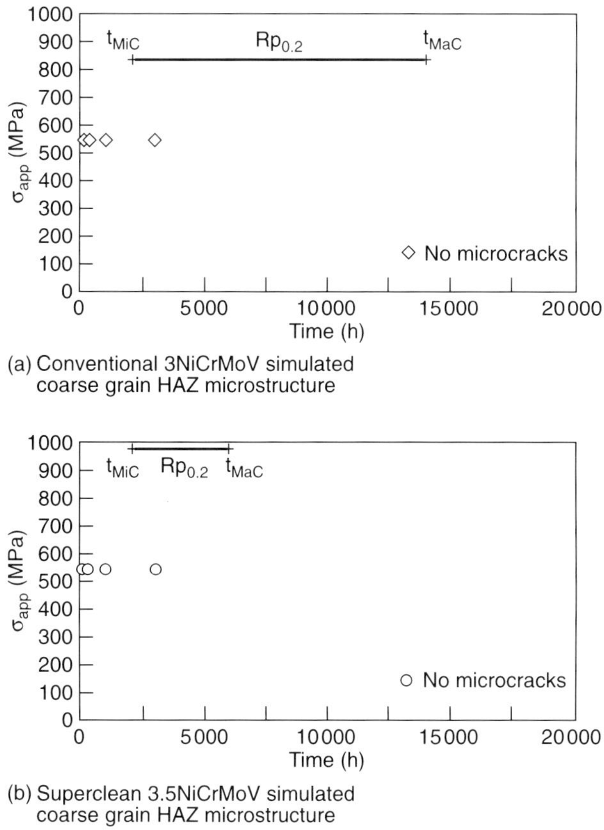

Fig. 13 SCC tests in pure flowing water with 100 ppb Na_2SO_4 at 90°C

weldment to SCC is the CGHAZ because of the high hardness of this metallurgical constituent. Tests have therefore been performed using specimens sectioned from homogeneous simulated CGHAZ material.

The results from the pure steam tests indicate that SCC macrocracking can occur in simulated CGHAZ microstructures for a welded conventional 700 MPa 3 NiCrMoV and a welded superclean 875 MPa 3.5 NiCrMoV steel after 10 000 h exposure at applied stresses of 400 to 550 MPa (Fig. 11). These microstructure are significantly more prone to SCC than their respective parent steels (Cf. Fig. 8).

There is no evidence of macrocracking or indeed microcracking at an applied stress of 550 MPa in pure water (Fig. 12) and water with 100 ppb Na_2SO_4 (Fig. 13) at 90°C.

5.2.2 Actual weldments

Cross weld tests using testpieces taken from a superclean weldment have been used to evaluate the SCC susceptibility of actual welded joints. Prior to

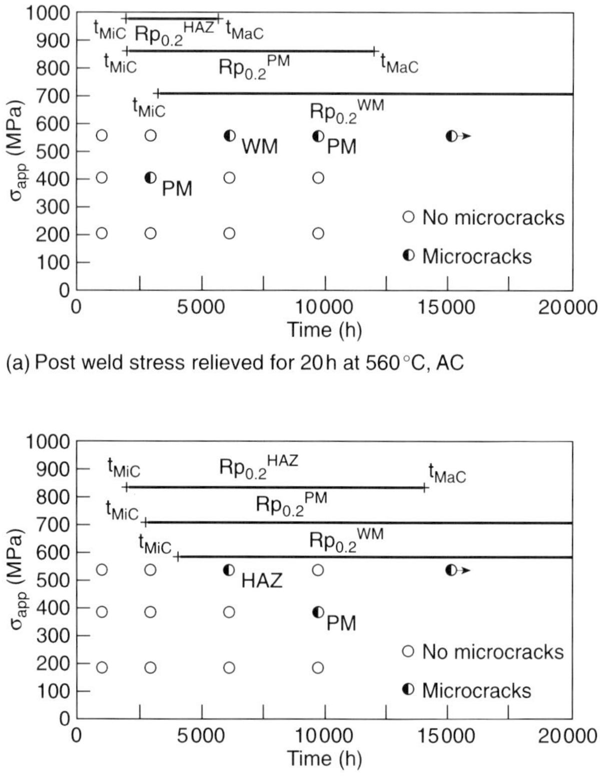

(a) Post weld stress relieved for 20 h at 560 °C, AC

(b) Post weld stress relieved for 20 h at 620 °C, AC

Fig. 14 Superclean 3.5 NiCrMoV weldment cross weld SCC tests in pure low oxygen condensing steam at 95°C

specimen removal, sections of the weldment were post weld stress relieved for (a) 20 h at 560°C (PWHT for 875 MPa parent material) and (b) 20 h at 620°C (PWHT for 700 MPa parent material).

In pure condensing steam at 95°C, only evidence of microcracking is evident after 10 000 h at applied stresses of 400 to 550 MPa (Fig. 14). This may occur in all microstructural constituents of the joint, but most usually in the parent material. It is emphasised that the CGHAZ regions are not clearly more vulnerable than the parent material or the weld metal, as might have been predicted from the homogeneous simulated CGHAZ microstructure tests.

There is no evidence of microcracking in cross weld specimens loaded to 200 MPa (Fig. 14).

SCC microcracking is not observed in cross weld specimens loaded to 550 MPa after 3 000 h immersion in pure flowing water (Fig. 15) and flowing water with 100 ppb Na_2SO_4 (Fig. 16) at 90°C.

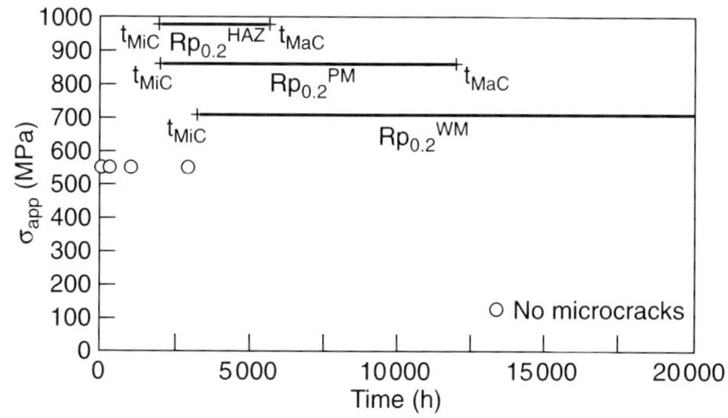

(a) Post weld stress relieved for 20 h at 560 °C, AC

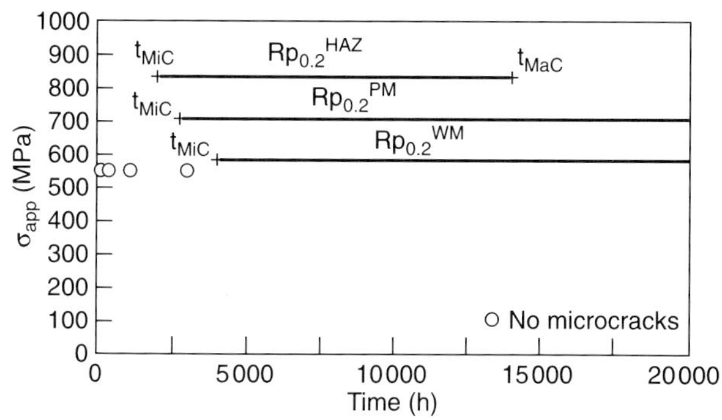

(b) Post weld stress relieved for 20 h at 620 °C, AC

Fig. 15 Superclean 3.5 NiCrMoV weldment cross weld SCC tests in pure flowing water at 90°C

6. CONCLUSIONS

A series of multi-specimen stress corrosion crack initiation tests have been performed in pure condensing steam at 95°C, and pure flowing water and water with 100 ppb Na_2SO_4 at 90°C, on conventional 700 MPa 3 NiCrMoV and superclean 875 MPa 3.5 NiCrMoV LP rotor steels in parent and simulated CGHAZ material conditions. In addition, cross weld tests have been conducted on specimens taken from a superclean 875 MPa 3.5 NiCrMoV weldment, post weld stress relieved for (a) 20 h at 580°C and (b) 20 h at 650°C. The following conclusions have been drawn during the course of post test metallographic inspection.

1 Superclean steel is more resistant to pitting than LP rotor steels of conventional purity in pure low oxygen condensing steam and simulated early condensate, particularly in terms of pit growth kinetics. The advantage is not clear in pure low oxygen flowing water at 90°C.

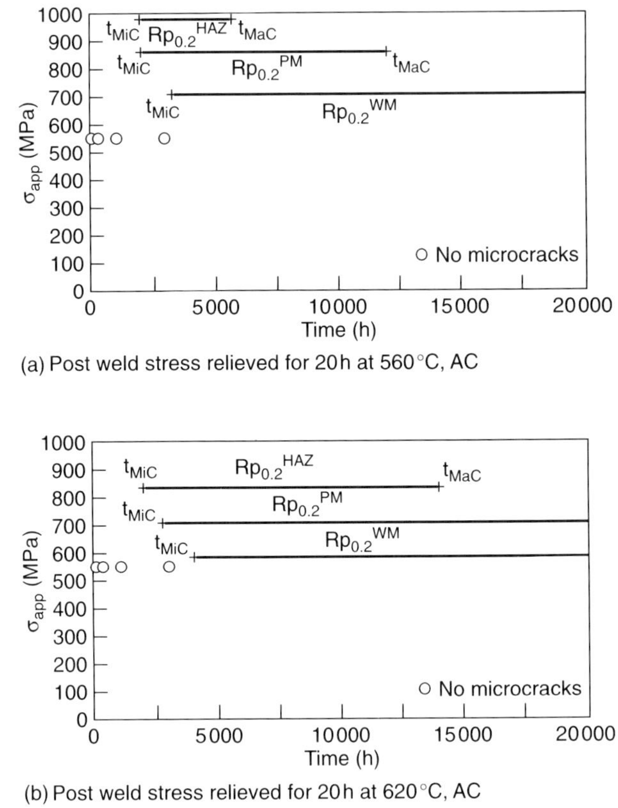

Fig. 16 Superclean 3.5 NiCrMoV weldment cross weld SCC tests in flowing water with 100 ppb Na_2SO_4 at 90°C

2 At applied stress levels close to $Rp_{0.2}$, the SCC initiation resistance of superclean 875 MPa 3.5 NiCrMoV is significantly less than that of conventional 700 MPa 3 NiCrMoV.
At lower applied stress levels, the SCC initiation resistance of the two steels both in the unwelded and PWHT welded conditions are similar.
3 The SCC initiation resistance of the 875 MPa superclean steel is significantly lower than that previously experienced for conventional steels of a similar strength level in pure low oxygen condensing steam when the applied stress is close to $Rp_{0.2}$.
4 The SCC initiation resistances exhibited by homogeneous simulated CGHAZ material for superclean 875 MPa 3.5 NiCrMoV and conventional 700 MPa 3 NiCrMoV at applied stresses of 400 and 550 MPa are less than the equivalent properties displayed by the respective parent steels. This lower SCC resistance of CGHAZ material is not evident in cross weld tests performed on material from an actual superclean weldment post weld heat treated to the same two strength conditions.

5 With stresses of 100 MPa in the vicinity of the joint, welded rotors have a high SCC safety margin.

REFERENCES

1. S. R. Holdsworth and G. Burnell: 'Stress corrosion crack initiation in LP turbine rotor steels', *COST Conf. Proc. High Temperature Materials for Power Engineering*, Liège, September 1990.
2. S. R. Holdsworth and D. V. Thornton: 'Resistance of superclean 3.5 NiCrMoV to stress corrosion cracking in pure steam', *Proc. Jaffee Memorial Symposium on Clean Materials*, TMS/ASM Materials Week, Chicago, 1992, November. On Clean Materials TMS/ASM Materials, TMS/ASM Materials Week, Chicago, 1992, November.
3. S. Bayard, M. Nougaret and A. Anis: 'Material Developments for large steam turbine rotors', Am. Power Conf., Nuclear Turbine Design and Operating Experience, US, 1986, April 1992.
4. C. Berger: '3.5%NiCrMoV – Superclean – EPRI test forgings', *Proc. Conf. on Superclean Rotor*, Dusseldorf, 1988, rep. 8.
5. J. Denk: 'Initiation of stress corrosion cracks LP steels', COST 501 WP8 (III) CH5 Final Report, July 1994.
6. M. Molière and Y. Verdier: 'Conception et realisation d'une boucle d'essais pour les études de corrosion en eau très pure', Métaux – Corrosion – Industrie, 984, 59, 81.
7. S. Inagki *et al*: 'Effects on material and environmental factors on SCC of NiCrMoV rotor steels', Proc. Fossil Steam Turbine Disc Cracking Workshop, Charlotte, 1990, October, GS-7250.

Susceptibility to Embrittlement of 4%Ni-CrMo High Strength Steels and 11%Cr-NiMoV Heat-Resistant Steel – Roles of Melting Methods and Residual Elements

J. BOURRAT and H. SCHAFF

Aubert et Duval, 41 rue de Villiers, 92202 Neuilly-sur-Seine, France

ABSTRACT

The security of turbine components critically depends on their fracture toughness after long time of service exposure at temperature, and hence on the brittle-ductile transition temperature after in service embrittlement. These characteristics depend at the same time on the initial quality of the steel, and on its susceptibility to embrittlement.

On two 4Ni-CrMo high strength low alloy steels (0.35C 3.8Ni 1.7Cr 0.25Mo and 0.30C 3.7Ni 1.25Cr 0.40Mo) and one heat-resistant 11%Cr-NiMoV grade, different levels of initial fracture toughness quality have been achieved by varying the method of melting (with and without consumable electrode remelting) and independently by varying the amount of residual elements. The embrittlement susceptibility is evaluated by testing after a Step-Cooling heat treatment. The results are discussed by trying to highlight the role of the initial quality, the intrinsic susceptibility of the grade to embrittlement, and the level of residual elements.

1. INTRODUCTION

The reliability of turbine components depends critically on their fracture toughness and thus on the brittle-ductile transition temperature after long term exposure to in service operating temperatures. The toughness after embrittlement depends both on the initial quality of the steel and its susceptibility to in service embrittlement.

The so called 'Superclean' steels containing very low levels of residual elements (phosphorus, arsenic, antimony, tin, etc.) and as low as possible in silicon and manganese are much less susceptible to in service embrittlement. It is difficult to know what level of impurities can be tolerated, for the fracture toughness of the component depends on numerous factors, among which can be cited:

The initial quality of the steel, that is, the K_{1c} transition curve of the component when new at the moment of entry into service. This does not

just depend on the level of impurities but also on the quality of the melting, thermomechanical transformation and heat treatment.
The inherent susceptibility of the steel, for we know that for the same level of impurities, some steels embrittle more than others.

The aim of this paper is to present some results obtained on production heats, on several steels and with several melting methods, to see how this problem of the permissible level of residual element content can be tackled.

2. EXAMPLE 1: HIGH STRENGTH STEEL – AFNOR GRADE 35 NCD 16

Four 'qualities' of this high strength 4%NiCrMo steel were melted as follows:

A: Electric arc melting with vacuum degassing.
B: As 'A' but with reduced levels of phosphorus and manganese and very low levels of Sn, As and Sb. We call this quality 'HP' (High Purity) for it is not strictly a true 'Superclean' steel which would require silicon and manganese contents of the order of 0.05%.
C: Vacuum arc remelting of an electrode of quality A.
D: Vacuum arc remelting of an electrode of quality B.

The resultant cast analyses are given below:

Table 1 Chemical analysis of four Afnor 35 NCD 16 heats (wt%)

	A	B	C	D
C	0.358	0.352	0.359	0.360
Si	0.36	0.15	0.36	0.17
Mn	0.42	0.42	0.45	0.33
S	0.002	0.001	0.001	0.001
P	0.0090	0.0035	0.0090	0.0030
Ni	3.75	3.73	3.75	3.74
Cr	1.68	1.69	1.70	1.70
Mo	0.29	0.26	0.28	0.27
Al	0.007	0.007	0.006	0.006
V	0.06	0.06	0.07	0.055
Co	0.09	0.013	0.09	0.013
Cu	0.05	0.02	0.03	0.02
Sn	30 ppm	20 ppm	29 ppm	19 ppm
As	44 ppm	31 ppm	48 ppm	30 ppm
Sb	8 ppm	5 ppm	7 ppm	4 ppm
O_2	21 ppm	20 ppm	12 ppm	10 ppm
N_2	45 ppm	44 ppm	37 ppm	20 ppm

From these ingots, bars of 150 mm diameter were rolled and heat treated to an U.T.S. of 1100/1250 MPa by austenitisation and hardening in oil from 875°C followed by a double temper (585°C/4 hours + 570°C/4 hours). Once heat treated the bars were characterised by tensile testing, determination of the transition temperature using Charpy V impact tests and room temperature K_{1c} fracture toughness tests using ASTM CT test pieces 50 mm thick, taken in the L-R direction.

To evaluate susceptibility to in service embrittlement test pieces from each melting route were subjected to a 'Step-Cooling' treatment:

Second tempering → cooling 30°C/h → 510°C/15 h → cooling 30°C/h →
482°C/24 h → cooling 30°C/h → 468°C/72 h → cooling 30°C/h →
427°C/100 h → cooling 30°C/h → 399°C/168 h → air cooling to room temperature.

The same tests (tensile, K_V, K_{1c}) were carried out after this embrittlement treatment and compared with the original results.

The tensile results are given in Table 2 and the transition curves in Figures 1 and 2.

Table 2 Tensile tests (transverse direction)

	As heat-treated				After Step-Cooling			
	UTS N/mm^2	0,2% YS N/mm^2	EI %	RA %	UTS N/mm^2	0,2% YS N/mm^2	EI %	RA %
A	1167	1051	12.7	56	1199	1087	15.0	55
B	1149	1022	14.4	59	1146	1039	13.9	54
C	1164	1038	14.0	60	1198	1080	15.8	56
D	1134	1009	15.9	66	1154	1038	16.0	65

The comparison of Figs 1 and 2 shows that the embrittlement is considerable: at the K_V = 40 Joules level for example, the difference is around 130°C for casts A and C, 100°C for B and 70°C for cast D. Nevertheless, it can be observed that at room temperature, the impact strength of bars from cast D remain very high (approx 90 J) and close to the upper shelf of the transition curve. This embrittlement test therefore shows that product D is quite satisfactory for those components exposed to service temperatures down to 0°C.

The room temperature K_{1c} tests confirm that products B and D maintain a very high toughness after embrittlement:

- After the standard heat treatment, all four steels give non-valid results according to the ASTM criteria, despite the section of the test blocks

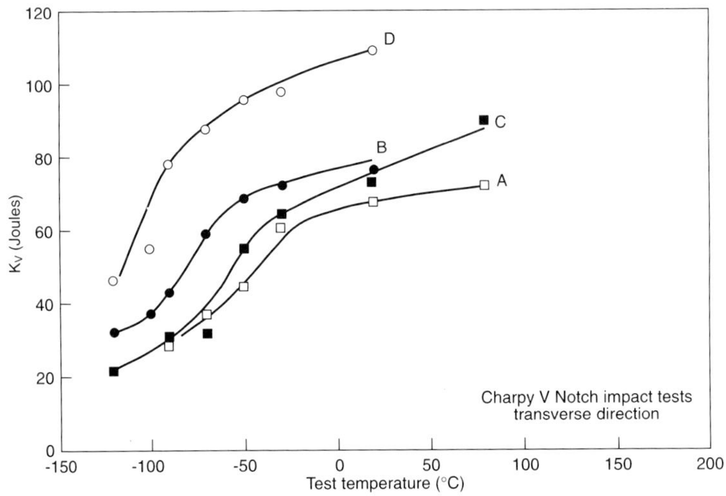

Fig. 1 CHARPY V transition curves in the as heat-treated condition (Afnor 35 NCD 16)

(50 mm): KQ being between 180 and 205 MPa$\sqrt{m}$, with the plasticity value Pmax/PQ between 1,18 and 1,40. The fracture faces are ductile with very pronounced plastic deformation dimpling.

- After Step-Cooling, the facets become brittle for products A and C. The lowest value obtained, corresponding to an impact value of around 30 Joules, is K_{1c} = 129 MPa$\sqrt{m}$ with Pmax/PQ = 1,00 and a totally flat

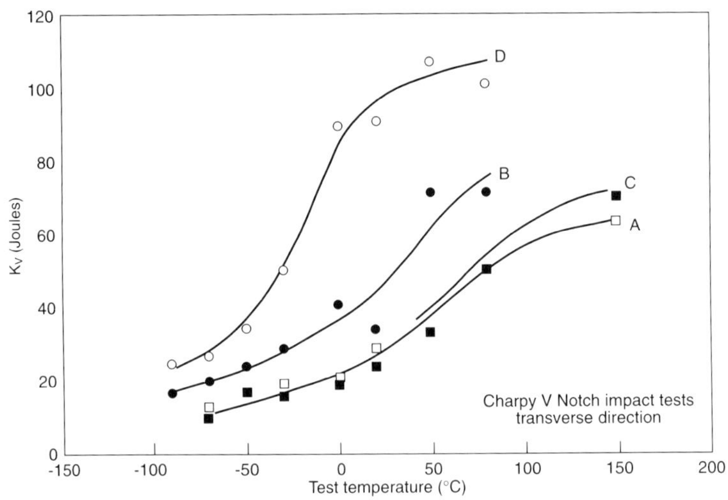

Fig. 2 CHARPY V transition curves after embrittlement by Step-Cooling (Afnor 35 NCD 16)

rupture surface. On the other hand, products B and D maintain a remarkably ductile behaviour: KQ = 206 and 218 MPa$\sqrt{m}$, Pmax/PQ = 1,24 and 1,28.

Collectively, these results show that product D (H.P. quality + remelting) exhibits a satisfactory behaviour vis-à-vis the Step-Cooling embrittlement test and a very great improvement compared to product C (classic quality + remelting).

A reduction in, but not complete elimination of silicon and manganese, and low residuals permits quite distinct improvements in this type of steel.

3. EXAMPLE 2: HIGH STRENGTH STEEL – AFNOR GRADE 30 NCD 16

A similar study has been conducted on 3 casts from a similar family:

E: Standard quality, without remelting
F: HP quality, without remelting
G: Standard quality, with remelting.

Table 3 Chemical analysis of three Afnor 30 NCD 16 heats (wt%)

	E	F	G
C	0.297	0.300	0.298
Si	0.39	0.15	0.37
Mn	0.46	0.27	0.53
S	0.001	0.001	0.001
P	0.0074	0.0026	0.0090
Ni	3.44	3.44	3.42
Cr	1.33	1. 33	1.29
Mo	0.43	0.43	0.43
Al	0.006	0.005	0.005
V	0.01	≤0.01	0.03
Co	0.07	≤0.02	0.07
Cu	0.08	0.02	0.03
Sn	60 ppm	18 ppm	25 ppm
As	90 ppm	15 ppm	50 ppm
Sb	16 ppm	4 ppm	15 ppm
O_2	13 ppm	16 ppm	15 ppm
N_2	55 ppm	40 ppm	45 ppm

The chemical compositions are of the same type as the preceding steel but with lower levels of carbon, nickel and chromium and a higher molybdenum content. The 150 mm diameter bars were heat treated to the same level of tensile properties.

Table 4 Tensile tests in the as heat-treated condition

	UTS N/mm^2	0,2% YS N/mm^2	EI %	RA %
E	1165	1039	17.3	64
F	1120	1010	18.4	72
G	1165	1049	17	68

Fracture toughness tests after Step-Cooling embrittlement revealed an important difference between the Afnor grades 30 NCD 16 en 35 NCD 16: Even for the standard quality casts E and G, the fracture mode remains ductile (Pmax/PQ > 1,10) which shows that for the same levels of impurities, this grade is intrinsically less susceptible to embrittlement.

4. EXAMPLE 3: HEAT RESISTANT STEEL 11%CrNiMoV

Three casts of a martensitic stainless steel of the type C = 0.12, Cr = 11.5, Ni = 2.8, Mo = 1.5, V = 0.30 were studied, in the form of 120 mm diameter rolled bars. The melting routes were as follows:

H: Standard quality
I: 'Superclean' quality with low levels of residual elements and very low level of silicon and manganese.
J: As 'I' followed by remelting under vacuum.

Table 5 Chemical analyses of three 11% Cr–NiMoV heats (wt%)

	H	I	J
C	0.116	0.122	0.114
Si	0.230	0.019	0.004
Mn	0.880	0.029	0.016
S	0.0010	0.0010	0.0010
P	0.0090	0.0025	0.0034
Ni	2.78	2.77	2.81
Cr	11.93	11.90	11.73
Mo	1.86	1.53	1.55
Al	0.013	<0.003	<0.003
V	0.288	0.370	0.340
Co	0.045	<0.020	<0.020
Cu	0.041	0.028	<0.020
B	13 ppm	<5 ppm	<5 ppm
As	44 ppm	44 ppm	43 ppm
Sb	6 ppm	3 ppm	2 ppm
Sn	21 ppm	13 ppm	10 ppm
N_2	0.026	0.022	0.010

After austenitisation at 1050°C and air cooling, the tempering temperatures were adjusted to obtain the same tensile properties across all three casts.

H: Tempered at 720°C/4 h + 650°C/4 h
I: Tempered at 675°C/4 h + 650°C/4 h
J: Tempered at 640°C/4 h + 620°C/4 h

It can be noted that the almost complete absence of silicon and manganese has resulted in a significant lowering of the tempering temperature. This was not the case with the casts of low alloy steel § 2 and 3 for which the same tempering temperature was used for the different qualities.

The tensile properties obtained in the transverse direction are as follows:

Table 6 Tensile tests (transverse direction)

	As heat-treated				After Step-Cooling			
	UTS N/mm²	0,2% YS N/mm²	El %	RA %	UTS N/mm²	0,2% YS N/mm²	El %	RA %
A	923	715	16.9	58	937	740	18.2	58
B	897	739	18.2	64	910	793	17.8	63
C	872	721	18.5	71	892	782	19.3	70

The transition curves of Charpy V impact values are given in Figs 3 and 4.

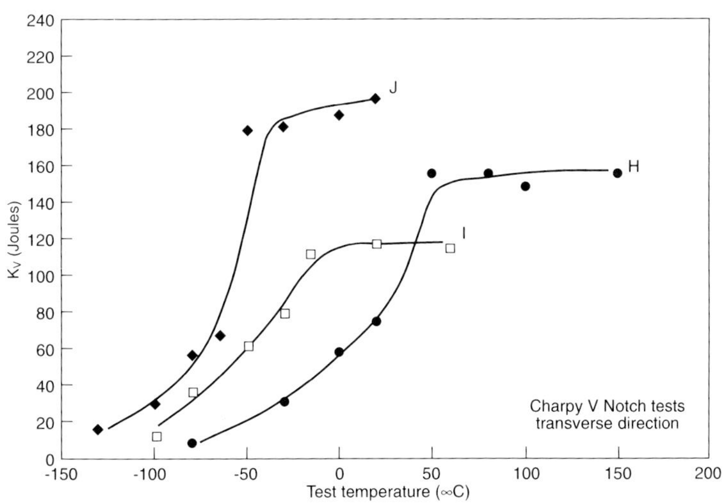

Fig. 3 CHARPY V transition curves in the as heat-treated condition (11% Cr–NiMoV steel)

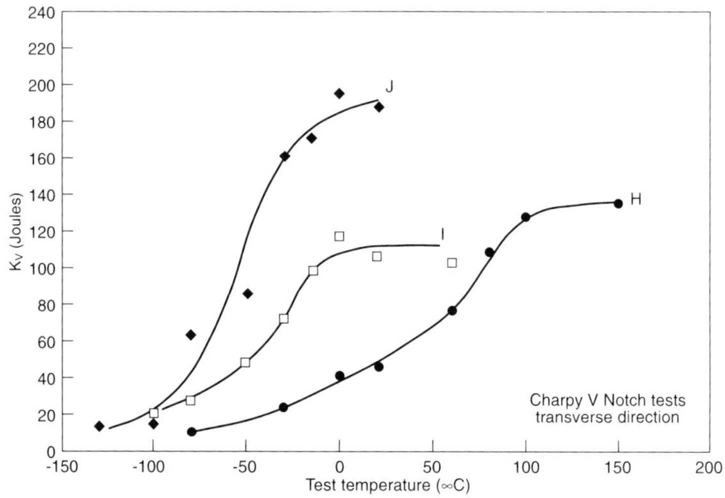

Fig. 4 CHARPY V transition curves after embrittlement by Step-Cooling

In the heat treated condition, there is a great divergence. The 'Superclean' cast I shows a clear improvement in the transition temperature, but also a drop in the ductile shelf. This drop is due to the mediocre metallurgical cleanness of this cast, this being linked to the difficulties of deoxydation which exist when melting some grades free of silicon and manganese. Remelting (cast J) allows a satisfactory metallurgical cleanness to be attained, resulting in an excellent impact strength transition curve.

Step-Cooling provokes a net embrittlement of the standard H cast (a differential of about 40°C) and has practically no effect on the 'Superclean' cast I and J. The transition curve of cast J remains very good.

5. DISCUSSION AND CONCLUSION

The following conclusions have been drawn from the three examples presented:

- For the Afnor steel grade 35 NCD 16, the remelted 'HP' quality (Si = 0.17, Mn = 0.33) exhibits significant embrittlement during step-cooling but retains an excellent impact transition curve and very high K_{1c} fracture toughness at room temperature.
- The Afnor steel grade 30 NCD 16 is less susceptible to embrittlement than 35 NCD 16. Within the same steel family therefore, relatively small variations in composition can result in significant improvement and optimisation, with regards to susceptibility to embrittlement.

- The Superclean 11%Cr (Si = 0.004, Mn = 0.016) steel is practically immune to embrittlement by Step-Cooling. Remelting by the consumable electrode method is necessary to achieve an optimum impact transition curve. This composition leads to a significant reduction in the tempering temperature. We think that it would be useful to study and compare the structural stability and the creep resistance of this type of steel during very long term exposure to temperature.

In order to define those levels of residual elements and silicon and manganese contents which are acceptable, it is necessary to answer the following questions:

What are the minimum K_V impact transition curves and K_{1c} values that will allow a turbine component to complete satisfactorily its required operational life?
How representative is the laboratory embrittlement test relative to real operational conditions?

The difficulty lies in choosing the embrittlement test. Knowledge of the physical phenomena responsible for embrittlement and the quantifying of their relative importance are necessary. Among the possible mechanisms we can cite:

The inter-granular segregation of 'tramp elements' such as phosphorus, arsenic, etc. leading to grain boundary embrittlement.
The precipitation of new phases, either intra or inter-granular.
The slow dissolution, or conversely, the slow growth, of existing precipitates.

It ought to be possible to develop a model to predict the kinetics of phase segregation or phase dissolution and through the resultant computation to define a representative accelerated embrittlement test giving meaningful results. For a given alloy type, such models could be based on the study of microstructural modifications observed on real components and validated by tests on laboratory scale casts with varying contents of those elements being studied.

Concerning, for example, that part of the embrittlement process linked to the segregation of 'tramp elements' at grain boundaries, one can ask whether the phenomenon of embrittlement does not stabilise after a certain time, at least in the case of steels of relatively high purity. If this 'saturation' exists, it should be possible to tune the initial quality of the alloy to the maximum intensity of embrittlement expected in service, thereby adjusting the initial quality and the strength level to achieve K_V and K_{1c} transition curves which exactly meet operational needs.

This type of approach would mean stipulating to the steel maker precise objectives, without prejudging the methods he should use to achieve them.

Pitting and Stress Corrosion Cracking of Conventional and High Purity LP Turbine Rotor Steel

J. DENK

ABB Power Generation Ltd., Materials Technology, CH-5401 Baden, Switzerland

ABSTRACT

Service experience and laboratory investigations showed that corrosion pits are often the initiation sites of stress corrosion cracks. Since non-metallic inclusions are the most likely sites of pit formation the behaviour of a high purity steel in terms of pitting and stress corrosion crack initiation was compared with conventional LP steels.

With electrochemical methods and with exposure tests at open circuit conditions the pit initiation and growth behaviour was investigated. In order to study the crack initiation loaded specimens were exposed to various environments. The spectrum of corrosion media ranged from very aggressive environments such as highly concentrated caustic solutions to mild media such as deaerated steam.

The high purity steel exhibited a significantly higher pitting resistance than the conventional steels. The role of pitting as nucleation sites of stress corrosion cracks in different media has been demonstrated. Consequently the high purity steel had a certain advantage if this type of crack initiation mechanism was predominant. In corrosion systems where a film rupture mechanism causes crack initiation the behaviour of HP and conventional steels is similar.

It was also shown that transient conditions can lead to early crack initiations.

1. INTRODUCTION

Low pressure steam turbine rotors are endangered by stress corrosion cracking. In the past decade significant efforts were undertaken to understand the mechanisms and to take this knowledge into account for the design and for the material selection.[1–10]

In many cases the initiation of stress corrosion cracks was associated with the presence of pitting. Non-metallic inclusions are the most likely sites of pit formation. Besides their stress concentration effects these structural defects which can develop local environments with higher aggressivity, especially when they are in contact with the bulk medium. The reduction of the content of steel impurities or the avoidance of the most sensitive types of inclusions should therefore result in a higher pitting resistance and so in a lower SCC susceptibility.

Hence the investigation of the pitting and stress corrosion behaviour of a high purity steel and the comparison with conventional steels seemed to be worthwhile. The work presented here was part of a COST 501 programme.[11]

2. MATERIALS

For LP rotors and disks a material is required which exhibits both a sufficient strength level and a high toughness at all locations of the component, i.e. a good forgability and hardenability are demanded. Low alloyed forged steels satisfy these demands. In order to increase the hardenability in the core of the forgings these steels are alloyed with nickel. The larger the dimensions of the forging the higher the nickel content has to be to obtain the same mechanical properties.

It is well-known that these steels are sensitive to embrittlement after long-term exposure at elevated temperatures due to the segregation of phosphorus and other impurities to the grain boundaries. This led to the development of high purity steels, which were interesting also from a corrosion point of view due to their low content of non-metallic inclusions.[12–13]

In the present programme the pitting and stress corrosion cracking behaviour of a high purity 26NiCrMoV145 steel (= 'HP 3.5NiCrMoV') was compared with a conventional 26NiCrMoV145 (= 'Conv. 3.5NiCrMoV steel'; this steel was investigated in two strength levels). The chemical analyses and the mechanical properties can be seen from the Tables 1 and 2.

Table 1 Chemical Composition (wt%)

Material	C	Si	Mn	S	P	Cr	Ni	Mo	V	Cu	Al
High purity 3.5NiCrMoV	0.27	0.04	0.03	0.003	0.004	1.51	3.61	0.32	0.11	0.06	0.001
Conventional 3.5NiCrMoV	0.28	0.17	0.26	0.006	0.008	1.59	3.70	0.42	0.11	0.04	0.001

The inclusion characteristics of the 3.5NiCrMoV types (conventional and high purity) were quanitified by a metallographic investigation. For comparison reasons an old rotor material with a high impurity content, designated as 'dirty steel', was included.

It appeared that the conventional steel had considerably more and larger inclusions than the HP steel (see Fig. 1 and Table 3). The larger inclusions of the conventional 3.5NiCrMoV were predominantly aluminium silicates with a rather inhomogeneous structure. These silicates were strongly extended in the forging direction and can have a length up to about 200 μm. Frequently MnS particles with a typical size between 10 to 40 μm were found. In the high purity steel small aluminium–magnesium oxide inclusions were observed; MnS particles were not detected.

Table 2 Mechanical Properties

Material	Temperature	Yield strength	Tensile strength	Elongation	Reduction of area	Absorbed Energy CharpyV
		$R_{p0.2}$	R_m	El	A	
	°C	MPa	MPa	%	%	J
High purity	RT	752	867	19.0	63	157
3.5NiCrMoV	100	722	825	18.0	64	175
steel	160	693	796	16.5	59	–
Conventional	RT	738	857	19.3	69	152
3.5NiCrMoV steel	100	709	821	19.0	70	–
lower strength	160	701	803	18.9	71	–
Conventional	RT	866	965	17.2	56	142
3.5NiCrMoV steel	100	844	924	17.0	56	141
higher strength	160	811	896	16.4	52	–

The density of small and mid-size particles of the 'dirty' steel was an order of magnitude higher than those of the conventional material; the maximum size however was somewhat lower. In contrast to the conventional 3.5NiCrMoV steel most of the inclusions of the 'dirty' 3Cr0.5Mo steel contained MnS.

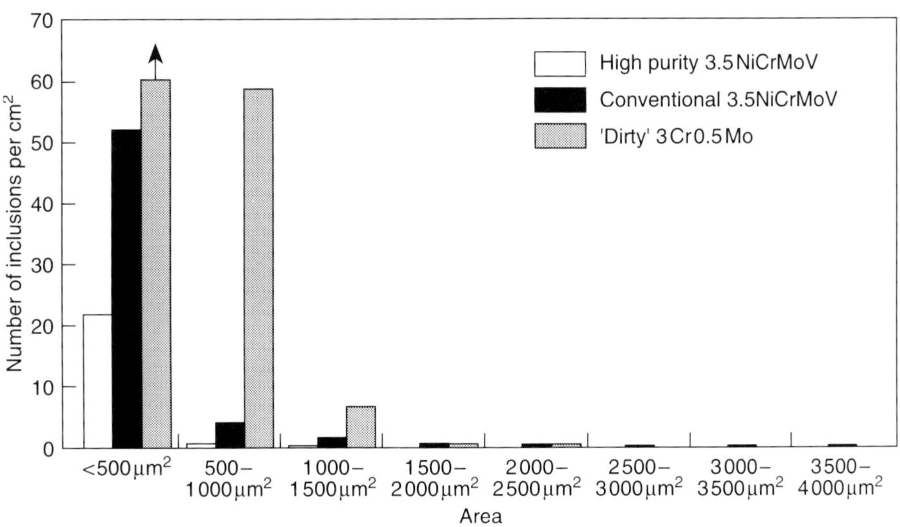

Fig. 1 Sizes (section areas) of non-metallic inclusions in LP steels

Table 3 Results of the Metallographic Investigations

	Materials		
	Conventional 3.5NiCrMoV steel	High purity 3.5NiCrMoV steel	'Dirty' 3Cr0.5Mo steel
Structure	Bainite	Bainite	Bainite
Grain size (ASTM)	7–8	8–9	3
Predominant inclusion types	Al-silicate Mn-sulphide	Al-oxide AlMg-oxide	Mn-sulphide Mn-silicate
Maximum length	190 μm	85 μm	148 μm
Maximum breadth	58 μm	28 μm	36 μm
Maximum area	3900 μm^2	1300 μm^2	2600 μm^2
Largest MnS inclusion	58 μm	none	148 μm

3. PITTING BEHAVIOUR

From the electrochemical point of view pitting occurs if the breakdown potential is exceeded, i.e. if the protecting passive film is locally destroyed. In the present work the activation behaviour of the conventional and high purity 3.5NiCrMoV steels was investigated by potentiodynamic measured current density – potential curves (polarisation rate 0.5 mV/min; 10 min prepolarisation at a cathodic potential).

Figure 2 shows the breakdown potentials (definition: U_B = potential, where the anodic current density exceeds 30 mA m^{-2}) versus the pH value in aerated and in deaerated 0.001 M NaCl solution. It can be seen that U_B increases with increasing pH value. At levels below about pH9 no real passivation occurred; i.e. the current density increases immediately after the transition from the cathodic to the anodic range. Above pH12 both types of steels are passive until oxygen development appears. If the lower bound curves are compared to those of the high purity 3.5NiCrMoV steel is shifted to lower pH values indicating a higher resistance to pit initiation.

A significant difference of the breakdown potentials in aerated and deaerated solutions was not observed. However, since the open circuit potential is considerably lower in deaerated media the difference between open circuit and breakdown potential is much larger if oxygen is removed.

The breakdown potential is reduced by the presence of aggressive ions such as chlorides. Figure 3 shows this influence for two pH levels. The HP steel exhibited again a larger range of passivity (characterised by the lower bound curves of the breakdown potentials) than the conventional steel.

The pit initiation and growth behaviour under open circuit conditions was investigated in a number of environments. At a free corroding metal surface

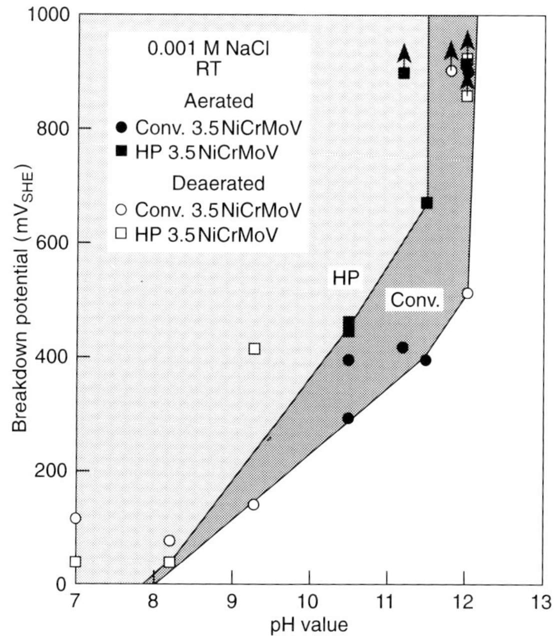

Fig. 2 Influence of the pH value on the breakdown potentials of conventional and high purity 3.5NiCrMoV steels in 0.001 M NaCl

anodic and cathodic areas form simultaneously. In contrast to the uniform corrosion, where these zones are unstable in terms of time and location, the types of local corrosion are characterised by stable anodic/cathodic areas.

If the specimens were exposed to stagnant aerated water, pits occurred in relatively short times. In refreshed water (circulation of the media through an ion exchanger) no macroscopic visible pit initiation took place. Even after long-term exposure the surfaces of specimens seemed not to be attacked. In the microscope however very small pits with typical diameters of about 20 μm and with depths of less than 5 μm were detected. Due to the circulation of the medium corrosion products were continuously removed and so the formed pits repassivated at a very early stage. This illustrates how important the deposition of corrosion products at the surface around the formed holes is for the development of stable anodic and cathodic zones and so for the pit growth kinetics. The pits of the conventional and of the HP steels had similar sizes under these refreshed conditions, the density however was found to be significantly higher in the conventional steel.

In contrast to aerated water at 80°C, where all steels showed similar pitting behaviour, a significant difference between conventional and HP steels was found in deaerated water and in water with pH9 (NH_3 addition). Under these milder conditions no pits occurred in the HP steel during the test periods while pits appeared in the conventional steels.

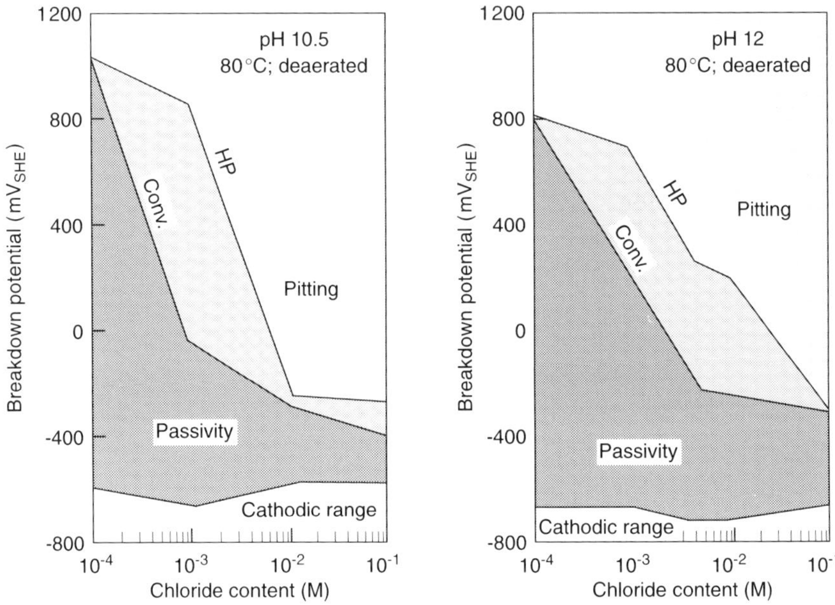

Fig. 3 Influence of the chloride concentration on the breakdown potentials of conventional and high purity 3.5NiCrMoV steels for two pH levels

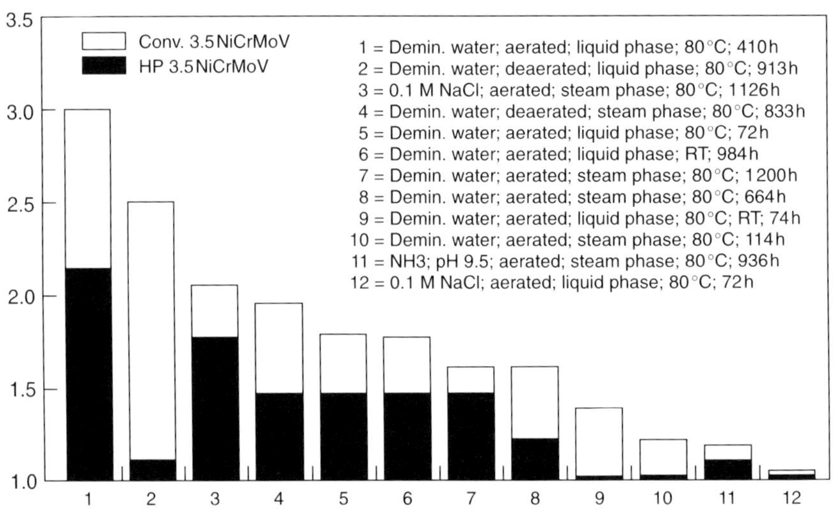

Fig. 4 Mechanical notch effect factor K_f of pits after exposure in various environments. K_f was measured by high cycle fatigue tests of precorroded specimens with stepwise increasing load amplitudes at $R = 0$

A good method to quantify the amount of damage due to corrosion is the determination of the notch effect factor in fatigue tests. Therefore HCF specimens were exposed to various media (steam and liquid phase). After the exposure the fatigue strength in air at room temperature of these specimens was measured using a computer controlled resonance fatigue machine. The load amplitude was increased stepwise until fracture appeared. The results of this method for uncorroded specimens were compared to conventionally determined HCF data and reasonable agreement was obtained. The notch effect factor K_f was calculated by relating the fatigue strengths of uncorroded and of precorroded specimens. The broken specimens were investigated in the scanning electron microscope in order to determine the size and the shape of the surface defect which had initiated the fatigue crack.

The measured notch effect factors are summarised in Fig. 4. The HP steel showed generally lower K_f values after all precorrosion conditions. The largest difference appeared in the liquid phase of deaerated water. It should be emphasised that this pure mechanical effect of pits can reduce the fatigue strength by more than a factor of 3.

4. STRESS CORROSION BEHAVIOUR

An essential part of the present work was the investigation of the initiation of stress corrosion cracks; i.e. the exposure of loaded specimens to various

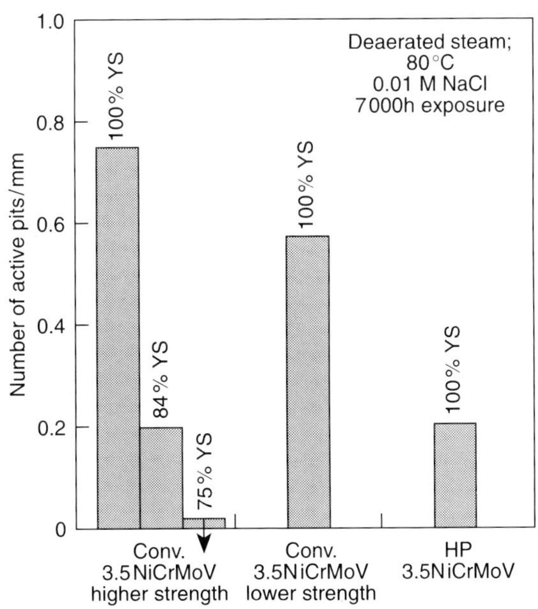

Fig. 5 Density of active pits after 7000 h exposure in steam of 0.01 M NaCl solution

environments. The spectrum of corrosion media reached from very aggressive environments such as highly concentrated caustic solutions to mild media such as deaerated steam. For the tests C-ring specimens according to ASTM G38 outer diameter : 40 mm; thickness 2.5 mm; breadth 10 mm) were used which were loaded in metallic or ceramic load frames.

Such C-ring specimens were loaded to their yield strength and exposed to various steam environments at 80°C. It appeared that crack-like defects were found only in the deaerated steam environments. In aerated steam the pits obviously repassivated and no active pits or microcracks developed. These crack-like extensions of pits were designated as 'active' pits; i.e. as stress-oriented pit growth. Slight indications of intergranular mode were observed only at a few pits of the conventional 3.5NiCrMoV steels. The depth of the active pits and the ratio of depth and width were higher in the conventional

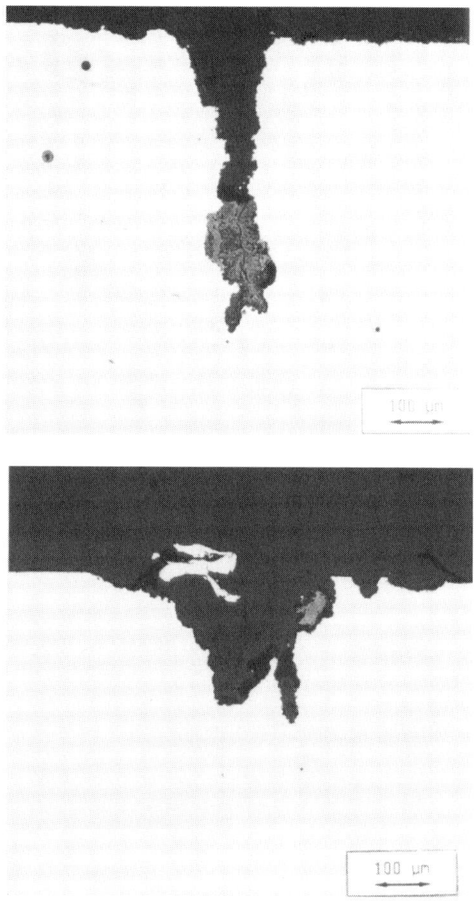

Fig. 6 Active pits (pits with crack-like extensions) in conventional (top picture) and in high purity (bottom picture) steels after 7000 h exposure in deaerated steam at 80°C

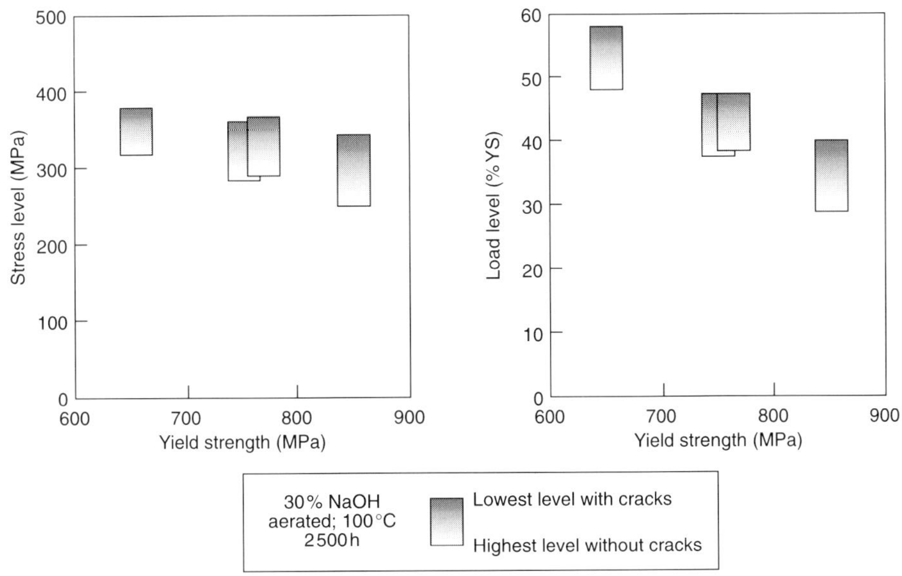

Fig. 7 Crack initiation in absolute and relative terms after 2500 h exposure in sodium hydroxide solution

steels. The most significant difference however was found in the density of these defects. In the chloride-containing steam a relatively homogeneous distribution of active pits appeared and allowed a quantitative determination of the density (Fig. 5). This diagram also indicates that stress levels in the range of the yield strength are necessary to form active pits. Figure 6 shows typical forms of active pits in conventional and in HP steels.

In corrosion systems were the stress corrosion cracks are not formed at corrosion pits no significant differences were observed between conventional and high purity steels.

The susceptibility of mild and low alloyed steels to stress corrosion cracking in strong alkaline media has been well-known since the beginning of the century. Such solutions are certainly not typical steam turbine environments, however they cannot be completely excluded during abnormal service conditions. Therefore a number of tests were performed in strong caustic media indicating the limit stresses under worst case conditions.

In 30% sodium hydroxide solution at 100°C intergranular stress corrosion cracks were observed in all types of the steels investigated if stresses above 300 to 400 MPa were applied. Figure 7 characterises the initiation behaviour for an exposure time of 2500 h in terms of the highest load level without SCC cracks and the lowest load level with SCC cracks and relates this transition range to the strength levels of the steels. The appearance of the intergranular crack mode was used as initiation criteria. If absolute stresses are considered the initiation of stress corrosion crack is almost independent of the strength

Fig. 8 Intergranular stress corrosion cracks in highly stressed specimens (110% YS) after exposure in high temperature steam (150°C; 6.5 bar) and transient test conditions (unloading cycles for intermediate inspections)

level. This means of course that higher strength materials have a lower resistance in terms of relative loading. Metallographic investigations show that the initiation sites of stress corrosion cracks are cracks in the relatively thick hydroxide surface film (50 to 10 μm). This observation and the fact that the limit stress is almost independent of the strength of the steel indicate a strain controlled crack initiation mechanism.

Typical intergranular cracks were also observed in highly stressed specimens of all types of steels after exposure in the steam phase of demineralised

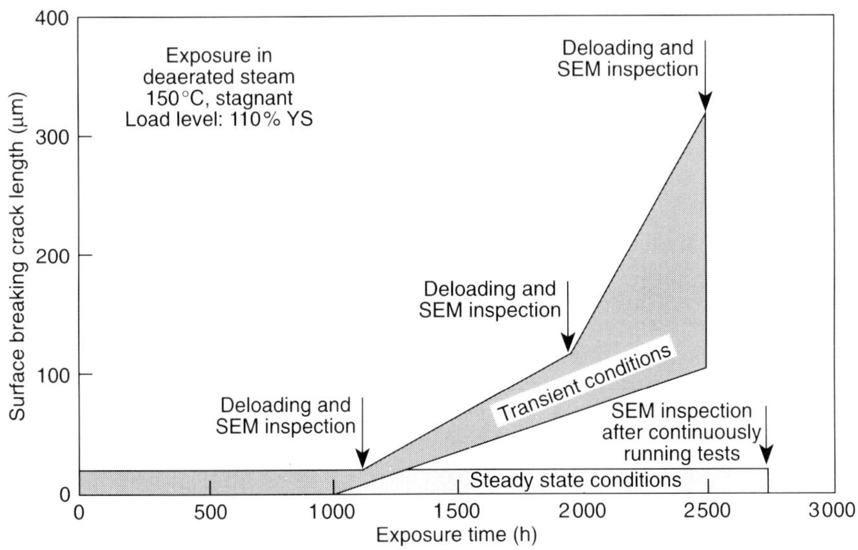

Fig. 9 Intergranular stress corrosion cracks appeared in high temperature steam if stress levels above the yield strength and transient test conditions were applied

water under autoclave conditions (150°C; 6.5 bar; see Fig. 8). However, these cracks appeared only if the tests were interrupted and the specimens were unloaded for intermediate inspections; no detectable cracking was found in specimens after a continuous exposure (Fig. 9). Under the applied conditions no pitting occurred but a dark magnetite film was formed on the surface. Obviously the cracks are caused by a film-rupture mechanism if transient conditions are present and if stresses above the yield strength are applied. These tests indicate that besides the applied stress level and the environmental conditions additional parameters (such as the loading history) influence the crack initiation.

In specimens, which were precorroded in aerated water prior to the loading and the exposure in the high temperature steam, intergranular cracks were formed at the pits. Due to the higher pitting resistance the depths of these cracks were significantly lower in the HP steel (Fig. 10).

5. DISCUSSION

The electrochemical measurements showed that a certain alkalisation of the environment is necessary in order to achieve stable passivity of low pressure rotor steels. As a consequence in many environments local attack in the form of pits or trough corrosion appeared under open-circuit conditions. The precipitation of corrosion products in the surroundings of the surface defect plays an important role for the stabilisation of anodic and cathodic areas and

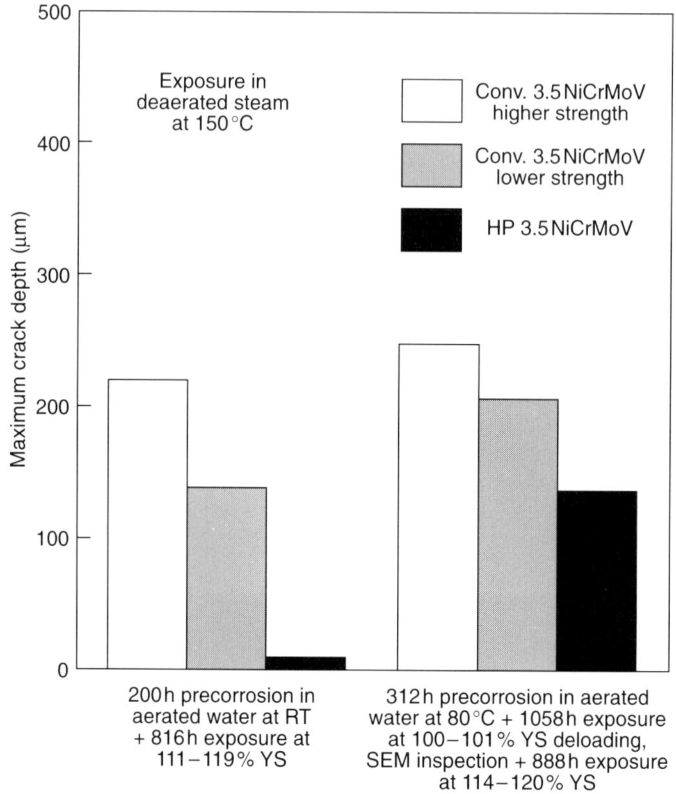

Fig. 10 Initiation of intergranular stress corrosion cracks after in high-temperature steam after pre-corrosion prior to the loading and exposure in the autoclave

so for the growth of the pit. If these deposits are avoided by the circulation of refreshed water the initiated pits repassivate in an early stage and no macroscopic visible attack appears.

Pits are mechanical notches with considerable stress concentration. In high cycle fatigue tests of precorroded specimens maximum notch effect factors of about 3 were observed (i.e. the fatigue strength in air at ambient temperature was reduced by a factor of 3).

The high purity steel showed, in a number of test media, an improved pitting resistance. Differences were observed especially in environments with intermediate aggressivity. This tendency was consistent with results of electrochemical measurements.

In long-term exposure tests of loaded specimens in different steam environments crack-like extensions were found at a number of pits. This stress-oriented growth of active pits is considered as the transition phase to the intergranular growth mode of stress corrosion cracks. It was surprising that these active pits appeared only in deaerated steam environments; in aerated media only hemispherical pits were observed. The usual growth kinetics of pits with decreasing penetration rates is obviously restrained in low-oxygen environments. This may be an effect of the different cathodic reaction and the different composition and structure of the corrosion products.

Due to the higher pitting resistance the high purity steel exhibited a considerably lower density of active (crack-like) pits. Also the depths and the ratios depth/width of these defects were lower. Consequently high purity steels with improved pitting resistance have a certain advantage if this type of crack initiation mechanism is predominant. In corrosion systems where a film rupture mechanism causes crack initiation the behaviour of HP and conventional steels is similar.

It was also demonstrated that transient conditions can lead to early crack initiation. Since real service conditions, with alternating stress/strain levels in start/stop cycles and steady state phases and different environmental conditions during service and shutdown, are relatively complex, more detailed work should be performed on this aspect.

REFERENCES

1. Lindinger and Curren: *Materials Performance*, Feb. 1982, 22–26.
2. Lyle and Burghard: *Materials Performance*, Nov. 1982, 35–44.
3. Lyle McMinn and Leverant: *Proc. Instn. Mech. Engrs.*, 1985, **199**, 59–67.
4. Czajkowski and Weeks: *Materials Performance*, March 1983, 21–25.
5. Speidel and Magdowski: *2nd Int. Symp. on Environmental Degradation of Materials in Nuclear Power Plants*, Montery, Sept. 9–12, 1985.
6. Curran: *EPRI Turbine Disc Integrity Seminar*, Minneapolis, September 21–23, 1982.
7. Speidel: *EPRI Seminar on Low Pressure Turbine Disc Integrity*, San Antonio, December 1–2, 1983.
8. Aubry and Duchateau: *Technology of Turbine Plants Operating with Wet Steam*, London, 1988.
9. Schleitthoff, David and Schmitz: *DVM-Arbeitskreis Bruchvorgänge*, 1988, 167–180.
10. Speidel, Denk and Scarlin: *Review of COST 505; ECI COST-0010-CH; EUR 13186 EN*, ISBN 92-826-2017-4, 1991.
11. Denk: *Final Report COST 501, WP8, Project CH5*, July 1994.
12. Jaffee: 'Super Clean 3.5%NiCrMoV – Steels and improve CrMoV – Steels', *Proc. VDEH Special Meeting*, Düsseldorf, 31 May–1 June 1988.
13. *EPRI Superclean Steel Workshop*, Sapporo, Japan, August 30–31, 1989.

100 000H Isothermal Ageing Test Results of NiCrMoV Steels for Low Pressure Steam Turbines

Y. TANAKA, T. AZUMA, N. YAEGASHI and Y. IKEDA

The Japan Steel Works, Ltd 4 Chatsu, Muroran 051, Japan

ABSTRACT

Currently, NiCrMoV steels have been widely used for low pressure steam turbine rotor shafts in fossil power generation plants. However, NiCrMoV steels have a strong potential to temper embrittlement and the temper embrittlement problem of the turbine material has been one of the largest concerns for the safe operation of power plants. On the other hand, the inlet steam temperature of LP turbines is progressively increasing in the newly designed power plants such as an ultra super critical unit. In these applications, the temper embrittlement susceptibility of the LP turbine material is a critical problem. As a solution to the temper embrittlement problem, superclean material and its production technologies have been established and several superclean rotors are already in operation in Japan. In the present paper, the results of isothermal ageing tests up to 100,000 h, by using LP turbine materials made around 15 years ago, are introduced. The effect of chemical and impurity contents on the temper embrittlement behaviour is discussed based on the data. The results show that the comparatively small J-factor material still has a significant sensitivity to temper embrittlement and this may suggest that the superclean material should be used for NiCrMoV rotor forgings for higher temperature application.

1. INTRODUCTION

NiCrMoV steels are extensively used for low pressure steam turbine rotor shafts in power generation plants since the material has good strength versus toughness balance. However, the service temperature of the NiCrMoV low pressure steam turbine is generally controlled below 343°C (650 °F), due to the strong potential of the material to temper embrittlement, which may cause appreciable degradation of fracture toughness and loss of reliability of the turbine. Due to the strong potential of NiCrMoV steel to temper embrittlement, attention has been paid to the long term embrittlement susceptibility of the steel in order to assess the safety and reliability of steam turbines. The mechanism of temper embrittlement has been investigated in parallel with the long term embrittlement tests.[1–7]

Recently the operating temperature of steam turbines has progressively increased and is expected to reach 400°C in the advanced type of power generation plants. The temperature falls just in the range where the material may suffer from significant temper embrittlement. Under these circumstances, the test results of long-term isothermal ageing of NiCrMoV steel must be useful to assess the reliability of low pressure steam turbines, but also to discuss the required purity in order to eliminate the temper embrittlement susceptibility.

The Japan Steel Works, Ltd initiated long time ageing tests of NiCrMoV forging steel around 15 years ago. The forgings at this time were produced by using basic electric furnace and cast via the vacuum carbon deoxidation (VCD) process. Recently, exposure times have reached 100,000 h and impact tests were performed on these materials.

These materials have comparatively high contents of residual elements which may raise the temper embrittlement susceptibility. However, the isothermal ageing test data of such a material may contribute to predicting the temper embrittlement sensitivity of recently produced high purity materials. This paper summarises the embrittlement test data on NiCrMoV steel, aged up to 100,000 h. The data will be analysed by focusing on the effects of residual elements. The ageing test data of superclean NiCrMoV steel will be introduced.

2. MATERIALS TESTED AND EXPERIMENTAL PROCEDURE

The centre cores of seventeen heats of NiCrMoV steel rotor forgings were aged in the present investigation. These rotor forgings were produced from ingots weighing from 75 tonnes to 500 tonnes. All the materials were cast via the vacuum carbon deoxidation (VCD) process. The diameter of the forgings were from 1,200 mm to 1,800 mm. The chemical compositions of these test materials are summarised in Table 1. These NiCrMoV steels includes four heats of 3.8%Ni, nine heats of 3.5%Ni and 2.8%Ni grade steels. These materials were removed after a quenching and tempering heat treatment. Table 2 summarises the mechanical and impact properties of the steels tested.

The test materials were aged isothermally in specially designed electric furnaces controlled at 343, 399 and 454°C for 3000, 10,000, 30,000 and 100,000 h. The ageing temperatures were controlled within 2°C of the aim temperature. After ageing, Charpy impact test specimens in the longitudinal direction were machined and tested. The test materials were also subjected to the step cooling heat treatment[8] as shown in Fig. 1 or de-embrittlement heat treatment by water quenching from 595°C. After Charpy impact testing, metallurgical investigations were performed.

Table 1 Chemical composition of the materials tested

Steel (wt%)	C	Si	Mn	P	S	Ni	Cr	Mo	V	As	Sn	Sb	J-factor
A	0.26	0.07	0.29	0.008	0.010	3.36	1.73	0.41	0.12	0.010	0.008	0.0019	57.6
B	0.23	0.02	0.31	0.009	0.007	3.35	1.76	0.40	0.13	0.012	0.013	0.0034	72.6
C	0.24	0.02	0.26	0.009	0.008	3.40	1.74	0.43	0.10	0.010	0.010	0.0027	53.2
D	0.24	0.05	0.30	0.010	0.009	3.59	1.87	0.46	0.14	0.014	0.014	0.0027	84.0
E	0.29	0.06	0.36	0.011	0.009	3.56	1.92	0.45	0.15	0.016	0.016	0.0030	113.0
F	0.26	0.01	0.27	0.009	0.012	3.41	1.80	0.42	0.12	0.012	0.013	0.0025	61.6
G	0.33	0.05	0.31	0.009	0.009	3.09	1.89	0.41	0.14	0.012	0.014	0.0028	82.8
H	0.29	0.05	0.33	0.008	0.007	3.51	1.86	0.39	0.13	0.014	0.014	0.0022	83.6
I	0.32	0.01	0.22	0.006	0.010	3.80	1.87	0.41	0.10	0.011	0.009	0.0014	34.5
J	0.29	0.03	0.31	0.008	0.007	3.57	1.79	0.41	0.13	0.012	0.013	0.0021	71.4
K	0.26	0.01	0.27	0.007	0.006	3.39	1.80	0.39	0.09	0.010	0.011	0.0013	50.4
L	0.23	0.05	0.27	0.007	0.008	2.96	1.66	0.35	0.09	0.011	0.011	0.0024	57.6
M	0.25	0.02	0.28	0.007	0.009	3.76	1.81	0.39	0.10	0.010	0.010	0.0020	51.0
N	0.24	0.01	0.28	0.008	0.005	3.79	1.88	0.42	0.12	0.010	0.009	0.0020	49.3
O	0.24	0.03	0.28	0.008	0.008	3.05	1.56	0.32	0.11	0.011	0.010	0.0030	55.8
P	0.30	0.05	0.29	0.008	0.008	2.79	1.76	0.39	0.10	0.013	0.011	0.0028	64.6
Q	0.29	0.01	0.21	0.005	0.010	3.86	1.81	0.37	0.10	0.010	0.007	0.0014	26.4

Table 2 Mechanical properties of the materials tested

Steel	0.02%YS (MPa)	0.2%YS (MPa)	TS (MPa)	EL (%)	RA (%)	FATT		
						As-Temp. (°C)	De-emb. (°C)	Step cool (°C)
A	705	754	879	19	62	−65	−82	9
B	702	746	875	21	67	−12	−33	26
C	670	710	841	21	68	−33	−50	5
D	720	745	845	21	68	−10	−50	9
E	704	734	873	23	67	−14	−59	32
F	714	768	885	22	65	−25	−55	18
G	600	627	783	27	69	−50	−54	−21
H	618	646	782	26	71	−63	−77	−229
I	748	822	959	20	60	−53	−87	−33
J	716	752	892	21	64	−2	−23	44
K	690	717	849	22	71	−46	−59	−24
L	–	602	714	21	72	−28	−44	−20
M	–	797	925	21	62	−15	−18	18
N	–	777	903	19	67	8	−14	23
O	–	600	717	24	74	−46	−48	−32
P	–	620	760	22	70	−22	−28	14
Q	748	822	963	20	60	12	−12	37

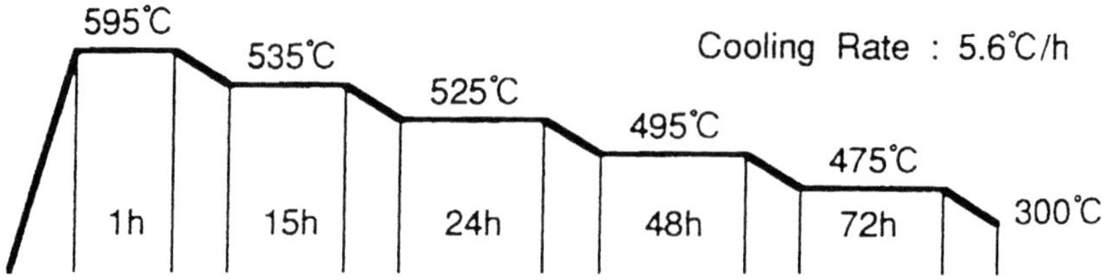

Fig. 1 Step cool heat treatment

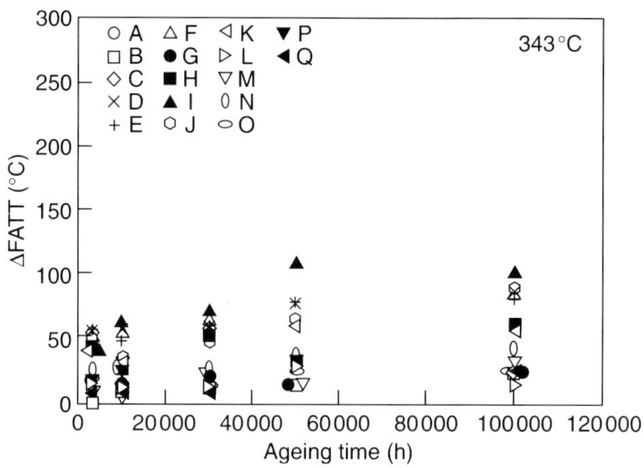

Fig. 2 Change in FATT by ageing at 343°C

3. 100,000 H ISOTHERMAL AGEING TEST RESULTS

Figures 2 to 4 show the change in FATT (Δ FATT) of all the steels aged at 343°C, 399°C and 454°C up to 100,000 h. Δ FATT is defined to be the increase of FATT from the FATT of the de-embrittled condition.

The Δ FATT at 343°C, which is the typical operating temperature for the majority of the low pressure steam turbines of conventional power plants, is a maximum of 100°C, and the average embrittlement is around 50°C, which is similar to those obtained by Newhouse.[9] On the other hand, materials aged at 399°C and 454°C suffered the maximum embrittlement of 250°C. The data spread of the large scatterband is due to the differences in chemistry, including the amount of residual elements, strength and microstructure.

Figure 5 shows a typical example of embrittlement behaviour at three ageing temperatures. The progress of embrittlement can be characterised as follows. At the higher temperature, 454°C, embrittlement proceeds rapidly and levels off after 50,000 h whereas the embrittlement proceeds up to

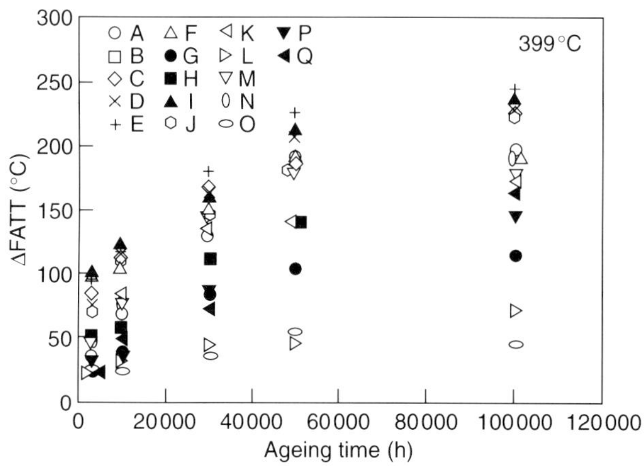

Fig. 3 Change in FATT by ageing at 399°C

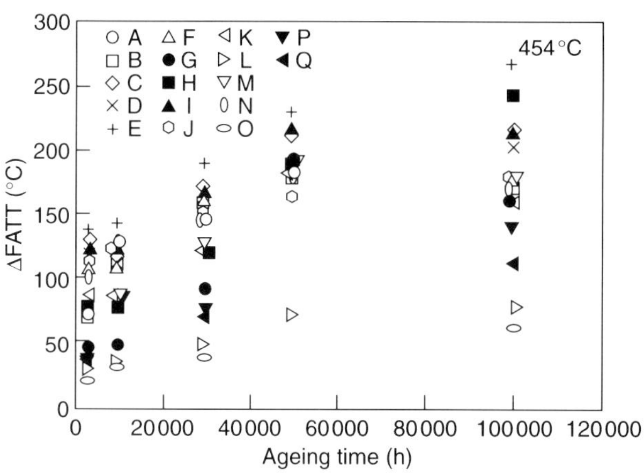

Fig. 4 Change in FATT by ageing at 454°C

100,000 h at 399°C. At the lower temperature, 343°C, embrittlement proceeds comparatively slowly due to the low diffusion rate at this temperature.

4. EFFECT OF IMPURITY ELEMENTS

It is well known that the segregation of residual elements such as P, Sn, As and Sb to the grain boundary causes the embrittlement of materials. This type of embrittlement occurs at a specific temperature range, between 350°C and

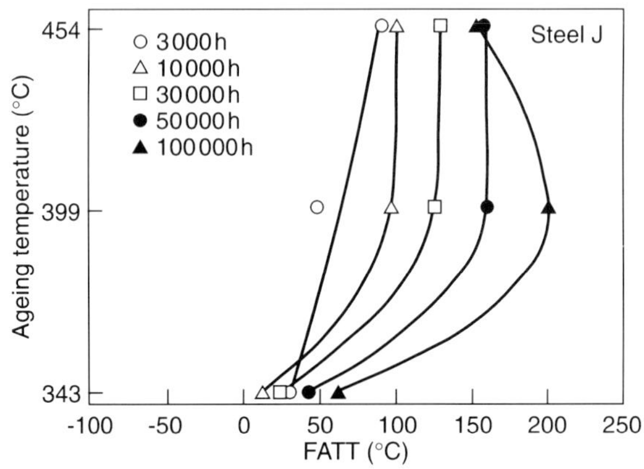

Fig. 5 Change in FATT at three ageing temperatures for steel J

550°C. Besides the residual elements mentioned above, Si and Mn exert a significant effect on the temper embrittlement susceptibility of the material. As a measure of temper embrittlement susceptibility of the steels, several chemical temper embrittlement parameters were presented. Among them, the J-factor proposed by Watanabe *et al*[6] and $\bar{X}$ by Bruscato[4] are widely used. These parameters are given by:

$$J = (Si + Mn)(P + Sn)\ 10^4\ (wt\%)$$
$$\bar{X} = (10P + 5Sb + 4Sn + As)\ 10^2\ (wt\%)$$

In the present test materials, the J-factor ranges from 26.4 to 113 and $\bar{X}$ ranges from 9.6 to 20.5. It should be noted that these values are typical of the VCD NiCrMoV steels produced 15 to 20 years ago while the J-factor of recent material is around 25 for low-Si VCD materials and below 5 for superclean steels. The present test data shows that embrittlement increases in accordance with the increase of these temper embrittlement parameters. The dependence on these parameters appears more clearly when the data are classified according to the Ni content.

4.1 3.5%NiCrMoV Steel

Figure 6 shows the relationship between the embrittlement (ΔFATT) at 399°C and J-factor of materials aged for 10,000 h, 30,000 h and 100,000 h. As mentioned earlier, embrittlement proceeds with increasing exposure time. The embrittlement seems to depend on the J-factor. However, it can be said that the increase of embrittlement is comparatively insensitive to the increase of J-factor in the materials tested. Though the 100,000 h data has not been obtained, investigations on temper embrittlement sensitivity for superclean steels, showed that the material is almost immune to temper embrittlement.[10–12] These facts suggest that the strong dependence of temper embrittle-

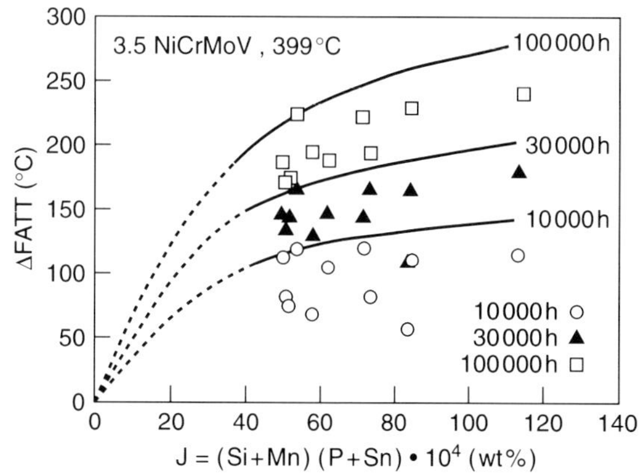

Fig. 6 Relationship between increase of FATT and J-factor at 399°C (3.5%NiCrMoV)

ment sensitivity may happen in the J-factor range, between 0 and approximately 40 in 3.5%NiCrMoV steel.

Figure 7 shows the correlation between ΔFATT and Bruscato factor $\bar{X}$ for 3.5%NiCrMoV steels aged at 399°C. This figure also shows that considerable temper embrittlement happens in the materials tested and this suggests the strong dependence of ΔFATT on $\bar{X}$ in the smaller $\bar{X}$ range.

4.2 3.8%NiCrMoV Steels

The increase of Ni content enhances the temper embrittlement susceptibility. Figure 8 shows the ΔFATT of 3.8%NiCrMoV steels aged at 399°C for 100,000 h in relation to the J-factor.

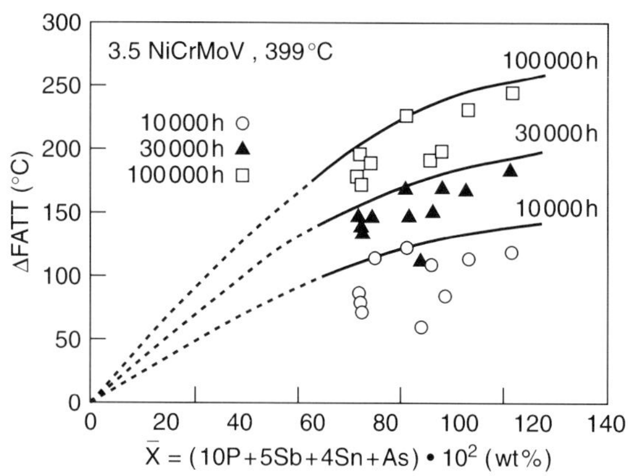

Fig. 7 Relationship between increase of FATT and $\bar{X}$

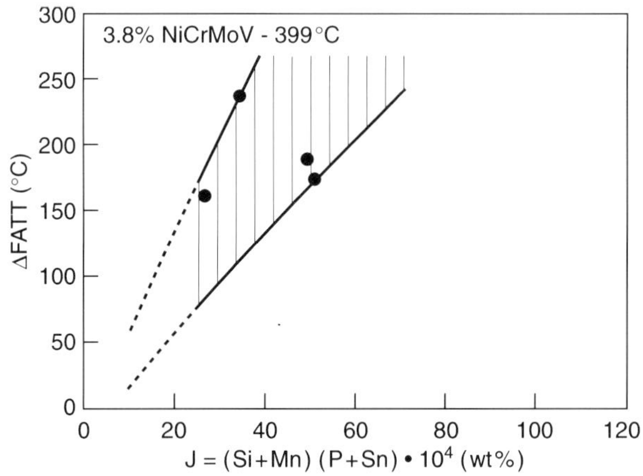

Fig. 8 Relationship between ΔFATT and J-factor (3.8%NiCrMoV)

4.3 2.8%NiCrMoV Steels

Lower Ni content materials showed comparatively small increases of FATT compared with the 3.5% and 3.8%NiCrMoV steels. Figure 9 shows the relationship between ΔFATT and J-factor for the 2.8%Ni materials aged at 399°C. The embrittlement is smaller than that of 3.8%Ni and 3.5%NiCrMoV steels. The increase of FATT by 100,000 h ageing at 343°C is less than 30°C, which is significantly smaller than those for 3.5%NiCrMoV and 3.8NiCrMoV steels.

It should be mentioned that the strength level of the higher Ni content materials are higher than that of lower Ni content materials. Therefore the

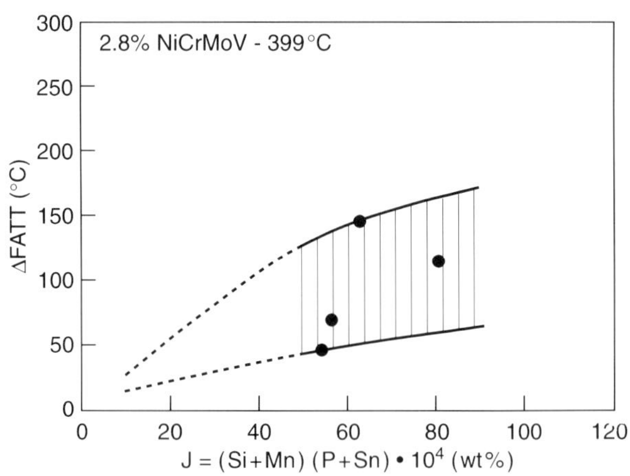

Fig. 9 Relationship between ΔFATT and J-factor (2.8%NiCrMoV)

above test results include not only the effect of residual elements but also the effect of strength level, though the magnitude of the effect of strength level is not clarified.

5. METALLURGICAL EXAMINATION

The change of microstructure by ageing was examined. Comparisons of the hardness between as-heat treated material and aged material showed softening of 3.5% and 3.8%NiCrMoV steels whereas the 2.8%NiCrMoV steels revealed no significant change of hardness. Softening is noticeable after ageing at 399°C and 454°C. This suggests that a change of microstructure proceeds during ageing. Transmission electron microscope observations showed that the coarsening of carbides proceeds in 3.5%Ni and 3.8%NiCrMoV steels. X-ray analysis of extracted residues showed increases of M_7C_3 type carbides in the aged materials. However, the influence of the effect of change in microstructure on embrittlement is not definite.

The amount of impurity elements on the grain boundaries was analysed by Auger electron spectroscopy. The general trend of increased P and Sn with increasing ageing time was observed.

6. CORRELATION BETWEEN THE STEP COOLING EMBRITTLEMENT AND 100,000 H AGEING EMBRITTLEMENT

The step cooling test[8] is one of the most widely used methods to assess the temper embrittlement susceptibility of steels. The step cooling treatment was devised to give a greater amount of temper embrittlement than a rotor shaft suffers during the tempering heat treatment. An advantage of the test is that it can be performed during a short time. When the correlation between the embrittlement due to 100,000 h ageing and that by step cooling test is found in the test results, it can be used to predict the long term ageing embrittlement from the step cooling test. Figure 10 shows the relationship between the embrittlement by 100,000 h isothermal ageing at 399°C and step cooling embrittlement The embrittlement due to step cooling is given by:

$$\Delta\text{FATT (step cool)} = \text{FATT (step cool)} - \text{FATT (de)embrittlement)}$$

Similarly, the embrittlement by 100,000 h ageing is given by:

$$\Delta\text{FATT (100,000 h)} = \text{FATT (100,000 h)} - \text{FATT(de)embrittlement)}$$

Fairly good correlations were found in the present test results. The correlations at 343, 399 and 454°C are as follows.

$$\Delta\text{FATT (100,000 h)} = 1.8 \times \Delta\text{FATT (step cool)} \quad \text{at } 343^\circ\text{C}$$
$$\Delta\text{FATT (100,000 h)} = 5.1 \times \Delta\text{FATT (step cool)} \quad \text{at } 399^\circ\text{C}$$
$$\Delta\text{FATT (100,000 h)} = 5.0 \times \Delta\text{FATT (step cool)} \quad \text{at } 454^\circ\text{C}$$

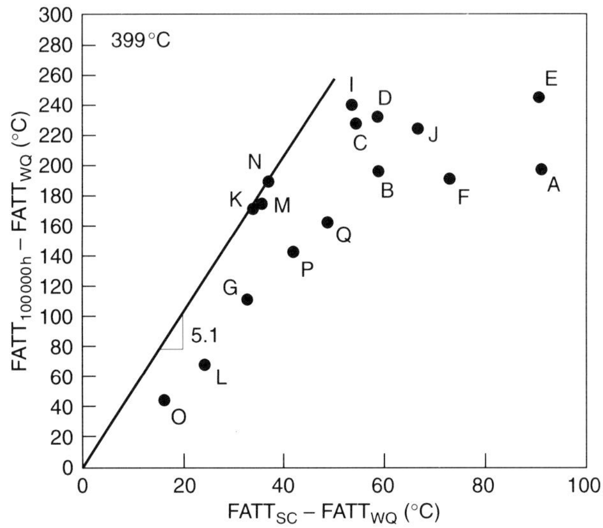

Fig. 10. Relationship between 100,000 h ageing embrittlement and embrittlement due to step cooling heat treatment

The 100,000 h embrittlement can be predicted by using these correlations at each temperature. Taking account of the trend that the embrittlement of the steel almost levels off at 100,000 at 399 and 454°C, the above estimation may predict the end of life embrittlement of NiCrMoV turbine material at these temperatures. When the above correlation is assumed to be valid for lower J-factor material, the long term embrittlement of recent clean steels can be predicted. In the case of materials which embrittled 10°C by step cooling, for example, the maximum embrittlement at 399°C after 100,000 h exposure will be 50°C. The validity of this should be verified by accumulating the clean steel data.

7. TEMPER EMBRITTLEMENT SUSCEPTIBILITY OF CLEAN AND SUPERCLEAN STEELS

The above test results suggested that even the comparatively small amount of residual elements, such as J-factor of less than 40, exerts a considerable amount of embrittlement. Several ageing test results appear in literature to support this trend.[13–15] Figure 11 shows the test results of 10,000 h ageing at 454°C for materials with comparatively low residual elements. The materials with a J-factor of around 20 exhibit some embrittlement susceptibility. Figure 12 shows the ageing test results of a superclean steel. The Charpy test results,

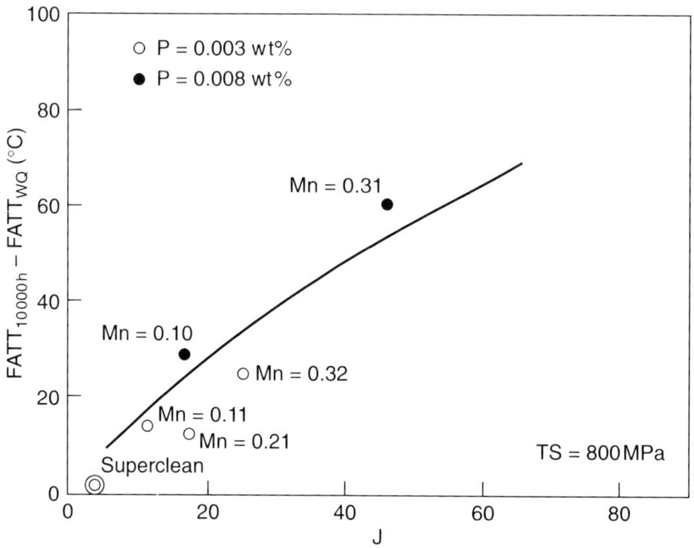

Fig. 11 Embrittlement behaviour of small J-factor 3.5%NiCrMoV steels

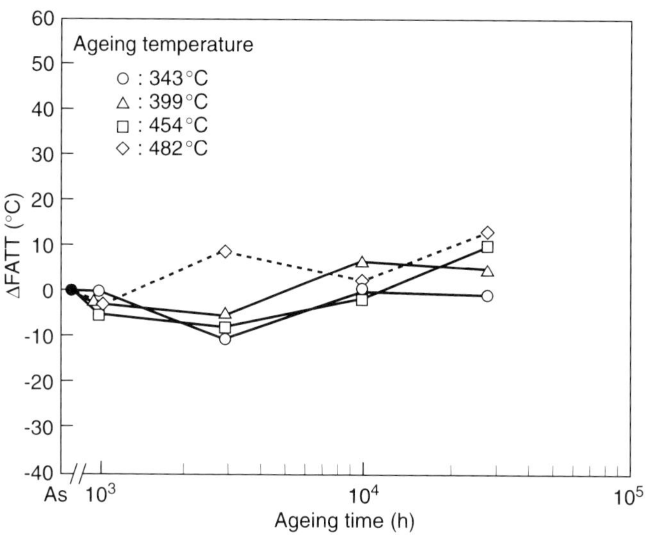

Fig. 12 Isothermal ageing test results of a superclean 3.5%NiCrMoV steel

after 30,000 h exposure, demonstrate that the superclean NiCrMoV steel is almost immune to temper embrittlement. These test results shows that the reduction of residual elements to the superclean version steel is essentially required to eliminate the temper embrittlement sensitivity.

CONCLUSION

The present isothermal ageing test results up to 100,000 h revealed embrittlement behaviour of NiCrMoV steels used turbines for low pressure steam turbines. The results can be summarised as follows.

1. The vacuum carbon deoxidised NiCrMoV steel produced around 20 years ago showed maximum degradation of FATT of 100°C at 343°C for 100,000 h ageing, while the embrittlement is enhanced to 250°C at 399°C and 454°C.
2. Ni content significantly affects the temper embrittlement susceptibility.
3. The embrittlement susceptibility increases with increasing J-factor and the results suggested a strong dependence of the temper embrittlement sensitivity on J factor even between 0 and 40.
4. Change of microstructure proceed when the material is aged at 399°C and over. Coarsening of carbides, precipitation of new types of carbides, cause the major changes of microstructure, hence the softening.
5. A relationship between the embrittlement by the step cooling heat treatment and 100,000 h isothermal ageing embrittlement was found.
6. It is necessary to reduce the residual elements to the superclean level in order to eliminate the temper embrittlement susceptibility.

REFERENCES

1. J. R. Low: *Proc. 1959 Conf. on Fracture of Engineering Materials*, ASM, Materials Park, OH, 1964, 127.
2. R. Narayan and M. C. Murphy: *J. Iron Steel Inst.*, 1973, **211**, 493.
3. McMahon Jr., *et al.*: *Elimination of Impurity-Induced Embrittlement in Steel (NP-1501)*, EPRI, Palo Alto, CA, 1980.
4. R. M. Bruscato: *Welding J.*, 1970, **35**, 148s.
5. J. Yu, C. J. McMahon Jr., and T. Wada: *Met Trans. A*, 1980, **11A**, 277.
6. J. Watanabe and Y. Murakami: *Prevention of Temper Embrittlement of CrMo Steel Vessels by the Use of Low Si Forged Shells*, American Petroleum Institute, Chicago, IL, 1981, 216.
7. C. J. McMahon Jr., A. K. Cianelli and C. Feng: *Met Trans., A*, 1977, **8A**, 1055.
8. G. C. Gould: *ASTM STP407*, ASTM, Philadelphia, PA, 1068, 59.
9. D. L. Newhouse: *Workshop Proceedings: Rotor Forging for Turbine and Generators (EPRI WS79-235)*, EPRI, Palo Alto, CA, 1981, 5–80.
10. R. I. Jaffee: *Workshop Proceedings: Superclean Rotor Steels*, Pergamon Press, Oxford, UK, 1989, 3.
11. W. P. McNaughton, R. H. Richiman and R. I. Jaffee: *J. Mater. Engng*, 1992, **13**, 9.

12. W. P. McNaughton, R. H. Richiman and R. I. Jaffee: *J. Mater. Engng*, 1992, **13**, 19.
13. Y. Yoshioka, O. Watanabe, M. Miyazaki and R. C. Schwant: *Proc. Second-ICPP Conf.*, Palo Alto, CA, November 2–4, 1988.
14. Y. Takeda, Y. Takano, T. Fujikawa, K. Sato, I. Kitagawa and S. Tanimoto: *Workshop Proceedings: Superclean Rotor Steels*, Pergamon Press, Oxford, UK, 1989, 71.
15. K. H. Mayer: *Workshop Proceedings: Superclean Rotor Steels*, Pergamon Press, Oxford, UK, 1989, 155.

Influence of Steel Cleanness on the Operational Properties of Turbine Components

K. H. MAYER* and C. BERGER†

* *MAN Energie GmbH, Nuremberg, Germany*

† *Siemens Power Generation (KWU), Mühlheim, Germany*

ABSTRACT

Long-term test results and operating experience with steam turbines have shown that steel cleanness has a decisive influence on the operational properties of highly stressed components, such as turbine rotors, discs, bolts and blades. Extensive knowledge is available on the positive influence of steel cleanness on component fatigue strength. Experience is also available on the negative effects of trace and deoxidation elements on the long-term toughness and ductility under conditions of creep stressing. For thick-walled components with large solidification cross sections the situation is intensified as a result of ingot segregation so that low toughness and ductility values can be present even when the affected component zones are in the new condition. Unacceptable ultrasonic indications in rotors owing to a local enrichment of non-metallic inclusions, such as Al_2O_3, SiO_2 and MnS, result even today in a scrapping rate of about 5%. Corrosion behaviour is less dominated by steel cleanness. For example, the influence on stress corrosion cracking is marginal. A higher cleanness distinctly improves the low-cycle fatigue strength in air and in corrosive environments.

Present knowledge of the effects of trace and deoxidation elements on the operational properties of highly stressed turbine components largely enables these elements to be specified according to requirements with due consideration being given to cost factors.

1. INTRODUCTION

Long-term test results and operating experience with steam turbines have shown that steel cleanness has a decisive influence on the operational properties of highly stressed components, such as turbine rotors, discs, bolts and blades. Figure 1 is a turbine longitudinal section with an overview of the relevant operational properties which must be taken into account in sizing, designing, manufacturing and operating the turbine, viz:

(i) Creep and thermal fatigue stresses in the steam inlet area of HP and IP turbines,

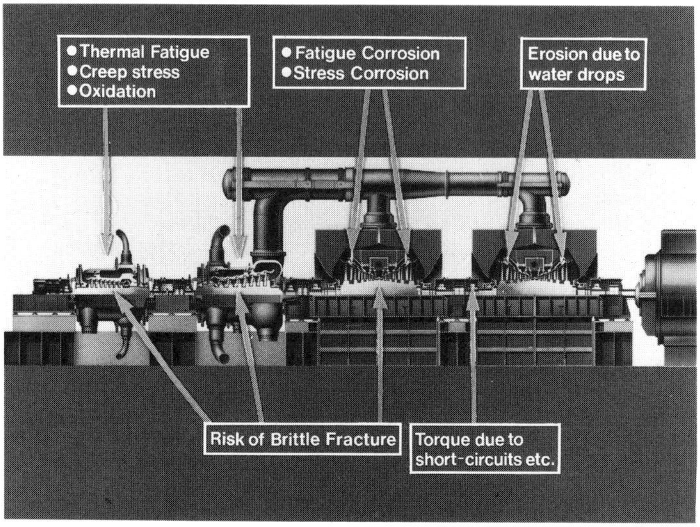

Fig. 1 Stresses, which limit the life expectancy of a steam turbine

(ii) brittle fracture stress of turbine rotors caused by centrifugal force and thermal stress,
(iii) torsion fatigue stresses of turbine rotors caused by short circuits and mal-synchronisation,
(iv) stress corrosion cracking and corrosion fatigue of rotors and blading in LP turbines, and
(v) wear of LP last-stage rotor blading due to water droplets as well as erosion/corrosion in the outlet area of the LP turbine.

For turbines in a fossil-fuel fired plant it has become important to consider the transient stresses occasioned by frequent starts (5,000–10,000) and load cycles (over 8,000), because these turbines, regardless of their unit size, are increasingly being operated in the medium-load and peak-load range of the electric grids.

As shown by the example of Fig. 3 for turbine rotors[1] as a function of the operating temperature, the essential *material characteristics* for the design and operation of turbines are:

- fracture toughness in the new condition and after long periods of operation
- 0.2% proof stress limit
- creep strength for 100,000 h and, in addition, low-cycle fatigue strength over the whole operating temperature range from RT up to steam admission temperature.

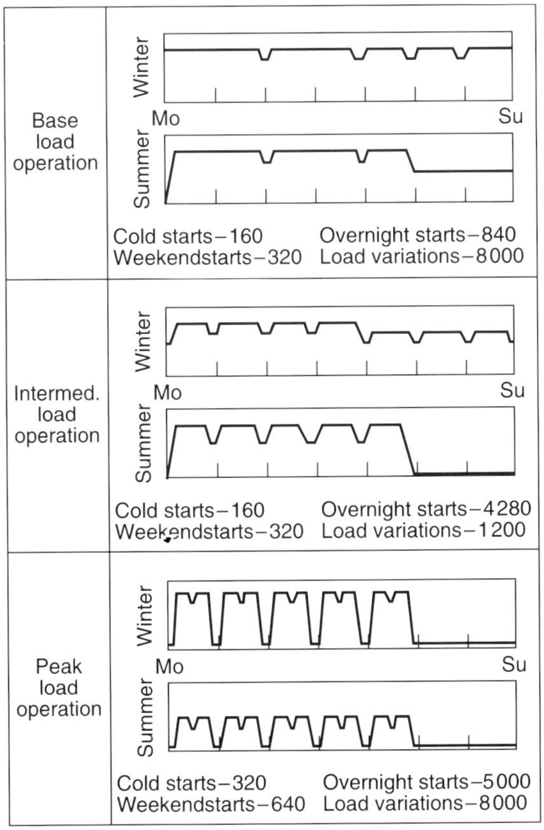

Fig. 2 Operational requirements in 40 years for different load regimes

Besides that, only materials are used which show an adequately high ductility under creep stressing in order to eliminate cracking in stress concentration zones and premature creep damage.

With regard to fatigue strength a wealth of knowledge is available on the influence of steel cleanness, Fig. 4a. Extensive experience is also available on the negative effects of the trace elements and deoxidation elements on long-term toughness, Fig. 4b. Likewise, ductility under creep stressing can be substantially reduced by the trace and deoxidation elements, Fig. 4c. Steel cleanness has a less dominating effect on stress corrosion cracking behaviour, Fig. 4d.

But again, any local enrichment of non-metallic inclusions, such as Al_2O_3, SiO_2 and MnS in large forgings tends to have a negative effect, Fig. 5. The present scrapping rate as a result of unacceptable ultrasonic indications is still in the order of 5%.

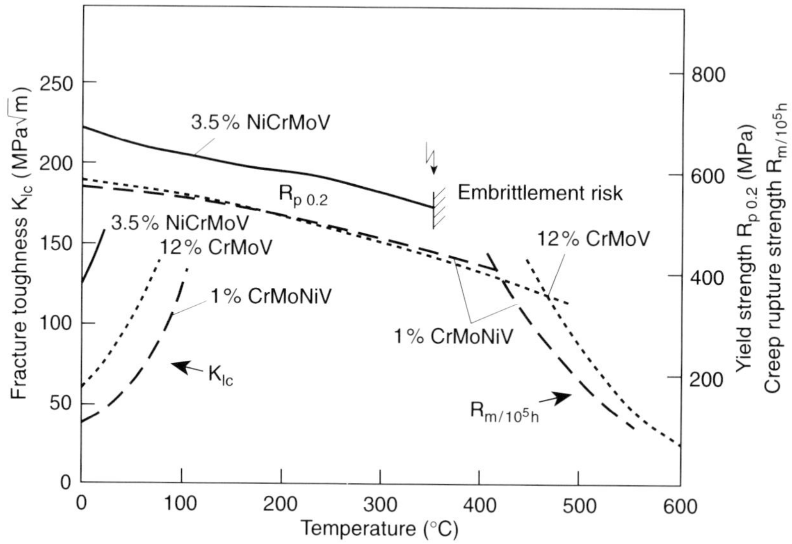

Fig. 3 Strength and toughness of rotor materials

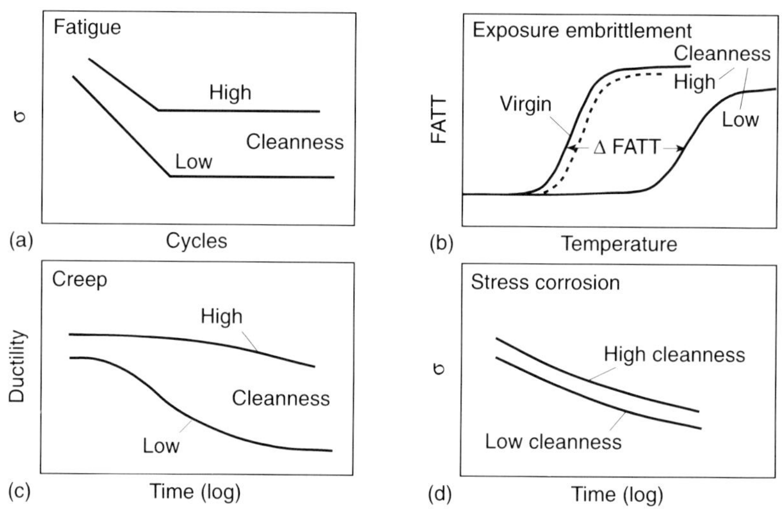

Fig. 4 Influence of steel cleanness on operational properties of steam turbine materials/components

2. TYPICAL TEST RESULTS

2.1 Long-term Embrittlement

In order to render the 3.5% NiCrMoV rotor steel of high strength and toughness suitable for optimised units operating at a temperature of up to

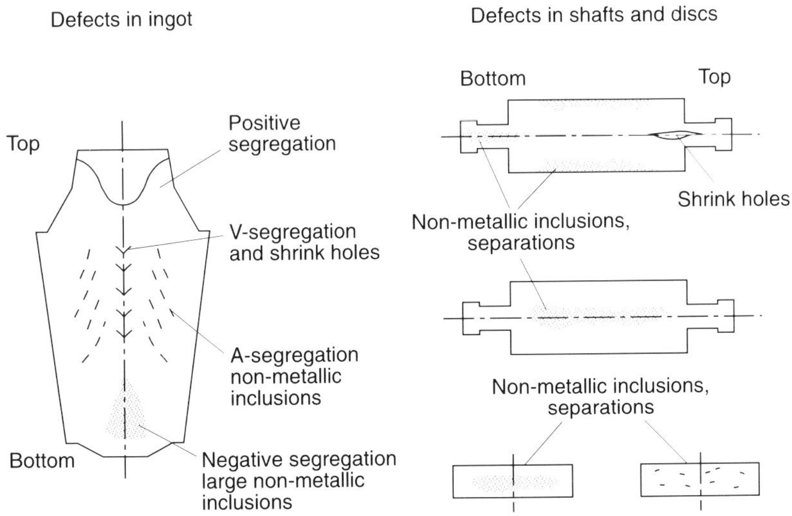

Fig. 5 Typical defects during manufacturing of forgings

about 500°C, many tests carried out by national and international institutes in the 1970s and 1980s served the purpose of investigating the causes of the tendency of this steel to embrittlement. A main cause for embrittlement was established to be an enrichment of phosphorus and tin at the grain boundaries after long-term exposure at temperatures over about 350°C,[2] Fig. 6. Elements tending to encourage embrittlement are also silicon and manganese. Figure 7

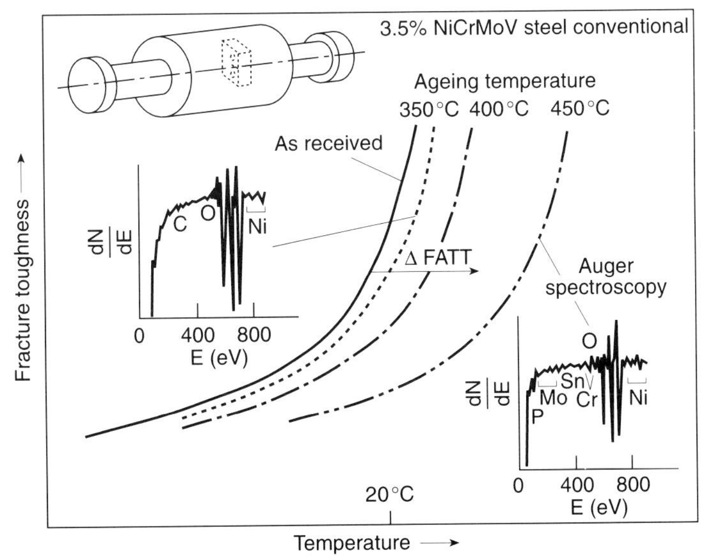

Fig. 6 Alteration of the material condition and fracture toughness due to long-time exposure (schematically)

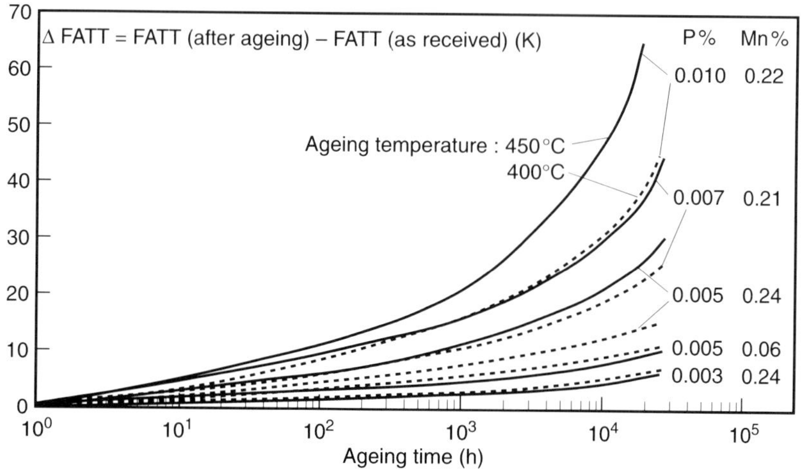

Fig. 7 Long-term toughness of 3.5% NiCrMoV: effect of P and Mn on the ΔFATT after ageing

illustrates the results of exposure tests up to 30,000 h at exposure temperatures of 450°C.[3] The change in the FATT brittle fracture transition temperature with 50% fibrous fracture after exposure versus the as-delivered condition is shown as the difference ΔFATT. A reduction in phosphorus and manganese results in a distinct reduction in the ΔFATT. Auger tests on intercrystalline broken specimens have verified the detrimental influence, especially of the phosphorus, on the reduction in toughness.[4] These tests, carried out on laboratory melts in 1987, were confirmed by the Electric Power Research Institute (EPRI, Palo Alto, USA).[5]

Thanks to an initiative by EPRI in the 1980s[6–8] a joint international programme verified that the tempering and long-term embrittlement of 3.5% NiCrMoV rotor steels can largely be suppressed by high-purity melting characterised by low contents of trace elements (S, P, Sn, Sb, As), gas contents (O_2, N_2, H_2) and silicon and manganese contents (inter alia 9). Figure 8 shows, by way of example, the $FATT_{50}$ values established after 24,000 and 36,000 h exposure at 480°C, in each case for the rim and core positions of pilot rotors.[10,11]

The highest long-term stability is shown by the specimens of rotor VEW1 which has the highest cleanness and the lowest X and K values. The bainite structure of the rotor core reveals practically no change in toughness and the martensitic rim structure which is more susceptible to embrittlement only has a FATT shift of about plus 8°C. The behaviour of the specimens of the JSW rotor show similar behaviour in spite of the distinctly higher J_{mod} and K values.

The specimens of rotor VEW2 with the highest content of trace elements revealed a distinct reduction in toughness. The ΔFATT of the bainite structure in the rotor interior is plus 17 or 18°C regardless of the 0.2% proof stress limit.

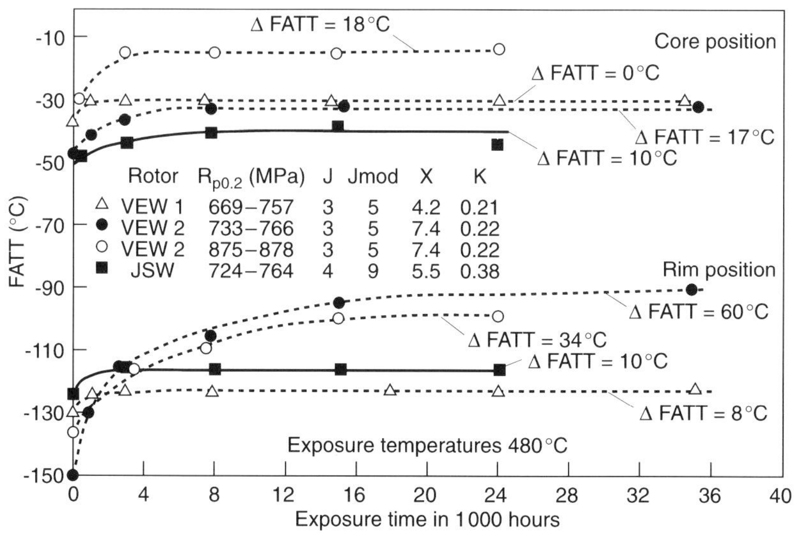

Fig. 8 Long-term toughness behaviour of superclean 3.5% NiCrMoV rotors as a function of chem. compostion, specimen position and yield strength

The change in the martensitic rim structure is even more distinct, i.e. plus 60°C after 36,000 h exposure for the lower 0.2% proof stress limit of roughly 750 MPa and plus 34°C after 24,000 h exposure for the higher 0.2% proof stress limit of 875 MPa.

To check the permissible limit values all hitherto available values of exposure tests up to 10,000 h versus the Mn content and the Bruscato factor x (Fig. 9) and versus the P-content (Fig. 10) were plotted. The plots impressively demonstrate that the P-content is the dominating factor for long-term embrittlement behaviour. According to Fig. 10 a P-content of less than 30 ppm would be required to suppress long-term embrittlement.

Exposure embrittlement versus cleanness, determined by means of the ISO-V-notch impact specimens, is confirmed on determining the fracture toughness by means of compact tension specimens, Fig. 11.[22]

2.2 *Creep Behaviour*

Creep tests on specimens of superclean 3.5% NiCrMoV pilot rotors have also verified that superclean melting improves the creep strength at temperatures around 500°C.[10] Figure 12 contains a summary of published results and our own test results on the basis of a Larson–Miller Parameter in comparison with the 3.5% NiCrMoV rotor steel produced by the conventional method according to SEW 555.[12]

However, for the specimens of the VEW2 rotor with the high 0.2% proof stress limit (850 MPa) at 500°C a distinct reduction in ductility and a notch-weakening effect ($\alpha_k = 4.3$) was established, Fig. 13. Possibly this reduction in

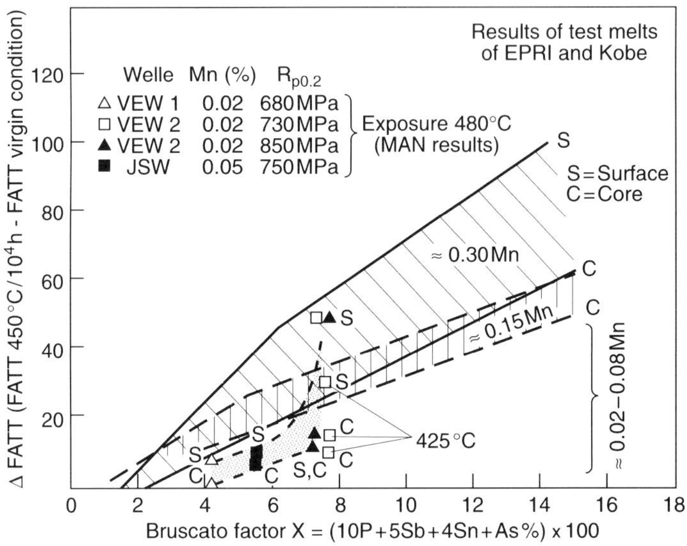

Fig. 9 ΔFATT as a function of Bruscato factor and Mn-content exposure time 10 000 h at 450°C and 480°C, 3.5NiCrMoV rotor steel

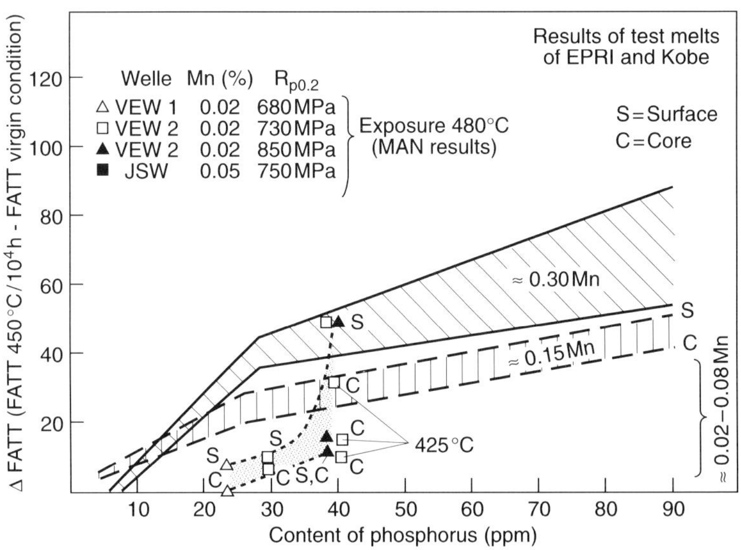

Fig. 10 ΔFATT as a function of P- and Mn-content and the specimen location exposure time 10 000 h at 450°C and 480°C, 3.5% NiCrMoV rotor steel

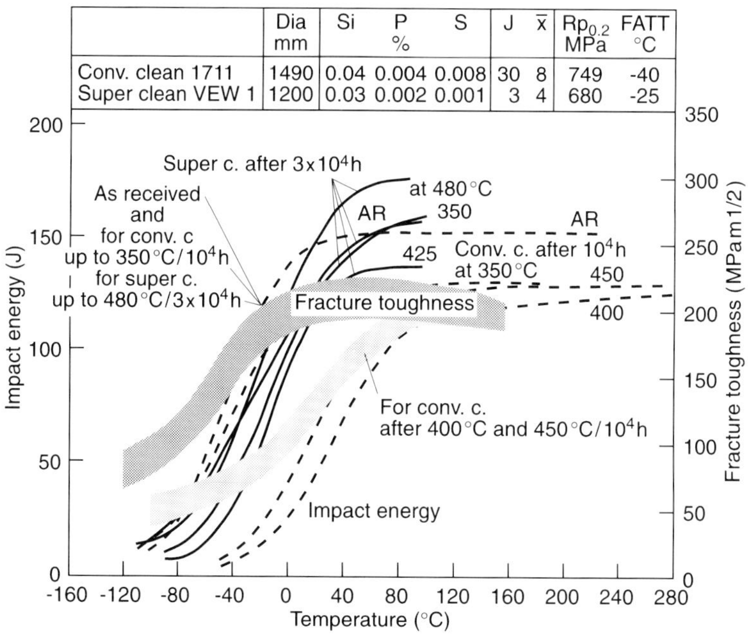

Fig. 11 Long-term toughness of conventional and superclean 3.5% NiCrMoV – rotors

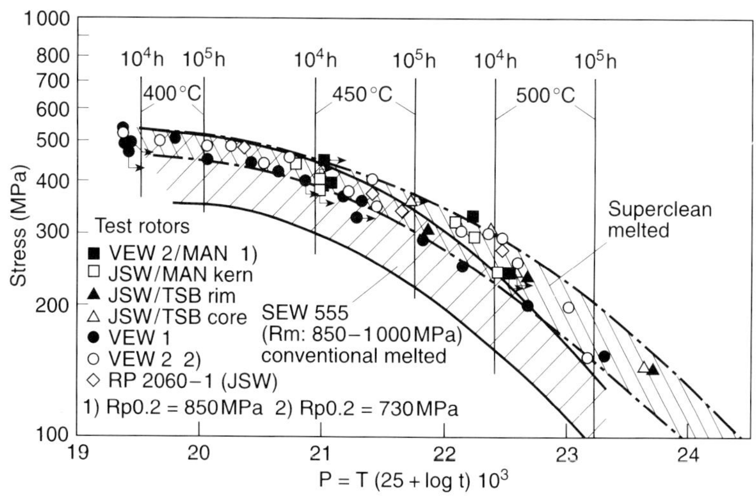

Fig. 12 Creep rupture strength of 3.5% NiCrMoV rotor steel in superclean and conventional melted condition

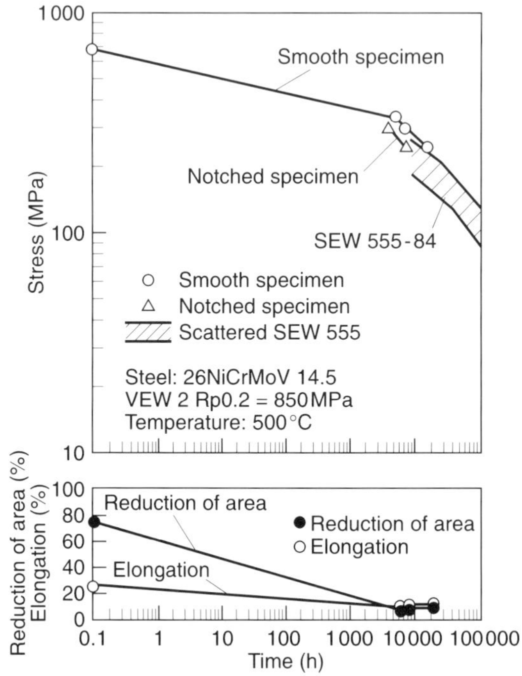

Fig. 13 Creep rupture behaviour of pilot rotor VEW 2 at 500°C (Rp0.2 = 850 MPa)

ductility is associated with the somewhat lower degree of cleanness. M. Kohne *et al.*[13] also established a loss of ductility in the creep test at 500°C, which was very pronounced at a very low Mn content of 0.02%, and was also present in the case of the VEW2 rotor.

In a further initiative by EPRI, which focused on improving the degree of cleanness of the 1% NiCrMoV rotor steel,[14] creep tests on specimens of three rotors melted by low sulphur silicon deoxidation, electroslag remelting (ESR) and vacuum carbon deoxidation (VCD), revealed that with VCD a higher creep fracture strength can be achieved.[15] These findings were in fact confirmed by the German Working Committee for Heat-resistant Steels after checking the influence of the melting process on the creep behaviour of the 1% CrMoV HP and IP turbine rotors.[16,17] Figure 14 shows the results of 15 different production rotors which were melted by CAB, ESR, BEST and VCD. As shown at the bottom of the figure, the rotors melted by VCD reveal a distinctly higher fracture deformation strength than the rotors manufactured by the other melting processes. However, there is no visible influence on creep fracture strength. An analysis of the results showed that the reduction of area after fracture shows higher values when the silicon content during deoxidation is less than 0.09%. If, after VCD treatment, the Si content is raised by further alloying to 0.20% the higher reduction of area is maintained. The best deformation strength was achieved by the melts which had the lowest

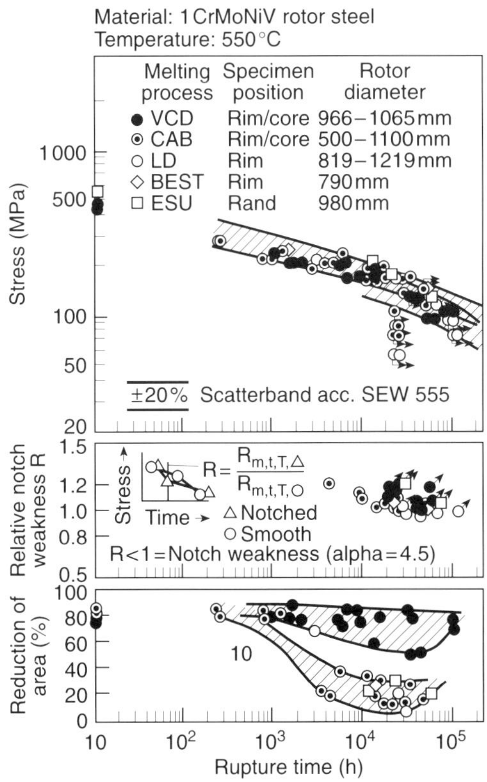

Fig. 14 Creep rupture behaviour of 1CrMoNiV rotor steel at 550°C

aluminium and silicon contents. No clear relationship can be established between the reduction of area after creep fracture and the trace elements or, similarly, between the Bruscato factor and the sulphur content. This finding corresponds with the results of the Auger tests of ruptured creep specimens. None of the specimens revealed a grain boundary segregation of trace elements.[16]

A distinct improvement in the reduction of area after creep fracture was also determined when castings of the modified 9% CrMo steel G–X12CrMoVNbN 9 1 were refined by secondary metallurgy. As shown by Fig. 15, the material which had been subjected to AOD treatment showed a distinctly higher creep fracture ductility in the 600°C creep tests than the melts from the induction furnace.[18] The figure also provides information on the quantity of deoxidation and trace elements of the melts. Whereas the aluminium, silicon and sulphur contents of the AOD melts are lower, the trace elements are roughly on the same level. It is possible that the better response occasioned by AOD melting is also achieved by a lower macroporosity on account of the lower gas contents.

Melting process	Si	Mn	P	S	Al	As	Sn	Sb
Induction furnace	0.42–0.49	0.41–0.50	0.006–0.008	0.004–0.005	0.009–0.037			
Argon Oxygen decarburisation	0.29–0.31	0.50–0.51	0.007–0.009	0.002	0.007	0.016	0.004	0.003

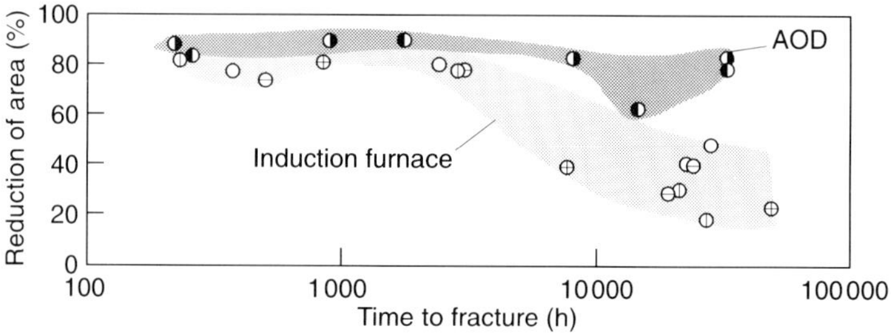

Fig. 15 Creep rupture ductility of modified 9% CrMo cast steel at 600°C influence of melting technology

In extensive creep tests by the German Working Committee for Heat-resistant Steels on specimens of the 11% Cr bolt and blade steel X19CrMoVNbN 11 1 the deoxidation element aluminium was clearly identified as a detrimental element with respect to creep fracture deformation behaviour, Fig. 16.[19] Al contents of max. 0.01% are recommended in order to obtain a higher reduction of area for this high-temperature-resistant steel.

2.3 *Low-cycle Fatigue Behaviour*

Figure 17 shows the values determined in air at room temperature in comparison with published results and our own results with specimens from

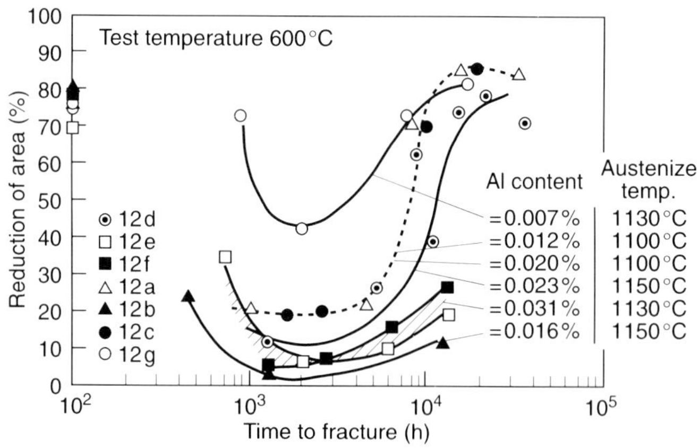

Fig. 16 Influence of Al-content and austenise temperature on creep rupture ductility of steel × 19 CrMoVNbN 11 1

1) Tests at 90 °C	Rp0.2 (MPa)	C	Si	Mn	P	S	Cr	Mo	Ni	V
○ EPRI/VEW 1	757	0.30	0.03	0.02	0.002	0.0012	1.68	0.41	3.44	0.09
△ EPRI/VEW 2	766	0.27	0.01	0.02	0.004	0.0010	1.71	0.43	3.73	0.10
1) ▲ EPRI/VEW 2	878	0.27	0.01	0.02	0.004	0.0010	0.71	0.43	3.37	0.10
1) ● EPRI/JSW	764	0.25	0.01	0.05	0.003	0.0010	0.70	0.40	3.70	0.12
□ LD-rotor	687	0.25	0.17	0.30	0.007	0.0060	1.61	0.37	3.57	0.10

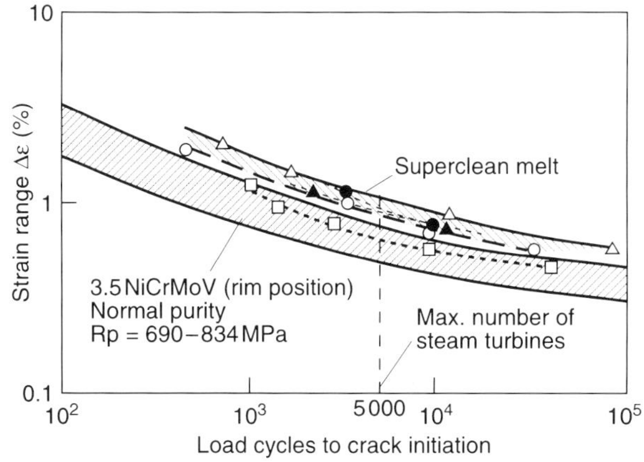

Fig. 17 Low-cycle fatigue test results of 3.5% NiCrMoV rotor steels determined in air and 20°C

conventionally melted 3.5% NiCrMoV rotors.[10] Over the whole low-cycle fatigue range of interest in the operation of steam turbines, the specimens of the superclean-melted EPRI test rotors feature a distinctly higher life expectancy. On average, the life expectancy is increased at least by a factor of 5. Another noticeable fact is the relatively narrow scatter band of the test values of the superclean-melted rotors in which it is no longer possible to differentiate between the different contents of trace elements and 0.2% proof stress limits.

The results determined in turbine condensate with pH values of 3, 5, 7 and 9 are given in Fig. 18. Even under these ambient conditions superclean melting offers a clear gain in life expectancy. As expected, the results obtained with a pH value of 3 are the lowest. This test condition simulates the worst case in the crevices of the blade root.

2.4 *Creep Crack Growth Behaviour*

Under a long-term research project of the German turbine manufacturers to determine the influence of natural flaws in turbine rotors on operating behaviour an investigation was also made into the crack initiation behaviour under creep and fatigue creep stressing on specimens from scrapped forgings of the 1% CrMoV rotor steel[20] with the following deoxidation and trace elements:

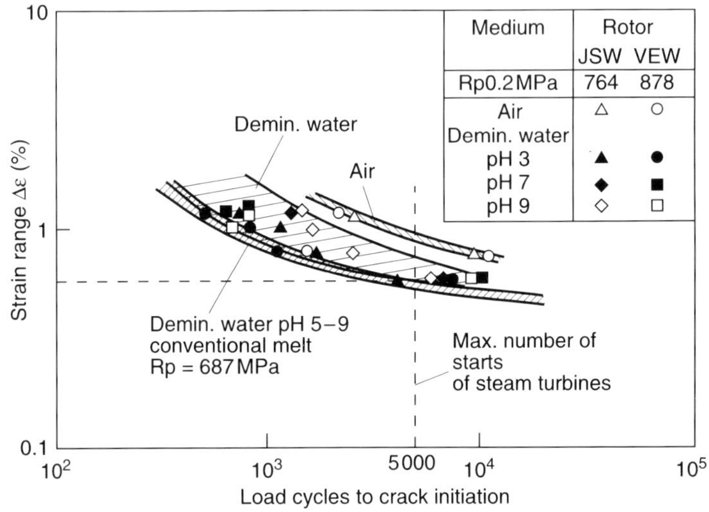

Fig. 18 Low-cycle fatigue test results of 3.5% NiCrMoV rotor steels determined in steam condensate with different pH values and at 90°C

Rotor (wt%)	Si	Mn	Al	P	S	Sb
K	.35	.43	.012	.010	.004	.002
T	.21	.76	.006	.008	.005	.001
U	.05	.71	.001	.008	.009	.001

Detrimental factors, independent of the amounts of the deoxidation and trace elements, have been found in local accumulations of oxide and MnS inclusions which, when subjected to creep stresses according to Fig. 19, are liable to link up into larger flaws. The fact that linkups of this kind occurred in material zones without recordable ultrasonic indications is a particular drawback. A linkup of embedded group flaws can occur under conditions of creep deformation strain above roughly 0.4%. An increase in echo height through the linkup of embedded group flaws has been observed from about 1% creep deformation upwards. However, this value is above the permissible global creep deformation of turbine rotors.

3. SUMMARY AND CONCLUSION

The insights and experience gained during the last 20–30 years regarding the influence of steel cleanness on the operational properties of high-temperature-resistant steels have culminated in developments in steel manufacture which

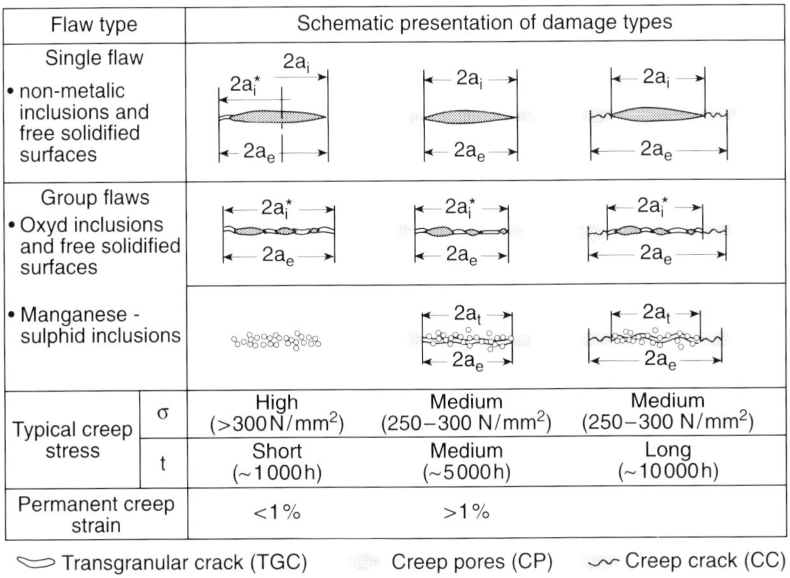

Fig. 19 Schematic presentation of type of creep damage as a function of type of natural flaw and creep stress

have not only resulted in the ability to cope with higher operational stresses, but have also brought forth new turbine concepts, e.g. combined-cycle power stations, and the possibility of using the steels in a larger temperature range. A good example is the superclean melting of the 3.5% NiCrMoV steel where, by reducing the trace and deoxidation elements, the tendency of this steel to long-term temper embrittlement is suppressed while the creep strength and low-cycle fatigue strength are distinctly increased for operation from RT to about 500°C.

Less spectacular, but also of great significance for the useful life of highly-stressed turbine components, is the increase in creep fracture deformation strength of temperature-resistant and high-temperature-resistant steels due to the markedly reduced silicon and aluminium contents, achieved by secondary metallurgy such as VCD and AOD.

No clear relationships have been discovered between unacceptable ultrasonic indications in turbine rotors due to local accumulation of non-metallic inclusions and the contents of trace and deoxidation elements. In the cases under review the positive influences of greater steel cleanness were possibly outweighed by the dominating influence of the melting method.

Investigations by fracture mechanics have demonstrated the negative influence of a local accumulation of non-metallic inclusions in respect of crack initiation under operational stresses, resulting in lower loadability and a shorter useful life of the turbine rotor.

REFERENCES

1. K. H. Mayer, W. Kachler and H. König: *Power Industry E-1'90 Technology Trends Conference and Exhibition*, Johannesburg, SA, Eskom College, Midrand, 21–23 May 1990.
2. M. Krause, J. Granacher, K. H. Keienburg and K. H. Mayer: *Dreißig Jahre Arbeitsgemeinschaft für warmfeste Stähle*, VGB Werkstofftagung, Düsseldorf 1980.
3. T. Hajdu, K. H. Keienburg, H. J. Grabke and W. Rohde: *11. Vortragsveranstaltung der AGW*, 25 November 1988, Düsseldorf.
4. R. Möller, H. Erhard and H. J. Grabke: *Arch. Eisenhüttenwes. 55"*, 1984, **11**, 543–548.
5. R. L. Bodnar, *et al*: *EPRI Research Project 2060-1, 2. Schlußbericht NP-5399*, September 1987.
6. R. I. Jaffee, P. Machner, W. Meyer and J. E. Steiner: *Ironmaking Steelmaking*, 1986, **13**(6), 322–326.
7. P. Machner, W. Meyer, A. Kucharz and R. I. Jaffee: *Berg- und Hüttenmännische Monatshefte*, October 1986, 433–442.
8. R. I. Jaffee, T. Ohhashi, O. Watanabe and P. Machner: Presented at Iron and Steel Society of ASME Toronto, October 1987.
9. R. Viswanathan: *Proceedings at the R. I. Jaffee Memorial Symposium*, 2–5 November 1992, Chicago, Illinois, USA.
10. K. H. Mayer, W. Kachler and H. Könit: *Proceedings at the R. I. Jaffee Memorial Symposium*, 2–5 November 1992, Chicago, Illinois, USA, 205–212.
11. C. Berger: *Proceedings of the R. I. Jaffee Memorial Symposium*, 2–5 November 1992, Chicago, Illinois, USA, 205–212.
12. 'Stähle für großere Schmiedestücke für Bauteile for Turbinen- und Generatoranlagen', *Stahl-Eisen-Werkstoffblatt (SEW) 555*, 1st issue, August 1984. Verlag Stahleisen mbH, Postfach 8229, 4000 Düsseldorf 1, West Germany.
13. H. Kohno *et al*: International ASME-EPRI Conference on *Advances in Materials Technology for Fossil Power Plants*, 1–3 September 1987, Chicago, Illinois, USA.
14. J. E. Steiner and R. I. Jaffee: ASTM Symposium Williamsburg, VA, USA, 28–30 November 1984, *ASTM STP 903*, 35–44.
15. V. P. Swaminathan, R. I. Jaffee, J. E. Steiner: ASTM Symposium Williamsburg, VA, USA, 28–30 November 1984, *ASTM STP 903*, 124–142.
16. H. König, K. H. Mayer, K. H. Keienburg, H. J. Grabke and E. Gärtner: *Steel Research 61*, 1990, **6**, 274–278.
17. Studiengesellschaft Stahlanwendung e.V. Project 195 Bewertung der Langzeitfestigkeit von Stählen für den Einsatz bei erhöhten Temperaturen, 166–178.
18. K. H. Mayer, W. Gysel: *International Symposium on Improved Technology for Fossil Power Plants*, 1–3 March 1993, Washington, DC, USA.

19. H. König: 16. Vortragsveranstaltung *Langzeitverhalten warmfester Stähle und Hochtemperaturwerkstoffe*, 26 November 1993, Düsseldorf.
20. H. Theofel, K. Maile, C. Berger, J. Ewald, M. Weiß and K. H. Mayer: *DVM-Arbeitskreis Bruchworgänge*, Magdeburg, 22–23 February 1994.
21. K. H. Mayer, C. Berger, C. Gerdes: *12th International Forgemaster Conference*, Chicago, IL, USA, 11–16 September 1994.
22. C. Berger: Unpublished data by Siemens/KWU.

Low Residuals in Bolting Steels

H. EVERSON* and J. ORR†

* *UES Steels, Stocksbridge, UK*

† *British Steel Technical, Rotherham, UK*

ABSTRACT

Bolts for use in high temperature steam turbines require good elevated temperature strength and ductility. The development and enhancement of 1%CrMoV based bolting alloys for use in UK designed steam turbines is described. Initial ductility problems encountered over 30 years ago were solved by the addition of Ti and B to the base alloy.

Subsequent developments have concentrated on reducing the residual levels (P, Sn, Sb, As, Cu). Current materials made to low residual practice show excellent ductility and tests in progress indicate that the low residual material shows notch strengthening.

1. INTRODUCTION

The development and improvement of high temperature bolting steels has been the subject of many investigations due to the critical nature of the component involved. The principal bolting material used in UK designed steam turbines, operating at steam temperatures in the range 500–570°C, is Durehete 1055.

Although a mature product, having been manufactured and used for over 30 years, Durehete 1055 has been the subject of continuous improvement programmes, the latest of which is the production of very low residual 'clean' Durehete 1055 in significant commercial quantities with an associated long term elevated temperature testing programme to generate property data.

2. INITIAL DEVELOPMENT

In order to appreciate the current developments it is necessary to review briefly historical factors which have led to the current enhancement of Durehete 1055.

When, in the 1950s, the UK decided to opt for a higher steam temperature of 565°C (1050°F), compared with the majority of designs throughout the rest of the world which operated at up to 538°C, appropriate materials with the necessary rupture strength were developed to deal with this higher temperature.

A 1%CrMoV steel known as Durehete 1050 was used as the bolting

Table 1 UK Ferritic Bolt Steel Designations

Trade Name	CEGB Code	BS 1506 Type No.	Nominal Compostion wt%				
			C	Cr	Mo	V	Others
Durehete 1050	GP 5 (1CrMoV)	–	0.2	1	1	0.75	–
Durehete 1055	GP 6 (1CrMoV)	681–820	0.2	1	1	0.75	Ti, B

material.[1] However test and service experience showed that the rupture ductility of Durehete 1055 deteriorated with time at temperature and a number of premature bolt failures resulted.[2]

Research and Development programmes by United Steel Companies (UES Steels' predecessor) explored ways of improving the ductility of the Durehete 1050 base. It was found that additions of boron and titanium, protected by prior aluminium additions, significantly improved the rupture ductility of the alloy, therefore the basic composition of Durehete 1055 was derived.

In contrast with recent trends in materials development and the theme of this conference, the problem was solved by selectively increasing the trace element content of the material. Table 1 lists the composition range of Durehete 1050 and the initial range of Durehete 1055. The ductility improvements shown by the boron and titanium additions are illustrated in Fig. 1.[3]

3. CONSOLIDATION OF EXPERIENCE

Commercial production of Durehete 1055 began in 1963 and all the Durehete 1050 bolting was progressively replaced.

Service and long term test experience showed Durehete 1055 to be a consistent and reliable bolting material for turbine use with a significant advantage over competitive 12%Cr and nickel based alloys in that its expansion characteristics matched those of the low alloy steel casing and valve body materials, thus minimising design problems and avoiding compensation devices and procedures.

In the early 1970s the operating utility (CEGB) experienced one or two bolting failures. Detailed investigations of the failed bolts and bolts that had parallel service history showed that there was no deficiency in the failed material with regard to conformance with published long term elevated temperature properties.[4] Nevertheless, it was considered that there would be advantages in improving the ductility of Durehete 1055 to accommodate unspecified features of design and operating conditions.

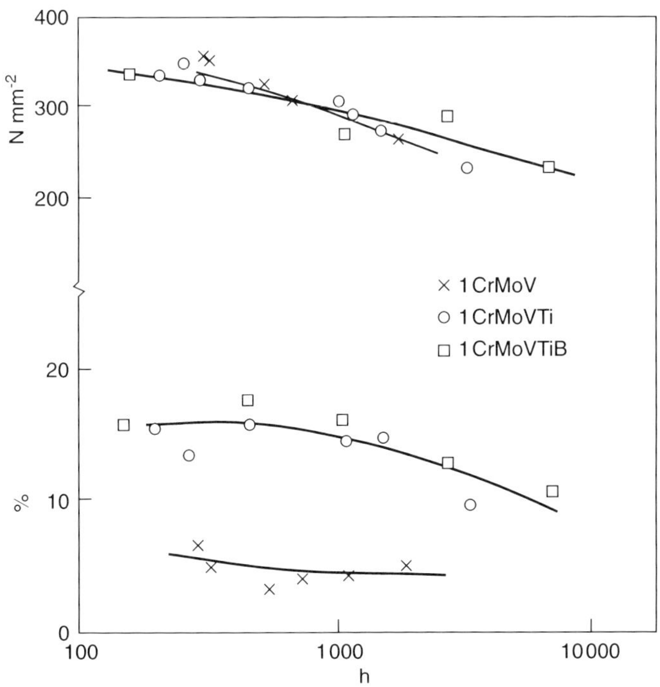

Fig. 1 Development of Durehete 1055 (Tests at 565°C)

Consequently a programme of work was commenced to look at the effect of residuals and steelmaking route on rupture ductility. Around the same time a number of workers explored the combined effects of residual elements and formulae were derived to express the relative embrittlement severity of selected trace elements.[5]

4. REVISION OF SPECIFICATIONS AND DEVELOPMENT PROGRAMME

The programme on D1055 initially used laboratory scale vacuum induction melts (VIM). These showed that by reducing the classical temper embrittling elements P, Sn, As and, to a lesser extent Cu, ductility improvements with consequent improvements in notched properties could be achieved. This led to further work using commercial low residual 15 tonne air melt casts, a vacuum arc remelted (VAR) ingot and a vacuum induction melted + vacuum arc remelted (VIM/VAR) ingot.

In parallel with this programme work was also being carried out within CEGB and a new specification CEGB, GDCD Standard 2, Issue 2 was agreed in 1982. The revised specification included restrictions on As, Cu and Sn in addition to tightening the ranges allowed for a number of alloying elements. Specified ranges are given in Table 2.

Results of the long term tests on the low residual materials produced from the experimental and trial casts have been published previously.[6]

Figure 2 shows the effect of the residual factor on 10 000 hours rupture strength at 550°C illustrating that reducing the residual content has not caused any reduction in rupture strength. Figure 3 shows the effect of the residual factor on rupture ductility at 550°C illustrating the improvement in reduction of area values as the residual factor reduces.

Figures 2 and 3 also show the differences between casts made before and casts made after the change in specification. The residual factor in these cases was calculated from the formula $R = P + 2.43\,As + 3.57\,Sn$. Sb and Cu levels were not available for all the materials investigated.

Figure 3 is particularly interesting in that although the laboratory melted VIM casts and the commercial sized Air VAR and VIM/VAR casts formed a discrete population with the highest reduction of area and lowest residual factors, the low residual commercial air melted casts showed a large increase in ductility compared with the early commercial material. In fact, rupture ductility proved to be more related to residual content rather than the melting route.

This was very encouraging as VIM and VAR materials are considerably more expensive than air melted materials and showed that, provided a reliable process route guaranteeing low residual air melt material could be developed at a realistic cost level, there was no necessity to use the VIM and VAR routes in order to achieve a significant increase in ductility.

It should be remembered that the Issue 2 material, with only modest restrictions on residual content and modifications to heat treatment, performs well. Thousands of tonnes are in service in UK designed sets throughout the world with no ductility problems.

5. CURRENT CLEAN STEEL DEVELOPMENTS (DUREHETE 1055, MK III)

In 1991, following discussions with a major European turbine builder, the further development of Durehete 1055 was commenced. UES Steels made the decision to reduce significantly the residual content of Durehete 1055.

By this time sufficient data on antimony levels had become available for this element to be included in the cumulative residual factor and the formula $R = P + 2.43\,As + 3.57\,Sn + 8.16\,Sb + 0.13\,Cu$ was adopted.

Re-analysis of available data using the full R factor indicated that a value of less than 0.10 was advisable to ensure that the material showed notch strengthening, i.e. has an average notched to plain rupture strength of $\geqslant$ unity. Figure 4 illustrates this point.[7]

Table 2 Development of Durehete 1055 Specifications

Specification	Year	Composition, Wt. %																
		C	Si	Mn	P	S	Cr	Mo	Ni	V	B	Ti	Al	As	Cu	Sb	Sn	R
CEGB	1969	0.15	0.10	0.35	–	–	0.09	0.85	–	0.60	0.001	0.05						
02596		0.25	0.35	0.75	0.04	0.04	1.30	1.10		0.80	0.010	0.20						
Issue 1																		
02596	1982	0.17	0.10	0.35	–	–	0.90	0.90	–	0.60	0.001	0.07	–	–	–	–		
GDCD-2		0.23	0.35	0.75	0.02	0.02	1.20	1.10	0.20	0.80	0.010	0.15	0.08	0.020	0.20	–	0.020	
Issue 2																		
BS1506	1986–	0.17	0.10	0.35	–	–	0.90	0.90	–	0.60	0.001	0.07	–	–	–	–	–	
681–820	1990	0.23	0.35	0.75	0.02	0.02	1.20	1.10	0.20	0.80	0.010	0.15	0.08	0.020	0.20	–	0.020	
Current	1995	0.17	0.10	0.35	–	–	0.90	0.90	–	0.60	0.001	0.07	–	–	–			
D1055		0.23	0.35	0.75	0.02	0.02	1.20	1.10	0.20	0.80	0.010	0.15	0.08	0.020	0.20	*	0.020	$\leqslant 0.14$

* Value reported to customer

$R = P + 2.43\,As + 3.57\,Sn + 8.16\,Sb + 0.13\,Cu$

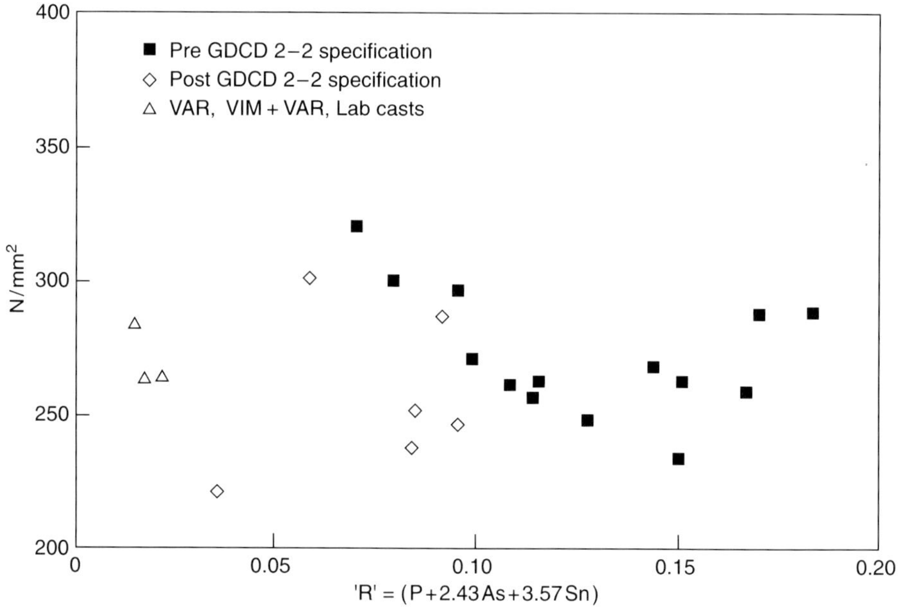

Fig. 2 Strength for 10 000 h at 550°C as a Function of Residual Element Content Factor 'R'

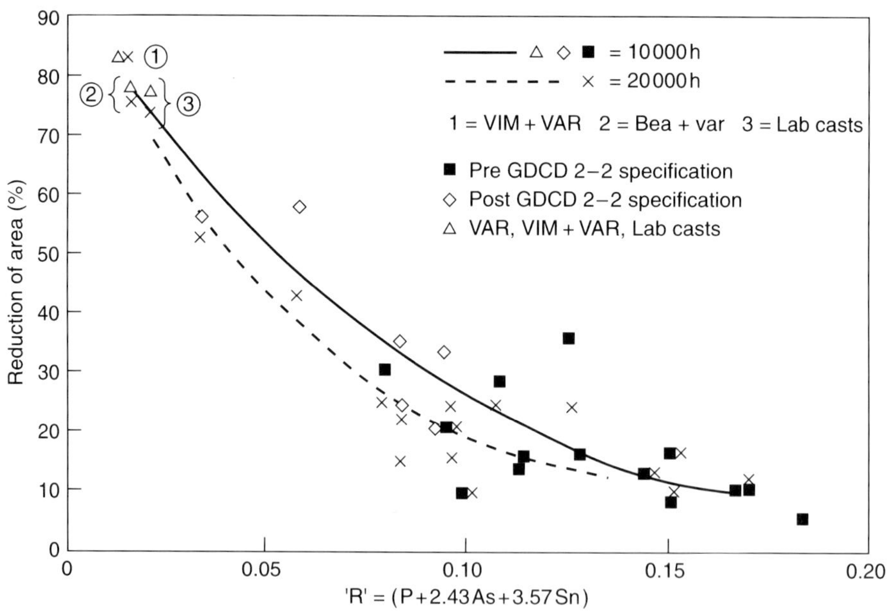

Fig. 3 Ductility at 550°C as a Function of Residual Element Content Factor 'R'

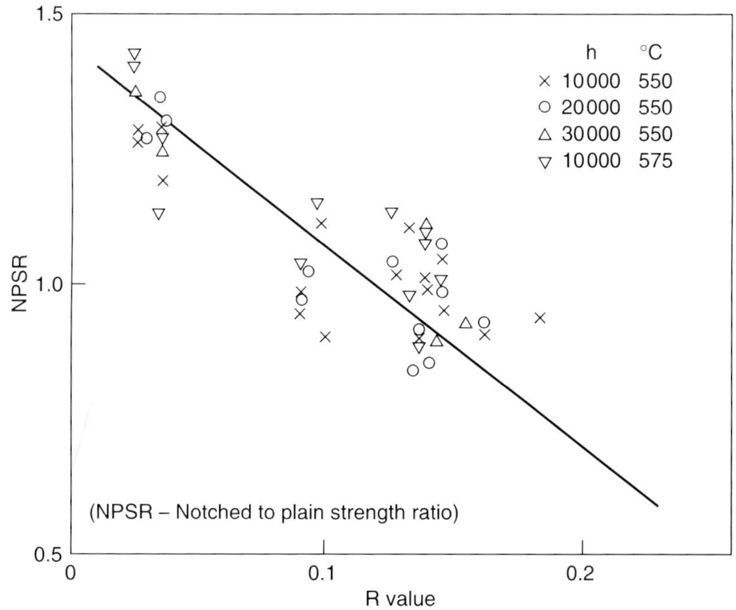

Fig. 4 NPSR Values for Durehete 1055

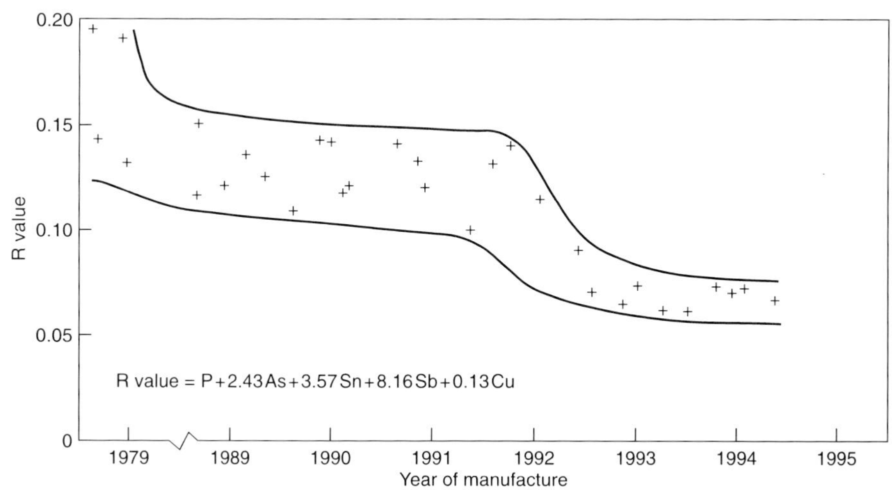

Fig. 5 R Values of Durehete 1055 Casts made by U.E.S.

By using carefully selected and segregated low residual scraps and ferro alloys, extended dephosphorisation procedures and selected steelmaking fluxes a steelmaking process has been developed.

Figure 5 shows the improvement in R values with time. The latest 1200

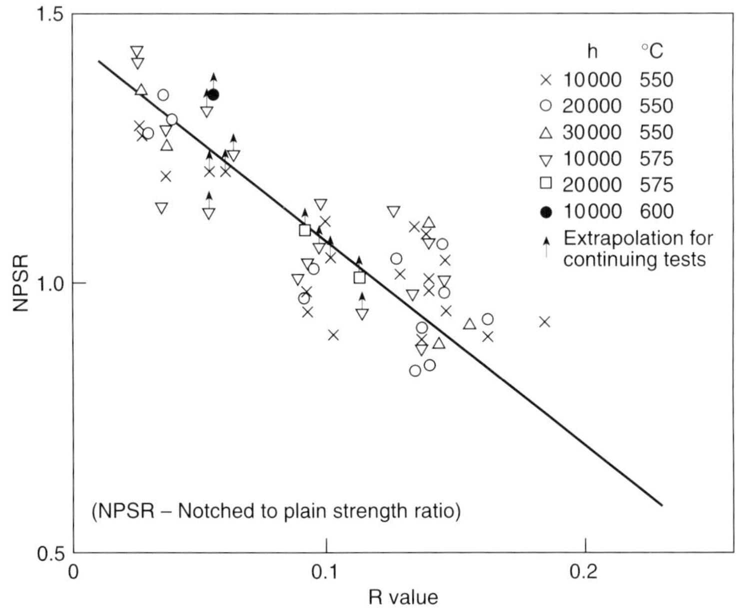

Fig. 6 NPSR Values for Durehete 1055

tonnes of Durehete 1055 made have met the aimed for R values of <0.1. With the most recent 1000 tonnes below 0.06 the practice is now well established.

The difficulty at present in accepting a specified R value of <0.1 is that during steelmaking rapid spectrographic analysis of residuals (especially Sb) to the accuracy required is difficult and subsequent confirmation by chemical methods is often required.

Materials from recent production material with R values <0.06 are on test. As it takes at least 3 years from making the cast to receiving any significant long term elevated temperature test results we do not have sufficient data to make a full comparison with the older material. However, the shorter term data (longest duration of broken tests about 9000 hours) indicates reduction of area values of the recent superclean casts generally exceed 70% and notch strengthening is evident. The values are in line with those predicted from the work on experimental and vacuum melted casts. Extrapolated values are given in Fig. 6.

6. CLEANNESS

To a steel maker 'cleanness' is a measure of the non-metallic inclusion content of a steel rather than residual element content. In the context where 'superclean' is defined as a very low residual content one residual that has an influence on non-metallic inclusion content is sulphur. It would be possible to manufacture Durehete 1055 with a low sulphur content – aiming for less than 50 ppm. However, this would have a very detrimental effect on machinability.

As all bolts undergo extensive machining operations, pressure from bolt manufacturers has deterred us from exploring very low sulphur levels. With the current low residual Durehete 1055 it is considered that exploring low sulphur levels would put us in the region of diminishing returns since the current product has more than adequate ductility for its purpose.

7. CONCLUSIONS

1. Although Durehete 1055 has been in use for over 30 years it has undergone a number of refinements since its introduction.
2. Current production made to low residual (clean steel) practices shows that significantly improved ductility and notch characteristics result from control of embrittling trace elements.
3. In the past Durehete 1055 bolting has been confined almost exclusively to UK designed turbine equipment, but it is now being specified for turbines designed in North America and Continental Europe where it will replace more expensive, less compatible alloys.

REFERENCES

1. M. Gemmill and J. D. Murray: *Engineering*, Dec. 1955.
2. G. D. Branch, J. H. M. Draper, N. W. Hodges, J. B. Marriott, M. C. Murphy, H. Smith and L. H. Toft: *International Conference on Creep and Fatigue in Elevated Temperature Applications*, Philadelphia, Sept. 1973 and Sheffield, April 1974.
3. H. Everson, J. Orr and D. Dulieu: 'Advances in Material Technology for Fossil Power Plant', *ASM/EPRI Conference*, Chicago, 1–3 Sept. 1987.
4. D. Hall and G. H. Whitley: *ASM/EPRI Conference*, Chicago, Sept. 1987.
5. B. L. King: *Phil. Trans. R. Soc.*, London, 1980, **A295**, 235.
6. H. Everson, J. Orr, D. Burton and D. Dulieu: *Rupture Ductility of Creep Resistant Steels*, Institute of Materials Conference, York, Dec. 1990.
7. J. Orr, H. Everson, D. Burton and J. Beardwood: *The Performance of Bolting Materials in High Temperature Plant*, Institute of Materials Conference, York, June 1994.

Zero Inclusion Steels

J. H. LOWE* and A. MITCHELL†
* *Axel Johnson Metals Corporation, Vallejo, CAL, USA*
† *Dept of Metals and Materials Engineering, University of British Columbia, Vancouver BC, Canada V6T 1Z4*

ABSTRACT

An examination of the influence of inclusions on steel properties indicates several general trends, although the specifics of each case can vary considerably. The first of the trends is that the most deleterious inclusions are the largest ones, which are almost invariably collections of small particles rather than one single precipitate. The second is that 'large' is a function of the property affected and of the alloy concerned, but non-metallic inclusions of less than one micron dimension, separated by distances which are greater than 10 microns, essentially have no effect on mechanical properties. Unfortunately, 'large' in most cases is not quite large enough to enable ultrasonic or other NDT methods to be used as a reliable quality filter. The third is that as melting processes become more refined, and in consequence the content of large inclusions in the product becomes smaller, our present methods of inclusion assessment become less and less effective in judging quality. In the cleanest alloys presently made by remelting techniques, the probability of metallographic detection of the small fraction of large inclusions which would affect properties is negligible and metallographic assessment is therefore not possible. Since this small fraction of large inclusions also constitutes a negligible fraction of the total oxygen content, any assessment based on chemical analysis is equally invalid.

The corollary to the above situation is that as inclusion quality assessment becomes more problematical, the reliance on process control becomes greater – a fact of production which in the aero-engine industry has led to a proliferation of specifications and frozen practices. The logical end-result of the quality improvement process has produced the idea of 'zero inclusion' steels. These are alloys in which the process of melting itself guarantees the absence of inclusions which are large enough to influence mechanical properties. It appears from research that the only process capable of this degree of cleanness is electron-beam hearth remelting, and it is in this context that we have carried out the developments described in this work.

The alloys which have been studied using EBCHM are austenitic stainless steels (for electronic applications) and maraging 250 steel (for aerospace applications). The compositions and procedures required for the zero inclusion condition were determined in laboratory experiments, and the alloys were subsequently melted using these techniques on a large scale. Results are reported for the stainless steel on the scale of 1 tonne ingot, and for the maraging steel at the scale of 5 tonne ingots.

It appears from the results of this programme that the zero inclusion condition can be achieved in appropriate alloy compositions i.e. those in which the evaporation experienced in EBCHM would not be prohibitive. (Two alloying elements which are a particular problem in this regard are Mn and Cu.) Since the EBCHM process is very amenable to on-line process control including image analysis of the liquid surface, we believe that it meets all of the quality control requirements of at least the near future in very high quality alloys. The cost attached to the product is significant, but not unreasonable in view of the quality attained and the critical nature of the end product. The EBCHM process is an accepted industrial tool in Japan, the USA and in Russia and the equipment required for the zero inclusion alloy production on the required scale is presently available.

1. INTRODUCTION

The role of inclusions in steels is a very wide one, depending principally on the physical parameters under study. In the context of the present applications in power generating equipment, the mechanical effects on crack nucleation and propagation together with the effects on corrosion behaviour are the most germane. Inclusions which affect these properties are generally in the size range of 10–500 μm and may be either exogenous or indigenous; they may be either sulphides or oxides or combinations of the two; in the smaller sizes, they may be single particles or in the larger sizes, agglomerations. The problems they pose to both producer and user are generally the same; expensive rejections and unpredictable machine life. As our mechanical designs have been progressively refined by the application of fracture mechanics concepts, the need for accurate information on 'probable defect distribution' has become critical, and with it the need for reliable inclusion assessment.

The detectability (by non-destructive testing (NDT)) of an inclusion is a function of the matrix dimensions, the inclusion geometry and orientation to the test direction, the ultrasonic (UST) penetrability of the matrix and finally, the inclusion's dimensions. Although these parameters combine to give each case its individual detection limit, in general we can observe that the lower limit of detectability by non-destructive testing is approximately the minimum dimension which is used to characterise the fracture behaviour (as illustrated in Fig. 1 for a range of nickel-base superalloys), a not unexpected result, but one which demonstrates the need for either further reduction in inclusion size distribution, or better testing methods. It is a fortunate fact of mechanical design that the most deleterious inclusions are those which intersect the free surface of a component and are therefore generally the most detectable. The thesis of this work is that an increase in detection reliability is unlikely and therefore we should aim towards process improvements which themselves will guarantee freedom from deleterious inclusion contents. An additional factor which supports this direction of development is that when any NDT method is used towards the limit of its capability, the statistical reliability of the result is reduced quite sharply.

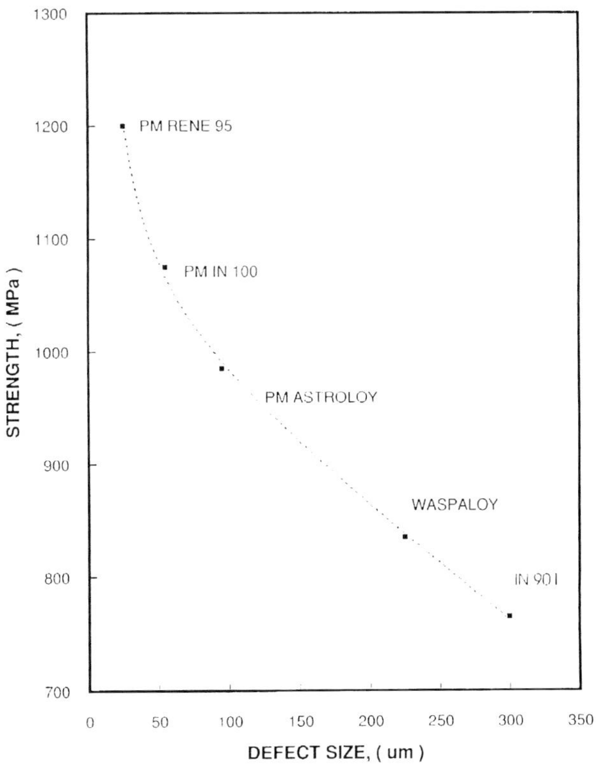

Fig. 1 Critical defect length required to initiate unstable crack growth in a range of superalloys at room temperature. (Courtesy Dr A. Lasalmonie, SNECMA)

The traditional method of assessment of inclusions in steels is that of optical metallography. Although this technique has been refined very extensively, it still suffers from the inherent disadvantage that it is a 2-D analysis of a very small portion of the component in question. For relatively dirty steels, it is probably reasonable to make the extrapolation of quality from a small surface to the bulk material, particularly when the inclusions assessed are in the same size range as those which are used to determine quality. However, when the steel is very clean, and when the inclusions which affect properties are bigger than the average of those used to make the metallographic assessment, the validity of this extrapolation becomes very questionable. This point may be illustrated by reference to the literature,[1] where we find that the 'clean' steels typically have an inclusion distribution, as measured optically, as shown in Fig. 2. The cut-off, at the large-size inclusions is 25 μm, with a concentration of 0.01 cm^{-2}. This concentration, representing a realistic limit in the optical technique (approx. 100 fields @ ×400 to detect one inclusion), corresponds to approximately 500 inclusions of 25 $\mu m\ kg^{-1}$, and is in fact far from negligible

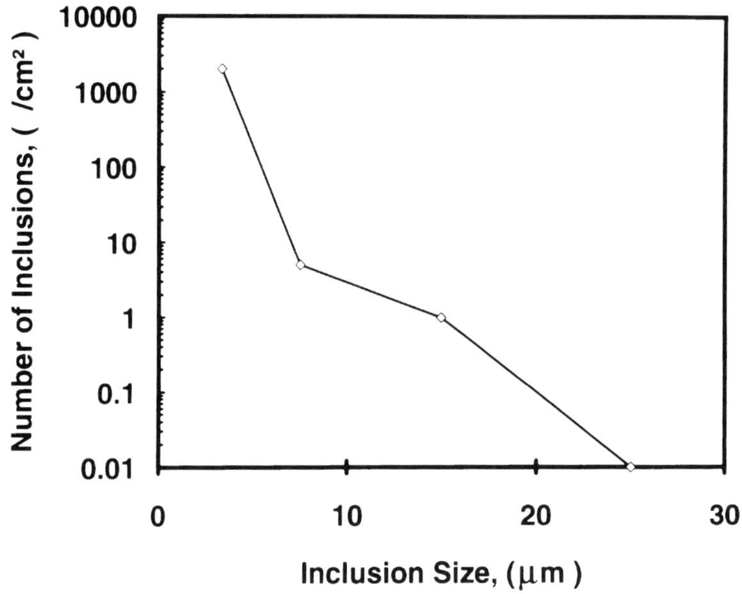

Fig. 2 Typical inclusion size distribution in a clean steel as assessed by an optical method at 400× on 100 random fields

in terms of mechanical behaviour. If the relation of Fig. 2 is extrapolated into the size range of 250 μm at which an effect on, for example, fatigue endurance limits would be observed, the concentration falls to approximately one inclusion per 2000T, at which the statistical probability of either non-detection by UST, or alternatively the probability that this inclusion will be in a position in the component where it will affect properties in service are both negligible. (By optical examination, it has been estimated that for a reliable assessment of concentration at this level of occurrence of 250 μm inclusions, it would be necessary to examine approximately 10^8 mm^2 of sample area). If, however, the extrapolation is valid, the inclusion distribution shown in Fig. 2 assures the user that the steel will be effectively free from property-affecting defects. Unfortunately, the reported incidence of large inclusions in this type of steel is considerably higher than that estimated above. Thomas *et al.*[2] report a frequency of detection for 500 μm inclusions at 10^{-3} cm^{-3} (or 1 inclusion per 7.8T) in steel which was optically assessed as having a cut-off of 0.1 cm^{-3} inclusions of 30 μm, by examination of 1000 mm^2. The incidence of large inclusions in this steel would be frequent enough to be a matter of concern to the user of, for example, large rotating parts. We are forced to conclude that the extrapolation of Fig. 2 into large size ranges is not valid; a reasonable conclusion, since the mechanism of formation of large inclusions is essentially different from that of the small inclusions typically counted by optical methods. The former arise by entrapment or by agglomeration; the latter as individual chemical precipitates.

The analytical oxygen content of the steel has been frequently used as a quality standard for inclusion contents. This method also suffers from the inherent disadvantages described above. In relatively dirty steel, the relationship between total inclusion content and the oxygen analysis is quite good, but in clean steels not only is the accuracy of the analysis in question, but also its relevance to the inclusion size range which is of mechanical concern. For example, in a steel which contains 1000 alumina inclusions of 50 μm diameter, per tonne, and has an oxygen content of 10 ppm, the large inclusions account for only 6% of the oxygen content, which is well within the normal analytical error in oxygen analysis at this level. Even gross variations in the large inclusion content would, therefore, remain unregistered by the oxygen analysis. This point is graphically illustrated by results reported on very clean bearing steels[3] where superior properties are reported for steels with a higher oxygen content, but made by a process which reduces the incidence of large inclusions.

In an effort to alleviate this problem, the aerospace industry has instituted the assessment by electron-beam button-melting.[4] This method has the advantage of sampling a larger volume of metal than the metallographic techniques, but has two serious drawbacks. First, the analysis estimates the total inclusion content and the size distribution with a good degree of accuracy, but does not determine the spatial distribution of the inclusions in the starting alloy. The effect of 'nearest neighbour' interactions, whose importance was stressed by Dijk *et al*,[5] cannot be defined. Second, the inclusions are necessarily heated strongly in vacuum and if they are of particular compositions which either volatilise or are reduced by the steel's carbon content, they may not be registered correctly in the eventual surface analysis of the button sample. Hence the method has been used on low-carbon alloy steels with success, but may not be applicable to forging grades with a higher carbon content. A method of induction levitation melting which may avoid the latter problem is under development,[6] but still suffers from the first of the difficulties described above.

It is clear from the above survey that there is a strong need in very high quality steels to produce components in which the manufacturing process itself assures freedom from large, random inclusions. We propose that this is best accomplished by a melting/casting procedure in which we remove the large inclusions by a solution process, so that the final content of sulphur, nitrogen and oxygen is completely in solution at temperatures close to the solidus. In this way, there are no inclusions present during cooling and solidification and, therefore, there can be no agglomeration leading to large inclusion particles. The final chemical analysis is hence an assurance of the absence of large inclusions. The following section details work directed to that end.

2. THEORY OF ZERO INCLUSION STEELS

The design of the process described above requires a computation or experimental determination of the solubility product of the inclusions concerned. The required thermochemical data is largely available in the literature and the saturation solubilities and their temperature dependence, for various inclusion compositions may be computed. The values obtained may also be checked experimentally in the temperature region above the liquidus by using the EB button technique itself, in cases where the alloy does not contain a large concentration of volatile elements. Both of these methods have been described in the literature,[7,8] as applied to nickel-base superalloys and stainless steels, with results which take into account the effects on the driving force for inclusion precipitation of segregation during freezing and also of interfacial tension. The results are summarised in Table 1 in terms of the chemical composition of the alloy which will not precipitate inclusions before complete solidification has taken place, under solidification conditions typical of remelt ingots.

Two points are to be noted from Table 1. First, the compositions indicated are quite close to those obtainable with existing melting practices and are hence realistic for industrial operations. Second, the role of calcium oxide must be taken into account. Calcium is very insoluble in liquid steel and its oxide is extremely stable at high temperature. These two factors combine to act against any solution mechanism for the removal of CaO–containing inclusions from the system and we must use melting and refining techniques which do not introduce them in prior processing. The use, for example, of CaO–containing refractories and slags must be minimised and their effects neutralised in the final processing stages by the application of alternative compositions based on the alumina–silica–magnesia system. This problem has been studied in relation to the refining of stainless steels and the appropriate solutions indicated.[7]

Since the final refining step in the production of the zero-inclusion composition is necessarily a vacuum process in which sulphur removal is not possible, the starting composition of the alloy must already have the required low sulphur content, obtained by a process which does not leave CaO–containing inclusions in the alloy. This may be done by a choice of raw materials; by refining with CaO–containing slags followed by ESR remelting in an acid slag; or by CaO ladle treatment followed by a second ladle

Table 1 Composition for Zero Inclusions

Al_2O_3	Al ppm	O ppm
2000 dyn. cm^{-1}	20	<10
1500	20	<2

treatment in an alumina-lined ladle with an acid slag. The first is the chosen route for superalloys,[8] the second for alloy steels and the third for stainless steels.[7] The input to the final refining process is therefore a cast ingot with the appropriate sulphur content, and with an inclusion content of CaO–free compositions which are to be removed by the refining step.

The final refining step must be able to remove all inclusions in the liquid by solution (and also flotation where appropriate) and produce an additional removal of O & N by volatilisation/reduction mechanisms. This must be accomplished at a temperature as close as possible to the liquidus (to ensure maximum rejection of any solid inclusions) and in a refractory-less environment. The only process which fulfills these conditions is electron-beam cold-hearth refining.[9] An additional advantage of this process is the high degree of visibility of the liquid metal surface being processed, which permits on-line image analysis and imaging pyrometry to be used as control tools. After processing, the zero-inclusion alloy is processed in the same manner as a conventionally-remelted ingot.

3. RESULTS OF REFINING STUDIES

3.1 Stainless Steels

As reported previously, with austenitic stainless steels, achievement of the required composition in Al and O resulted in a steel which contained no detectable inclusions, either by metallographic or EB button assessment techniques, in the size range greater than 1 μm in diameter. The evaluation of the steel's properties is difficult, since we must focus on properties which are likely to show differences within the statistical limits of the test chosen. For this reason, the tests selected are not the conventional ones of tensile or fracture behaviour, but instead relate to the production experience in the application in question. The first application is that of thin sheet for electronic purposes; the second is that of tubing for the transport of reactive gases used in microchip manufacture.

In the first case, the steel is rolled to a thickness of 50 μm and is then inspected for surface flaws. It is then used in making masks which have a complex pattern of holes which are precision drilled by EB. The finished product is assessed by optically registering the position of the holes and the absence of extraneous penetrations. Inclusions cause material to be rejected for surface flaws at the initial stage; for mis-shaped or misplaced holes and for unintentional penetrations. At the present time approximately 70 tonnes of steel have been produced by this method without any rejections for these causes. Steel masks produced by the conventional remelting routes typically have a rejection rate of approximately 10% for inclusion-related defects.

In the second case, inclusions cause pitting corrosion to occur (particularly around welded points) in the small-bore tubing. The corrosion products, including the inclusion particle, are subsequently deposited onto the chip in process, causing a rejection. The critical test in this case is the pitting potential of the EB-processed steel, as compared with the conventional alloy. Presently-available laboratory results indicate that the EB alloy does not show pitting corrosion in the potential range characteristic of the conventional alloy, and that exceptional conditions must be applied in order to make the alloy pit. Whilst this test demonstrates the value of the steel for the application chosen, it is more interesting to note that it also indicates the steel might be a substitute for the more expensive nickel-base alloys which are used to avoid the pitting problem in many chemical process reactors. At the present time a 1500 Kg heat has been produced to the required chemical specification for zero inclusions and corrosion testing is in progress to verify the absence of pitting corrosion both in the bulk alloy and in the heat-affected zone of welds.

It is interesting to note that results on the melting of both this alloy and the nickel-base superalloy IN713C indicate that the EB process has some capability to remove sulphur. This conclusion was reached on the basis of analysis of the condensate present on the furnace roof after a melting trial. For a feed alloy (304L) containing 350 ppm S, the condensate contained 2000 ppm S; for the same alloy with 98 ppm S, the condensate contained 200 ppm S. The corresponding alloy sulphur contents in the EB ingots were 275 ppm and 49 ppm. Presumably the removal mechanism is one of evaporation, but the detailed process is not yet defined.

3.2 *Maraging Steel*

The steel under test is a maraging 250 steel intended for aero-engine main-shaft application. The critical properties in this respect are the strength/LCF relationships in the temperature range −50/+250C. The fracture initiation sites in the conventionally remelted alloy are found to be the Ti–C–N–S particles formed on solidification. The precipitation sequence of this composition range is extremely complex, but in the present context can be analysed in terms of the alloy content in respect of the C, N and S below which no inclusions will precipitate at temperatures above the solidus of the alloy, as was done for the stainless steels. The critical analysis, for the range of freezing rates commonly encountered in practical remelt ingots, is in this case:

$$C < 40\,ppm : S < 5\,ppm : O < 8\,ppm: N < 4\,ppm$$

Conventionally-remelted alloy has contents of these elements which are greater than these values. In the present work, a starting alloy with:

$$C\ 40\,ppm\ ;\ S\ 5\,ppm\ ;\ 0\ 11\,ppm\ ;\ N\ 10\,ppm$$

was processed by EBCHM, using a proprietary technique to reduce the impurity content, to manufacture ingots of 450 mm dia and approximately 5 tonnes in weight. The ingots were chemically analysed and were found to have impurity contents which were below the above critical limits. Using optical techniques, at 1000x, extensive examination (with a resolution of approximately 1 μm) failed to reveal any inclusions. It may be concluded that the initial premise of composition control to eliminate the inclusions is valid. Mechanical testing is in progress to investigate the effect of removing all inclusions on the LCF life and fracture initiation mechanism.

4. CONCLUSIONS

1. In very clean materials, it is clear that we have reached a situation where conventional methods of inclusion assessment will not give a reliable validation of the alloy cleanness.
2. Since failures in the very clean materials are initiated more by random inclusions than by inclusions assessable by metallographic or related techniques, the only feasible method for their control is process control in the melting operation.
3. The EBCHM process offers the best alternative for on-line control of this problem.
4. The two alloy series tested on a large scale have shown considerable promise of being true 'zero inclusion' quality; the viability of the production process will depend on the extent to which the property improvements can be utilised.

ACKNOWLEDGEMENTS

The authors gratefully acknowledge the support of the Daido Steel Corp and the Axel Johnson Metals Corp in the course of this work together with the Nickel Development Institute.

REFERENCES

1. F. B. Pickering: 'Inclusions', *Institution of Metallurgists Monograph No 3*, 1979.
2. K. Thomas, P. J. Appleton, W. S. Wilson and M. J. Worrall: *Proc. 4th International Conference on Clean-Steels*, Balatonszeplak, Hungary, 1992, Institute of Materials, 308.
3. *Proc. 4th International Conference on Clean-Steels*, Balatonszeplak, Hungary, 1992, Institute of Materials, Fig 2, 315.
4. L. Barnard, R. F. Brooks, P. N. Quested and K. C. Mills: *Ironmaking Steelmaking*, 1993, **20**(5), 344.
5. J. Dijk, D. Manneveld and J. M. Rabenberg: *Proc. 4th International Conference*

on Clean-Steels, Balatonszeplak, Hungary, 1992, Institute of Materials, 179.

6. R. G. Menzies and C. B. Adasczik: *Proc. Vacuum Metallurgy Conference 1986*, L. W. Lherbier and G. K. Bhat eds, American Vacuum Soc, New York, 65.
7. S. Fukumoto, A. Mitchell and F. Reyes-Carmona: *Innovation in Stainless Steel*, W. Nicodemi ed., Assoc. Italiana di Metallugia, 1993, 2.27.
8. S. Cockcroft, T. Degawa, A. Mitchell, D. W. Tripp and A. Schmalz: *Superalloys 1992*, S. D. Antolovich *et al* eds., TMS-AIME, Warrendale, PA, 577.
9. W. H. Buttrill, C. E. Shamblen and G. B. Hunter: *Advanced Materials*, T. Kishi *et al* eds., SAMPE Japan, 1993, 1660.

Clean Steels for Fusion

D. S. GELLES
Pacific Northwest Laboratory
Richland, WA 99352 USA

ABSTRACT

Fusion energy production has an inherent advantage over fission: a fuel supply with reduced long term radioactivity. One of the leading candidate materials for structural applications in a fusion reactor is a tungsten stabilised 9% chromium martensitic steel. This alloy class is being considered because it offers the opportunity to maintain that advantage in the reactor structure as well as provide good high temperature strength and radiation induced swelling and embrittlement resistance. However, calculations indicate that to obtain acceptable radioactivity levels within 500 years after service, clean steel will be required because the niobium impurity levels must be kept below about 2 appm and nickel, molybdenum, nitrogen, copper, and aluminum must be intentionally restricted. International efforts are addressing the problems of clean steel production. Recently, a 5000 kg heat was vacuum induction melted in Japan using high purity commercial raw materials giving niobium levels less than 0.7 appm. This paper reviews the need for reduced long term radioactivity, defines the advantageous properties of the tungsten stabilised martensitic steel class, and describes the international efforts to produce acceptable clean steels.

1. INTRODUCTION

Ever since mankind has learned to duplicate the nuclear reaction controlling our sun, it has been a goal of society to harness that nuclear fusion process to create limitless quantities of energy. This goal is not yet achievable, and it is now apparent that immense finances will be needed because the machines required must be very large, and the technical problems to be overcome are very complicated. In an international effort to demonstrate the practicality of fusion energy power production, a device called the International Thermonuclear Experimental Reactor (ITER) is being designed. A recent partial assembly sketch of the ITER design is shown in Fig. 1.

In the quest to procure funds for the ITER and earlier machines, it was apparent that fusion power needed to demonstrate advantages over other sources of energy. One of the inherent advantages for fusion over fission energy production is a fuel supply with reduced long term radioactivity. This was first emphasised in 1983 with the publication of the *Report of the DOE*

* Operated for the U.S. Department of Energy by Battelle Memorial Institute under Contract DE-AC06–76RLO 1830.

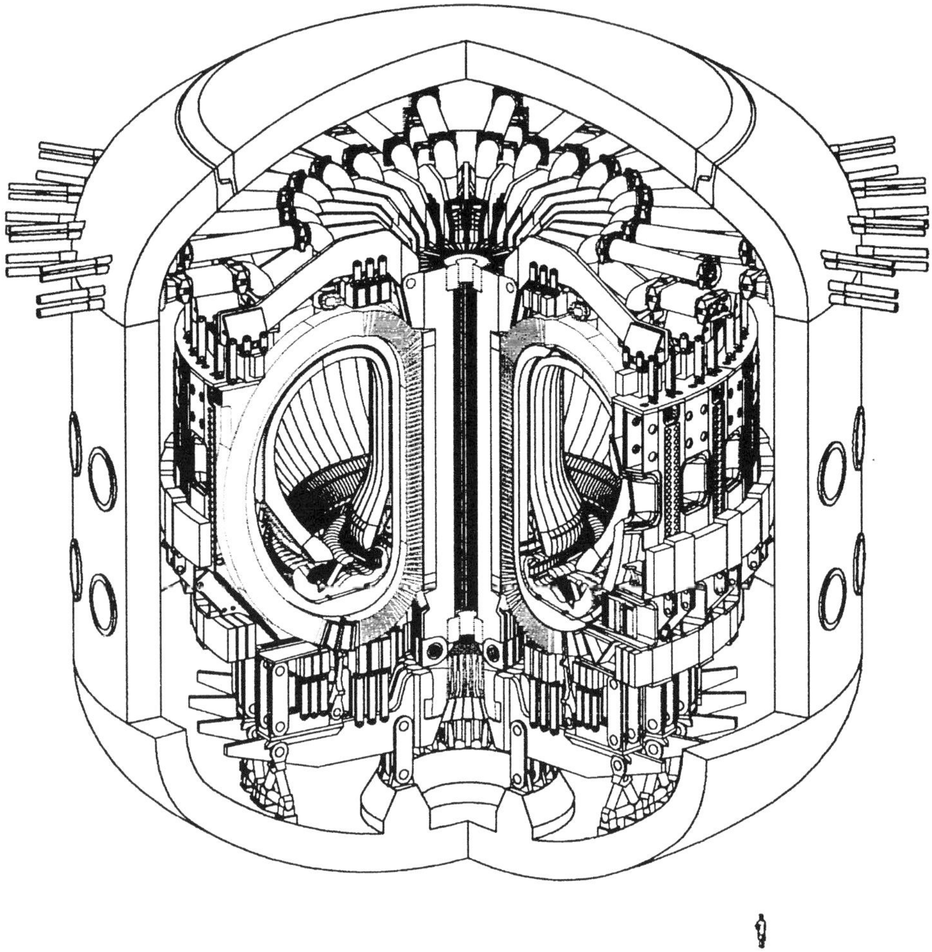

Fig. 1 ITER Partial Assembly Sketch as of July 18, 1994

Panel on Low Activation Materials for Fusion Applications.[1] The panel recommended that one of the goals of the fusion programme should be a reference reactor that meets the requirements for near surface disposal of radioactive waste. The recommendation was tied to a United States regulation, 10 CFR 61,[2] requiring that for shallow land burial of class C waste, radioactivity levels in waste must decay to safe levels after 500 years.

To meet the requirement of safe radioactivity levels within 500 years, the panel noted that reactor components that become radioactive must not contain certain deleterious elements that produce radioactive isotopes with very long half lives. The list included nickel, nitrogen, copper, molybdenum, and niobium, but the restriction on niobium was by far the most severe: 1 atom part per million in the first wall structure. An early example of the activation level

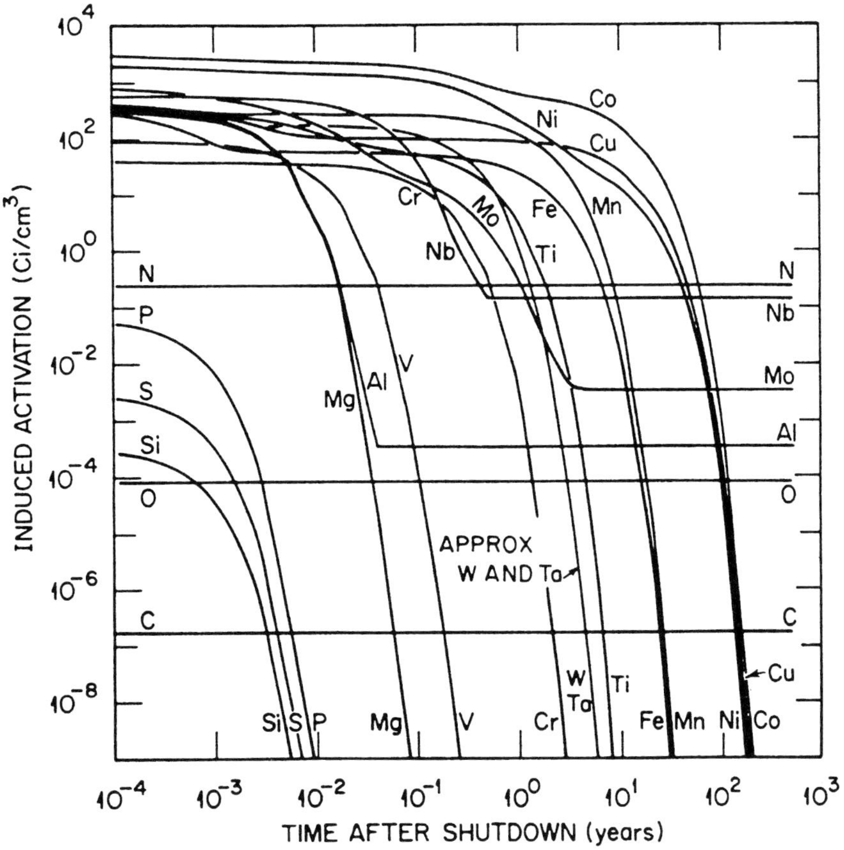

Fig. 2 Induced Radioactivity in the First Wall of a Fusion Reactor made from Austenitic Stainless Steel

and rate of decay for an austenitic stainless steel exposed for 9 MW-ym^{-2} in the neutron flux at a first wall of a fusion reactor is shown in Fig. 2.[3] Activity levels vary with time as a function of the half lives for the activated isotopes and daughter products of each element shown. High levels of radioactivity are maintained in the stainless steel first wall component for very long times due to nitrogen, niobium, and molybdenum. Therefore, impurities in reactor materials had to be strictly controlled to meet such a goal.

The materials development community took this suggestion to heart in subsequent years, and efforts are now proceeding to develop reduced activation structural material versions of the following alloy classes: silicon carbide/silicon carbide composites, vanadium alloys containing chromium and titanium, and tungsten stabilised martensitic steels. The purpose of this paper is to provide further details concerning the development of reduced activation tungsten stabilised martensitic steels.

2. LOW ACTIVATION GUIDELINES

Since 1983, the list of elements contributing to unacceptable long term radioactivity has come under further scrutiny.[4–7] The criterion has been expanded to consider not only shallow land burial, but also direct handling (or 'hands on' applications), and recycling of materials. The list of elements that can make an irradiated material unacceptable for shallow land burial is now considerably longer, as provided in Table 1. (Differences in the composition limits given in Table 1 are partly a measure of the assumed irradiation conditions used and partly calculational procedures.) Table 1 also includes two estimates of required composition limits for hands on applications. However, for the purposes of steel production, niobium levels on the order of 1 wppm for disposal, or 0.01 to 0.2 wppm for hands on applications, are expected to provide the greatest restriction.

3. DEVELOPMENT OF STEELS FOR FUSION

Super 12 chrome steels are found to be viable for fusion structural applications. Interest in this alloy class first evolved from materials development for liquid metal fast fission breeder reactors, where it was shown that ferritic steels had inherent resistance to irradiation induced swelling, a process whereby materials expand in volume due to the accumulation of vacancies in internal cavities.[8,9] The combination of high temperature strength, good thermal conductivity, and good resistance to irradiation induced swelling and creep in these martensitic steels has been demonstrated to provide an engineering window for fusion reactor design.[10–14] The high temperature limit may be restricted to temperatures as low as 450°C as a result of design specific details such as liquid lithium compatibility, but the swelling resistance of this class of materials appears to provide a stable structural material to very high irradiation doses, avoiding the need for component replacement. Precipitation of chromium rich α' results in irradiation induced hardening and embrittlement,[15] which can be avoided by lower chromium levels in the range 7 to 9%.

Composition specifications were possible for super 12 chrome low activation steel by restricting the additions of niobium, molybdenum, nitrogen, and nickel. A large number of alloys have been designed, manufactured, and tested at laboratories around the world,[16] and it has been shown that tungsten can be directly substituted for molybdenum on an atom for atom basis and that tantalum can be used for niobium, providing alloys with high temperature strength on a par with 9Cr-1Mo and 12Cr-1Mo steels. Lower chromium levels avoided the need for a nickel substitute or for higher carbon levels to maintain a fully martensitic structure, as well as preventing α' precipitation hardening. In fact, the tungsten stabilised martensitic steel class

Table 1 Low Activation Guidelines Based on Waste Disposal Criteria.

Element		Nb	Mo	Cu	N	Ni	Pb	Al	Ho	Tb	Ir	Gd	Bi	Re	Hf
Waste disposal criterion	1983[1] (appm)	1	3650	2400	3650	20000									
	1984[4] (wppm)	2.9	30	1200	3300	9100	31000	36000							
	1988[5] (wppm)	10	500	5000	1000	20000	500000	5000	1	2	10	20	100	1000	1000
Hands on criterion	1992[6] (appm)	.18	31	651		355	900	184	.057	.021	.32	.39	.13	72	.30
	1995[7] (wppm)	.0141	1.67			14800		5.24	.00608	0.0134	1.56	0.118	0.119	4.75	
Element		**Se**	**Cd**	**K**	**Sm**	**Tm**	**Pd**	**Ag**	**Eu**	**Lu**	**Dy**	**Yb**	**Os**	**Co**	**Er**
Waste disposal criterion	1988[5] (wppm)	1000	2000	2000	2000	3000	5000	20000	1000	100000	3000				100
Hands on criterion	1992[6] (appm)		4.9	137	.33		.72	.011	.074	.22	.54	.63	1.2	1.7	3.1
	1995[7] (wppm)		2.3	1652			0.212	.00494	.585		.0048		.517	119	0.196

is found to have excellent impact resistance, with ductile to brittle transition temperatures (DBTT) below −60°C, and excellent resistance to irradiation embrittlement.[16]

4. STUDIES ON CLEAN STEEL

Two papers have addressed the question of the practicality of producing clean low activation tungsten stabilised steels.[6,17] Butterworth and Keown[6] concluded in 1991 that state of the art residual levels are generally in the range 10 to 100 ppm for high-integrity steels used for aerospace, military, and other demanding applications and therefore did not meet the more exacting requirements for hands on limits. They proposed that the route used for producing nickel-based superalloys for aerospace components where specifications were possible for some impurities to the sub-ppm level might be more successful. Careful selection of starting materials was required to exclude specific high activation elements from the feedstock.

Feasibility has in fact been demonstrated in Japan. The Nippon Kokan Corporation (NKK) has been actively developing an alloy designated F-82H for fusion reactor applications.[17–20] This alloy's approximate composition is Fe-8Cr-2W-0.2V-0.04Ta (in wt%) with carbon levels of about 0.1%. As part of that effort, an attempt was made to reduce the content of activating elements below the guidelines of reduced activation.[17] A 5000 kg ingot was vacuum induction melted from high purity commercial raw materials using commercial practices. The chemical analysis for the starting materials is given in Table 2, and the composition of the mill product is given in Table 3. Given the tight requirement needed to satisfy activation requirements, only niobium was

Table 2 Chemical analysis of raw materials[a]

Element	Raw materials	Nb [ppm]	Mass of melting [kg]	Nominal content [wt.%]	Calculated Nb content in F-82H [ppm]
C	carbonet	–	5	0.1	–
Si	metallic silicon	<1.0	5	0.1	<0.001
Mn	metallic manganese	<1.0	5	0.1	<0.001
Cr	high purity metallic chromium	10	385	7.7	0.15
W	metallic tungsten	2.0	100	2.0	0.04
V	metallic vanadium	10.0	10	0.2	0.02
Ta	metallic tantalum	85.0	1	0.04	0.034
Ti	sponge titanium	1.1	1	0.02	<0.0004
Fe	converter steel	<0.5	4487	bal.	<0.45
total			5000	100	<0.7

[a] The dash means not analysed.

Table 3 Chemical Composition of Mill Product F-82H (wt.%).[a]

Impurity	Ladle	Plates			Impurity	Ladle	Plates		
		top	middle	bottom			top	middle	bottom
C	0.096	0.097	0.110	0.094	Ti	0.008	0.007	0.007	0.007
Si	0.10	0.09	0.09	0.09	Nb	<0.0005	0.00007	0.00005	0.00007
Mn	0.15	0.07	0.07	0.07	Sol.Al	0.005	0.007	0.007	0.008
P	0.003	0.002	0.003	0.003	W	2.1	2.1	2.1	2.1
S	0.003	0.0032	0.0029	0.0027	Ta	0.04	0.03	0.03	0.03
Cu	0.01	<0.01	<0.01	<0.01	B	–	0.0004	0.0003	0.0004
Ni	<0.1	0.03	0.03	0.03	T.N	0.0043	0.0043	0.0044	0.0047
Cr	7.71	7.46	7.46	7.46	T.O	0.0028	0.0033	0.0049	0.0037
Mo	<0.01	<0.0 01	<0.001	<001	Ag –	–	<0.005	–	
V	0.18	0.18	0.18	0.18	Co	–	–	0.005	–

[a] The dash means not analysed.

analysed in the raw materials. From Table 2, it can be shown that the major source of niobium is from iron and chromium, with lesser amounts from tungsten, tantalum, and vanadium. Mill product chemical analysis shows that the niobium content is quite uniform, varying from 0.5 wppm in the centre to 0.07 wppm at both ends, and that silver and cobalt levels are 50 wppm. It can therefore be concluded from comparison with Table 1 that the NKK 5000 kg heat would satisfy the waste disposal criterion, but would not meet hands on criteria based on silver and niobium impurities, and cobalt levels may be too high, depending on the irradiation conditions and calculation procedures used to define the hands on low activation limit.

DISCUSSION

The ITER development and testing programme is expected to continue well into the next century and is presently not being designed within low activation guidelines. The first commercial fusion reactor is not expected to be constructed before 2025. A commercial reactor would be expected to follow low activation guidelines, but it is not yet clear that a martensitic steel will be selected as the structural material. Therefore, commercial production of a clean steel for fusion will not be needed in the immediate future; we have 30 years in which to develop the technology.

6. CONCLUSIONS

Fusion power systems will be required to use low activation structural materials to allow shallow land burial of waste and possibly hands on or recycling options. This will mean that certain elements will be restricted from structural materials specifications. In particular, niobium must be kept below about 2 appm for shallow land burial and 0.01 to 0.1 wppm for hands on and recycling options. A tungsten stabilised martensitic steel appears to be a good choice for fusion structural materials applications, and therefore clean steels may be selected. However, commercial production of such steels will not be needed for another thirty years, giving ample time to advance steel making to the levels needed.

REFERENCES

1. *Report of the DOE Panel on Low Activation Materials for Fusion Applications*, UCLA/PPF-728, June 1983.
2. *Federal Register*, Monday, December 27, 1982, **47**(248), 57446–57482.
3. F. W. Wiffen and R. T. Santoro: *Proceedings of Topical Conference on Ferritic Alloys for Use in Nuclear Energy Technologies*, J. W. Davis and D. J. Michel, eds., AIME, Warrendale, PA, 1984, 195.
4. F. M. Mann: *Fusion Tech.*, 1984, **6**, 273; or *J. Nucl. Mater.*, 1984, **122** & **123**, 1053.
5. C. Ponti: *Fusion Tech.*, 1988, **13**, 157.
6. G. J. Butterworth and S. R. Keown, *J. Nucl. Mater.*, 1992, **186**, 283.
7. K. Ehrlich, S. W. Cierjacks, S. Kelzenberg and A. Möslang: 'The Development of Structural Materials for Reduce Long-Term Activation' submitted for publication in *Effects of Radiation on Materials: 17th International Symposium*, ASTM STP 1270, ASTM, Philadelphia, PA 1996, to be published.
8. S. D. Harkness, B. J. Kestel and P. Okamoto: *Radiation-Induced Voids in Metals*, J. W. Corbett and L. C. Ianniello, eds., CONF-710601, USAEC, Oak Ridge, TN 1972, 334.
9. D. S. Gelles: *Ferritic Steels for High Temperature Applications*, A. K. Khare, ed., ASM, Metals Park, OH 1982, 197.
10. *Starfire, A Commercial Tokamak Fusion Power Plant Study*, ANL/FPP-80–1, V1, Sept. 1980, 10–232.
11. *A Demonstration Tokamak Power Plant Study*, ANL/Fpp/82–1, Sept. 1982, 2–3.
12. *MARS Mirror Advanced Reactor Study*, Final Report, Volume 1–B, UCRL-53480, July 1984, 12–9 to 12–24.
13. *Blanket Comparison and Selection Study*, Final Report, ANL/FPP-84–1, Sept. 1984, 7.3–4.
14. *Modeling, Analysis & Experiments for Fusion Nuclear Technology*, FNT Progress Report: Modeling and FINESSE, Jan. 1987, 1021.
15. D. S. Gelles and L. E. Thomas: *Proceedings of Topical Conference on Ferritic*

Alloys for Use in Nuclear Energy Technologies, J. W. Davis and D. J. Michel, eds., AIME, Warrendale, PA 1984, 559.
16. D. S. Gelles: *Plasma Devices and Operations*, 1994, **3**, 31.
17. N. Yamanouchi, M. Tamura, H. Hayakawa, A. Hishinuma and T. Kondo: *J. Nucl. Mater.*, 1992, **191–194**, 822.
18. M. Tamura, H. Hayakawa, M. Tanimura, A. Hishinuma and T. Kondo: *J. Nucl. Mater.*, 1986, **141–143**, 1067.
19. M. Tamura, H. Hayakawa, A. Yoshitake, A. Hishinuma and T. Kondo: *J. Nucl. Mater.*, 1988, **155–157**, 620.
20. H. Hayakawa, A. Yoshitake, M. Tamura, S. Natsume, A. Gotoh and A. Hishinuma: *J. Nucl. Mater.*, 1991, **179–181**, 693.

Temper Embrittlement in High Strength Pressure Vessel Steels for Polyethylene Manufacture

G. A. HONEYMAN and J. E. HEESOM

Forgemasters Steel & Engineering Ltd, P.O. Box 287, Brightside Lane, Sheffield, United Kingdom

1. INTRODUCTION

The safe and reliable performance of pressure vessels used for polymerisation reactions is essential in the chemical industry. This need motivates the continual development of improved production routes and materials for pressure vessel technology.

Forgemasters Steel & Engineering Ltd manufactures reaction vessels for a wide range of operational conditions. The company's ability to assemble and proof test these items has been instrumental in the adoption of its equipment in Imperial Chemical Industries (ICI) designed plant throughout the world. Such plants produce low density polyethylene from ethylene in a reaction which proceeds under conditions of elevated temperature and pressure. The optimum conditions for polymerisation occur at approximately 200 MPa pressure at temperatures approaching 300°C. The containment systems, which range in capacity from 250 to 1500 litres, must, therefore, be absolutely reliable to maintain a high degree of safety and process efficiency. A schematic diagram of a typical reaction vessel is shown in Fig. 1.

2. MATERIAL DEVELOPMENT

The choice of steel used for the pressure vessels is limited by ductility and toughness requirements and also the need for adequate hardenability to achieve a uniform strength through the wall thickness. For the larger, more highly stressed vessels 3.5% NiCrMoV steels are preferred. This steel type was initially developed primarily for the electrical power industry for the highly stressed rotating parts of generators and low pressure turbines. The chemical analysis was modified to increase the strength to the level demanded by vessel design. Principally the carbon level was raised from 0.25% to 0.30%–0.35%.

The phenomenon of embrittlement is probably the most hazardous event in reaction vessel operation. For this reason, it is an ICI regulation that vessels are taken out of service after every 6 years of working life for reconditioning.

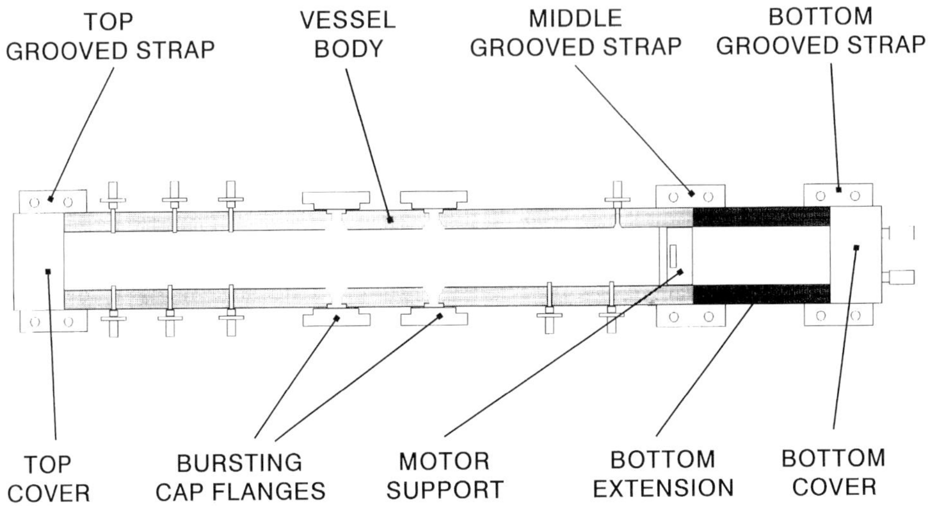

Fig. 1 Typical reaction vessel configuration

This procedure involves the dismantling, inspection and pressure testing of the assembly. The vessel body and the bottom cover are also given a de-embrittling stress relieving treatment at 550°C by oil quenching. Investigations by ICI (Fig. 2) optimised the treatment to return the impact properties close to the pre-operational values.

Several years ago, the specification for 3.5% NiCrMoV pressure vessel steels

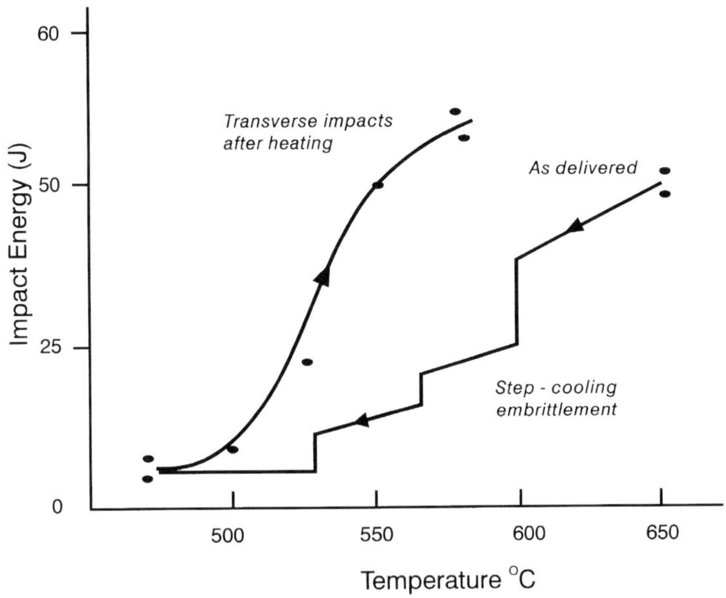

Fig. 2 Impact energy recovery after reconditioning

Table 1 Comparison of specifications

wt%	C	Si	Mn	S	P	Ni	Cr	Mo	V	As	Cu	Sn	Sb	J
Original Specification	0.30/ 0.45	0.35 max.	0.20 0.60	0.025 max.	0.025 max.	3.0/ 4.0	1.0/ 2.0	0.50/ 0.70	0.05/ 0.25	NS	NS	NS	NS	NS
Modified Specification	0.30/ 0.40	0.05/ 0.15	0.20/ 0.40	0.010 max.	0.010 max.	3.0/ 4.0	1.4/ 2.0	0.30/ 0.50	0.10/ 0.15	0.014 max.	0.15 max.	0.009 max.	0.009 max.	75 max.
Superclean Specification	0.30/ 0.35	0.02/ 0.05	0.02/ 0.05	0.002 max.	0.003 max.	3.50 3.75	1.65/ 2.00	0.45/ 0.50	0.10/ 0.15	0.005 max.	NS	0.005 max.	0.002 max.	10 max.

NS indicates not specified

Table 2 Mechanical property requirements

Specification	UTS MPa	0.2% PS MPa	R of A %	Charpy J	FATT °C
Original Specification	1112 to 1266	988 min.	30 min.	33 33	NS NS
Modified Specification	988 to 1080	834 min.	30 min.	53 min.	−50 max.

NS indicates not specified

did not impose restrictions on residual elements, however, after development work carried out by ICI and Imperial College, London, modifications were made to reduce the tendency of the material to embrittle in service.

The resultant specification is compared with the original in Table 1. The mechanical properties demanded for the two grades are listed in Table 2. The main difference in chemical composition is not only the level of residual elements specified but also the inclusion of a J factor restriction in the new specification. The need for superior toughness and ductility is reflected in the modified mechanical properties.

Worldwide investigations have resulted in the development of superclean NiCrMoV grades for steam turbine rotor forgings which significantly reduce temper embrittlement. FSEL considered that the superclean steel technology might improve the embrittlement characteristics of the 3.5% NiCrMoV pressure vessel steels still further and a trial vessel was manufactured for critical appraisal.

3. MANUFACTURE

Steel for pressure vessel components is melted in a Basic Electric Arc (BEA) furnace followed by refining in a vacuum arc degassing (VAD) unit. Further degassing is carried out by teeming into an ingot mould under vacuum using the Vacuum Stream Degassing (VSD) method ensuring hydrogen levels of 1 ppm or less. The move to a lower residual quality by ICI does not fundamentally modify the steelmaking procedure but high grade scrap is required to meet the J factor requirement.

In order to obtain a superclean steel to, typically, EPRI specification standards (Table 1), a more stringent approach to the steelmaking process is adopted and has been detailed in a previous publication.[1] The main differences as compared with conventionally cast 3.5% NiCrMoV grades are as follows:

- Basic oxygen furnace steel scrap of known high purity.
- Slag free tapping using sliding gate technology to the BEA furnace.
- New lining of basic refractory (dolomite) fitted to the refining ladle.
- Synthetic reducing slag during refining.

The superclean cast used for the manufacture of an experimental pressure vessel originated from an 81 tonne ingot of dimensions 1830 mm diameter × 3760 mm length. The remainder of the ingot was used for contract purposes and to manufacture a trial high carbon superclean rotor forging.

Pressure vessels are usually directly forged from the ingot to the required diameter before undergoing a preliminary heat treatment cycle which includes degassing and grain refining treatments.

Prior to quality heat treatment, forgings are rough machined and bored to maximise the through thickness tensile and toughness properties. The pressure vessels are austenitised at 850°C, water quenched and tempered between 600°C and 640°C.

4. RESULTS

The purposes of the investigation was to establish if superclean steel is less susceptible to embrittlement under pressure vessel operating conditions than conventional material. Both the low residual 3.5% NiCrMoV and the original higher residual 3.5% NiCrMoV were included in the evaluation so that the improvement in embrittlement characteristics already being achieved could be quantified. The third steel evaluated was the superclean cast made by FSEL.

The main aim of the investigation was to compare the toughness properties of each material type before and after a step cooling embrittlement treatment, this being illustrated schematically in Fig. 3. Some of the embrittlement samples were subsequently de-embrittled by a heat treatment of 10 hours at 550°C followed by oil quenching. The test results from this material facilitated an assessment of the efficiency of the reconditioning procedure.

Two casts of high residual 3.5% NiCrMoV were supplied to FSEL by ICI (HR1 and HR2). These test blocks originated from vessels that had been in

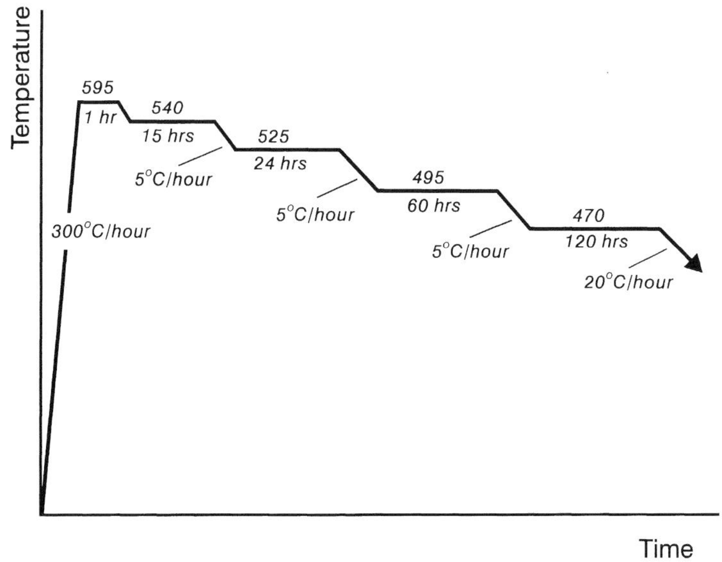

Fig. 3 Embrittling heat treatment

service for many years. For this reason both were given an additional de-embrittling treatment before being tested. Surplus test material from more recent contracts representing 2 casts of modified 3.5% NiCrMoV were also available (LR1 and LR2).

Prior to mechanical testing, chemical analysis was conducted on one sample per cast. Material properties were determined in the tangential direction for each cast in the as received, the embrittled and the de-embrittled conditions from a sub-surface location. Tensile tests were carried out at room temperature and fracture appearance transition temperatures (FATT) were derived from impact tests at a range of temperatures.

The chemical analyses for the five casts are listed in Table 3. In each case the requirements of the specification were met. The difference between the original and modified steels is demonstrated by the J factor. The residual levels of sulphur, phosphorus, arsenic, copper, tin and antimony are sig-

Table 3 Cast analyses (wt%)

Grade	C	Si	Mn	S	P	Ni	Cr	Mo	V	As	Cu	Al	Sn	Sb	J
HR1	0.32	0.19	0.19	.007	.008	3.52	1.54	0.54	0.12	.012	.09	.004	.007	.004	57
HR2	0.33	0.21	0.30	.008	.007	3.45	1.55	0.61	0.12	.023	.14	.004	.013	.0025	102
LR1	0.32	0.10	0.24	.003	.004	3.53	1.49	0.45	0.12	.006	.07	.005	.003	.002	24
LR2	0.31	0.12	0.23	.002	.006	3.53	1.47	0.45	0.12	.005	.05	.003	.003	.002	32
SC	0.35	0.02	0.05	.002	.003	3.58	1.69	0.48	0.12	.005	.04	.003	.004	.0015	5

Table 4 Mechanical properties

Condition	Material	UTS MPa	0.2% PS MPa	R of A %	Charpy (20°C) J	FATT °C
As Received	HR1	1127	987	59	39	+20
	HR	1211	1087	58	48	−52
	LR1	1072	961	55	88	−121
	LR2	1053	950	57	75	−125
	SC	1134	1007	60	88	−108
Embrittled	HR1	1080	945	60	28	+37
	HR2	1119	1001	52	12	+150
	LR1	1078	956	66	88	−115
	LR2	1054	934	60	71	−96
	SC	1109	976	57	88	−114
De-embrittled	HR1	1057	918	65	76	−13
	HR2	1113	986	56	43	−80
	LR1	1075	951	64	92	−125
	LR2	1047	928	61	72	−114
	SC	1082	934	61	85	−128

nificantly lower for the more recent casts. This, in conjunction with a reduced silicon content, has reduced the J factor to less than 40. For the superclean steel (SC), the residual and silicon contents have been further depressed as a result of the steelmaking practice. Additionally the manganese level is much lower.

The mechanical properties are presented in Table 4 for the materials in each condition. The tensile properties show little variation between the three conditions. It is particularly noteworthy that the ductility, as measured by the reduction of area at fracture, has not been adversely affected by the embrittlement procedure. Both the room temperature impact energies and the FATT values deteriorated after embrittlement for the original steels. However, this treatment did not influence the toughness of the modified and superclean casts. The effect of the de-embrittling treatment was to revert the properties to the superior levels measured prior to embrittlement.

The microstructure of all material types as evaluated by optical metallography was fully tempered martensite.

5. DISCUSSION

Recent advances in power generation design have precipitated the investigation of new alloys for steam turbine rotor forgings.[2] Turbines operating in advanced steam conditions of high temperature and pressure result in increased operating temperatures in the low pressure region. These higher temperatures are associated with the embrittlement range of conventional NiCrMoV steels used for low pressure rotor forgings. Extensive worldwide investigations[3,4] have resulted in the development of NiCrMoV grades with lower Mn, Si, P, As, Sn and Sb contents which significantly reduce the embrittlement problem. Temper embrittlement is also a fundamental problem for reaction vessels operating in the chemical industry. The propensity to embrittle during service effectively demands that the vessels are de-embrittled at periodic intervals. This results in considerable down-time and any extension to the reconditioning intervals is economically advantageous. The elimination of temper embrittlement has therefore been foremost in pressure vessel alloy development.

Low residual 3.5% NiCrMoV pressure vessel steel has already shown substantial improvement in toughness and resistance to temper embrittlement. The investigations into superclean technology suggested that superclean steel might offer further benefits.

The impact test results show that the toughness properties of the original composition are severely impaired by the embrittling treatment. This is demonstrated in Fig. 4 where the change in FATT, ΔFATT, for as received to embrittled material is plotted as a function of J factor. Above a J factor of approximately 60, ΔFATT increases very rapidly. An analogous plot for the

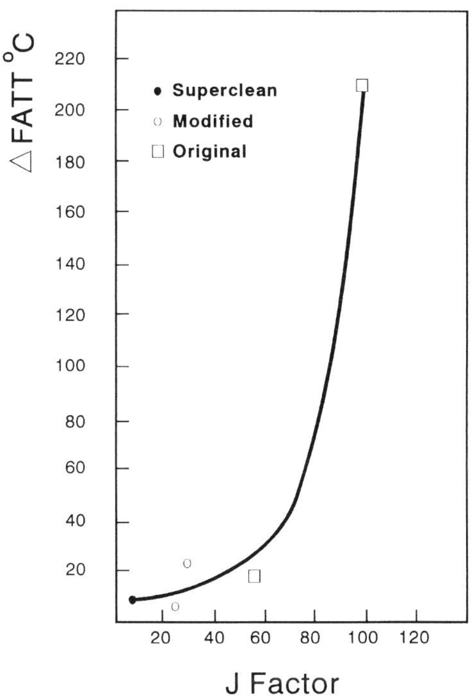

Fig. 4 ΔFATT for as received to embrittled

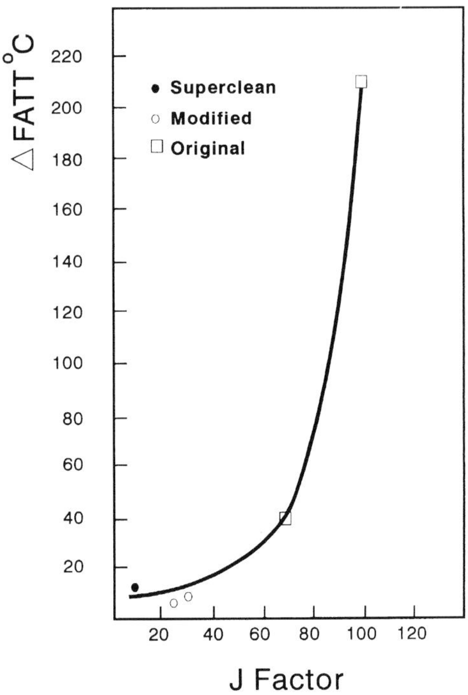

Fig. 5 ΔFATT for as embrittled to de-embrittled

embrittled to de-embrittled conditions, Fig. 5, follows a similar trend. However, in this case, the increase in ΔFATT becomes pronounced at a J factor of more than 40. Recovery of toughness after reconditioning justifies the belief that the embrittlement observed is, in fact, a reversible temper embrittlement phenomenon. Although pressure vessels operate at a temperature below the range traditionally recognised as the embrittlement region for NiCrMoV steels, the basic mechanism of the procedure is the same, albeit at a reduced rate. To identify further the critical role of the J factor, additional casts of 3.5% NiCrMoV pressure vessel steel have been included in Fig. 6. This shows the range of FATT measured for steels with various J factors. The inference of this data is that there is no major benefit to be gained by demanding a J factor of less than 40. The deterioration in FATT is extremely pronounced if J is allowed to exceed 60.

The deleterious effect of sulphur on upper shelf energy has been well documented.[5] It is perhaps more appropriate to consider the combined effect of sulphur and manganese since manganese sulphide inclusions markedly reduce toughness. This is illustrated in Fig. 7 by a sharp decrease in upper shelf energy when (S + Mn) exceeds 0.25%. With the low manganese levels required by superclean specifications, the (S + Mn) content is substantially lower than this critical value. For the modified 3.5% NiCrMoV standard a

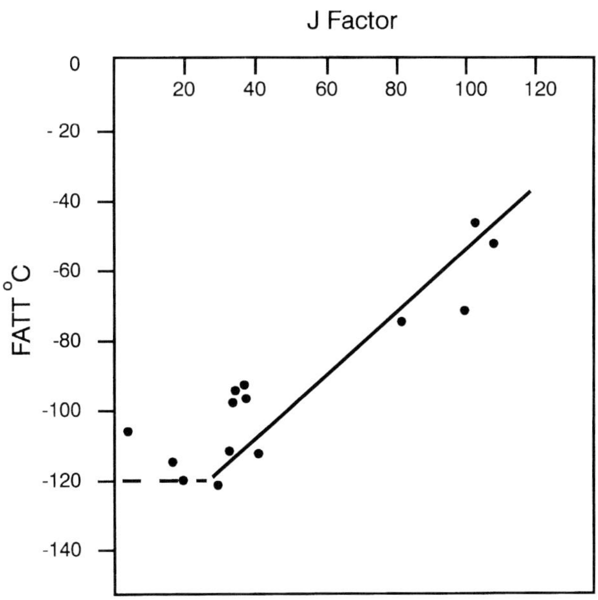

Fig. 6 Influence of J factor on FATT

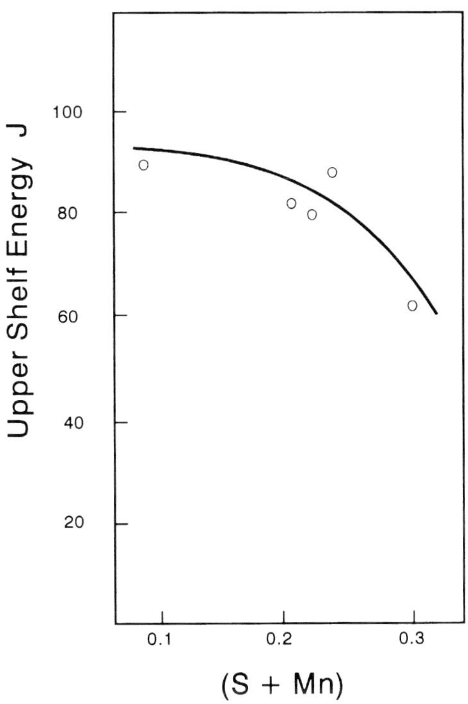

Fig. 7 Effect of (S + Mn) on upper shelf energy

minimum of 0.20% manganese is specified. It is thus important that manganese is kept towards the bottom end of the stipulated range (Table 1), and that sulphur is eliminated, as far as possible to maintain an acceptable upper shelf energy.

Although the benefits of a low residual content have been clearly demonstrated, this investigation has indicated that no improvement can be gained by further restricting the chemistry of the current ICI standard to an EPRI type superclean specification.

The difference in carbon content between pressure vessel and rotor forging 3.5% NiCrMoV must be implicated in the influence of residual elements on temper embrittlement at low J factors. Typically J factors of less than 10 are preferred in the Power Generation Industry to minimise embrittlement, however, it would appear that a more relaxed J factor of 40 maximum can be tolerated for the pressure vessel steels without influencing the propensity to significant in-service embrittlement. The presence of additional carbon produces an inherently less ductile steel. Grain boundary effects such as temper embrittlement are most prevalent when a material is in its toughest condition. The fracture mode is related to the relative strengths of the matrix and the grain boundaries.

If toughness is maximised, a small amount of weakening is sufficient to cause intergranular fracture; this is the case for 0.25% carbon 3.5% NiCrMoV. When the carbon content is increased to 0.30%–0.35%, much higher levels of the residual elements phosphorus, tin, arsenic and antimony are required to cause measurable embrittlement.

6. CONCLUSION

1. The presence of residual elements such as phosphorus, tin, arsenic and antimony significantly reduce the toughness and temper embrittlement resistance of high carbon 3.5% NiCrMoV pressure vessel steels as previously demonstrated by the lower carbon grades used in the Power Generation Industry.
2. The effect of residual elements on temper embrittlement at very low J factor values is not so pronounced in the pressure vessel steels when compared with rotor forging steels, which has been attributed to the presence of a higher carbon content.

ACKNOWLEDGEMENTS

Forgemasters Steel and Engineering Ltd would like to thank Mr R. J. Cooper of Imperial Chemical Industries PLC and Dr P. S. Crofton of Imperial College, London, for their contribution to the paper and Imperial Chemical Industries PLC for permission to publish the data.

REFERENCES

1. G. A. Honeyman and E. M. Lowe: 'Manufacturing Experience of Superclean Steels at Sheffield Forgemasters', *EPRI Superclean Steel Review*, Sapporo, Japan, September 1989.
2. R. I. Jaffee, T. Ohhashi, O. Watanabe and P. Machner: *Development of Superclean 3.5% NiCrMoV Low Pressure Steam Turbine Rotor Forging Steel*, Special Meeting VDEh, Dusseldorf, May/June 1988.
3. C. Berger, K. H. Mayer and R. I. Jaffee: 'Superclean 3.5% NiCrMoV Rotor Steel for Advanced Steam and Gas Turbines', *Third International Conference on Improved Coal-Fired Power Plants*, California, April 1991.
4. O. Watanabe, Y. Yoshioka and M. Miyazaki: 'Overview of Metallurgical and Mechanical Characteristics of Superclean 3.5% NiCrMoV Steel Rotor Forgings', *EPRI Superclean Steel Review*, Sapporo, Japan, September 1989.
5. F. B. Pickering: 'Towards Improved Toughness and Ductility', *Climax Molybdenum Company Symposium*, Kyoto, Japan, 1971.

Improvements of Manufacturing and In Service Mechanical Properties of Nuclear Components in Mn Ni Mo Steel through the Lowering of Contents in Residual Elements

L. DUNAND-ROUX, B. HOUSSIN and T. BERGER

Framatome
Site industriel de Chalon/Saint-Marcel, B.P. 13, 71380 Saint-Marcel, France

ABSTRACT

Reactor Pressure Vessels (R.P.V.) and Steam Generators (S.G.) forming part of Pressurised Water Reactor Nuclear Power Plants are manufactured using forgings and plates in manganese – nickel – molybdenum low alloy steel designated according to the French standard 16 MND 5 which is equivalent to the ASME/ASTM standards SA 508 Class 3 or SA 533 Type B Class 1.

The importance of having low level of residuals elements such as copper, phosphorus, sulphur, arsenic, tin, vanadium, is widely recognised for improving the reliability for long time service of heavy forgings and plates used for R.P.V. and S.G.

The objective of this paper is to show the change in the contents of residual elements in order to improve the following properties:

1. neutron irradiation embrittlement,
2. in service degradation by thermal ageing,
3. weld cracking sensitivity,
4. toughness properties in the base metal and heat affected zone.

1. INTRODUCTION

In Pressurised Water Reactor Nuclear Power Plants, the Reactor Pressure Vessels and the Steam Generators operate under severe conditions (neutron irradiation, temperature, pressure . . .).

These conditions lead FRAMATOME and their suppliers to have constant awareness of the requirements and for the means to be implemented to lower the level of the residuals elements.

These developments are linked with permanent research to improve the reliability for long time service of heavy forgings and plates in manganese, nickel, molybdenum low alloy steel used in the manufacture of the R.P.V. and the S.G.

2. MATERIALS USED FOR R.P.V. AND S.G.

2.1 General

The reactor Pressure Vessels and the Steam Generators are manufactured from heavy forgings and/or plates (see Fig. 1).

Several grade of steel can be used for the R.P.V. and the S.G., however due to extensive experience in material testing and the need to demonstrate the ability to operate under similar conditions in the USA, FRAMATOME selected the material, designated according to the French standards, 16 or 18 MND 5 which is equivalent to the ASME/ASTM standards SA 508 Class 3 for the forgings and SA 533 type B Class 1 for the plates.

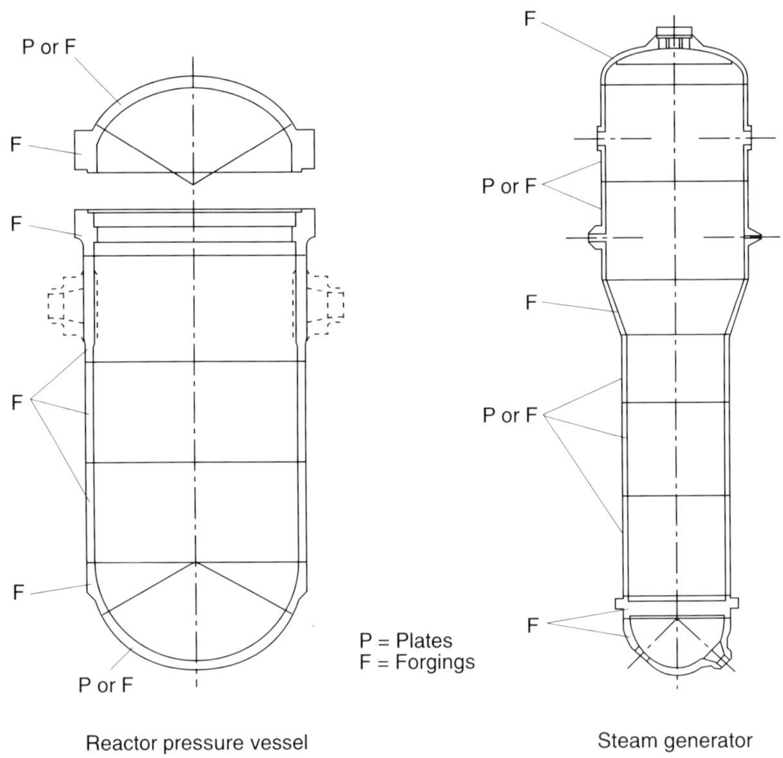

Fig. 1 Parts used for R.P.V. and S.G. manufacturing

2.2 *Requirements for the Chemical Composition*

The chemical composition was primarily based on those given for the ASME/ ASTM grade SA 508 Class 3 and SA 533 Type B Class 1, it has been optimised to improve the material properties and also its weldability.

Table 1 shows the typical specification for the parts of the components.

Table 1 Chemical Compositions Required

	Parts in R.P.V Core Region		Other parts	
Elements	Ladle in wt%	Product in wt%	Ladle in wt%	Product in wt%
Carbon max	0.20	0.22	0.20	0.22
Manganese	1.15 to 1.55	1.15 to 1.60	1.15 to 1.60	1.15 to 1.60
Phosphorus max	0.008	0.008	0.015	0.020
Sulphur max	0.008	0.008	0.012	0.015
Silicon	0.10 to 0.30	0.10 to 0.30	0.10 to 0.30	0.10 to 0.30
Nickel	0.50 to 0.80	0.50 to 0.80	0.50 to 0.80	0.50 to 0.80
Chromium max	0.25	0.25	0.25	0.25
Molybdenum	0.45 to 0.55	0.43 to 0.57	0.45 to 0.55	0.43 to 0.57
Vanadium max	0.01	0.01	0.01	0.01
Copper max	0.08	0.08	0.20	0.20
Aluminium max	0.04	0.04	0.04	0.04
Cobalt max	0.03	0.03	-/-	-/-
Arsenic (1) Tin (1)	The content shall be reported for information			

Note 1: Regarding the elements such as arsenic or tin, the French Code RCC-M[1] requires that no voluntary addition of these elements is allowed.

2.3 *Materials Properties*

The behaviour of the materials used in the R.P.V. and S.G. to resist the degradation due to the in-service conditions are continuously studied and there is permanent research to improve this resistance.

The main causes of degradation are the neutron irradiation embrittlement for the shells of R.P.V and the in-service ageing (thermal and strain ageing).

Some studies were made to understand the effect of the low residual elements and to demonstrate the improvement resulting from the lowering of the contents of these elements.

2.3.1 *Irradiation embrittlement*

One of the limitations of the life-time of the R.P.V. can arise from the core area embrittlement, more especially for the core shells which are located in front of the fuel elements and which therefore suffer of an important neutron irradiation embrittlement.

The sensitivity of the material to irradiation is mainly a function of the flux and the chemical composition of the steel particularly of the copper and the phosphorus contents.

Extensive studies have been carried out on the effect of the copper and phosphorus contents. The influence of these elements is taken into account in the evaluation of the shift in RT_{NDT} which is one of the means used to evaluate the evolution of the material toughness under irradiation.

The reference transition temperature RT_{NDT} shift can be determined using one of the following formulae:[2,3]

$$\Delta RT_{NDT} = [22 + 556\,(Cu - 0.08) + 2778\,(P - 0.008)]\,[f/10^{19}]^{1/2}$$

$$\Delta RT_{NDT} = 8 + [24 + 1537\,(P - 0.008) + 238\,(Cu - 0.08) + 191\,Ni^2 \times Cu]\,[f/10^{19}]^{0.33}$$

Where:

copper, phosphorus, nickel weight % with:
$\%P - 0.008 = 0$ if $P \leq 0.008\%$,
$\%Cu - 0.08 = 0$ if $Cu \leq 0.08\%$,

flux f: n cm^2

neutron energy $E \geq 1$ MeV

2.3.2 In-service ageing

The holding at service temperature during some ten years of the Reactor Pressure Vessels and Steam Generators is likely to induce a degradation of their toughness.

Two types of ageing can occur:

thermal ageing by segregation of elements that embrittle the grain boundaries,
strain ageing

The embrittlement due to these two phenomena are totally different and are considered to be non-cumulatives when their respective effect is moderate.

2.4 Thermal Ageing

The embrittlement of the manganese nickel molybdenum low alloy steel, by segregation of the phosphorus (and eventually other residuals like arsenic and tin) at the grain boundaries, was found[4] in the temperature range 400–500°C after 2.10^4 h.

This effect can also occur at lower temperatures, taking into consideration the ageing defined by the design life-time (3.10^5 h) of the pressure vessels of the primary loop of the PWR nuclear power plant.

The susceptibility of the 16 MND 5 (SA 508 Class 3/SA 533 Type B Class 1) to this type of embrittlement was extensively studied by FRAMATOME. It has been found in these studies that the embrittlement by service ageing of the forgings and plates is mainly a function of the following parameters:

metallurgical structure, notably grain size,
some residual element contents, mainly phosphorus,
time and temperature of ageing.

2.5 Strain Ageing

The sensitivity of 16 MND 5 (SA 508 Class 3/SA 533 Type B Class 1) to strain ageing has been evaluated by FRAMATOME.[5]

Strain ageing is mainly a function of the free interstitial contents of the material (nitrogen and carbon) and the material condition after welding (stress relieved or not).

The consequences of the strain ageing phenomena are:

increase of the yield and tensile strength,
material embrittlement involving a decrease of the rupture elongation and a shift of the Charpy-V transition temperature towards higher values.

The parameters decreasing the susceptibility to strain ageing are the elements precipitating with nitrogen and carbon such as aluminium, silicon and vanadium.

3. MATERIALS PROPERTIES IMPROVEMENTS

Table 2 summarises the elements having an action on the specific material properties of interest.

Table 2 Elements Influencing the Material Properties

Material Properties	Elements
Irradiation embrittlement	Copper, Phosphorus and Nickel contents
Thermal ageing	Phosphorus, Arsenic and Tin contents and Grain size
Strain ageing	Carbon and free Nitrogen contents

The influence of the chemical elements and the microstructure on the materials properties can be classified into 4 types:

1 the residual elements: sulphur, phosphorus, copper, arsenic and tin,
2 the grain size,
3 the carbon and nickel contents,
4 the free nitrogen content.

3.1 Evolution of the Residual Element Contents

FRAMATOME requests to the suppliers that the residual element contents shall be as low as possible. The suppliers were obliged to implement changes in their manufacturing processes in order to fulfill the requirements. The main change was the introduction of the refining of the steel in Ladle Refining Furnace (L.R.F). In addition to the L.R.F, another important parameter implemented by the suppliers to lower the level of the residual elements, is the stringent control of the raw materials by using well selected scrap, ferro-alloys and virgin materials. The steelmaking process currently used by all the suppliers can be summarised as follows: The steel is melted primarily in a basic electric arc furnace, then refined in Ladle Refining Furnace (L.R.F). The steel is mandatorily vacuum degassed either in L.R.F, or in D.H (Dortmund Hoerder) equipment or during vacuum steam degassing.

The introduction of the refining in L.R.F and the stringent selection of the raw materials greatly improve the reduction of the impurity levels of elements such as sulphur, phosphorus, copper, arsenic and tin contents and in a reduction of the scatter of the chemical analyses. The important improvement is illustrated by Figs 2 and 3 which show that the change of the sulphur and phosphorus contents achieved on parts delivered to FRAMATOME by one of the major suppliers.[6] Linked to the change of the sulphur content[7] Figs 4 and 5 illustrate respectively the improvement of the Fracture Appearance Transition Temperature (FATT) and the Upper Shelf Energy of the Charpy-V notch transition curve.

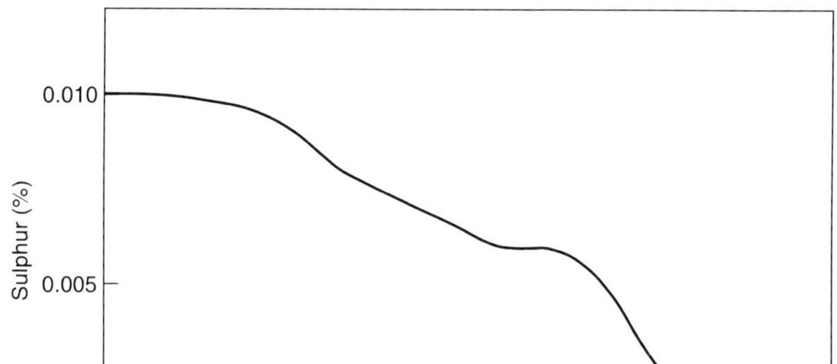

Fig. 2 Change of Sulphur content with time

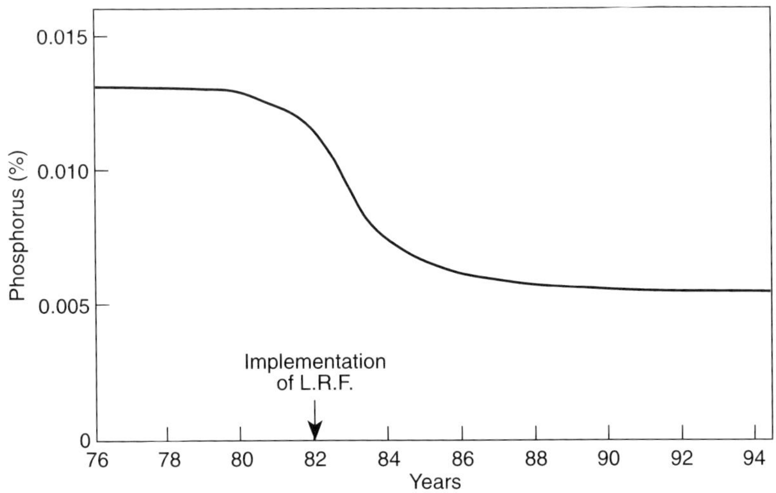

Fig. 3 Change of the Phosphorus content with time

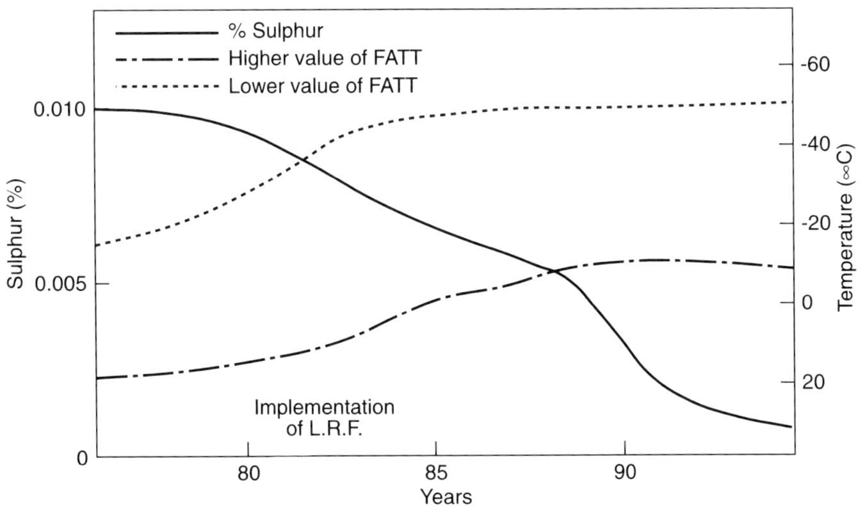

Fig. 4 Variation of FATT with S content

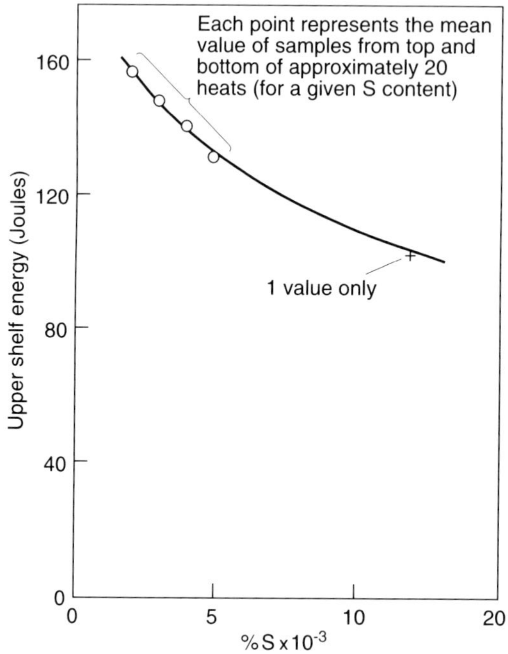

Fig. 5 Correlation between S and Charpy-V notch Upper Shelf Energy for plates in 16 MND 5

Table 3 gives the average and maximum contents of the residual elements achieved during recent production.

Table 3 Actual Residual Element Contents

	Forgings		Plates	
Elements (wt%)	Average	Maximum	Average	Maximum
Sulphur	0.001	0.0024	0.003	0.005
Phosphorus	0.005	0.0085	0.006	0.008
Copper for core region part	0.06	0.08	-/-	-/-
Copper for the other part	0.07	0.10	0.07	0.12
Arsenic	0.012	0.020	0.012	0.016
Tin	0.005	0.010	0.007	0.010

3.2 Grain Size

The control of the grain size is performed by optimisation of the austenitising heat treatment. For the austenitising cycle the supplier has to state:

a minimum and a maximum austenitising temperature,
a minimum and maximum holding time.

Another parameter for the control of the grain size is the aluminium or silicon killing of the steel. Homogenous ASTM grain size of about 8 is currently observed on products presently delivered.

3.3 Carbon and Nickel Contents

The carbon and nickel contents are well defined and controlled resulting in a low variation of these elements. A sufficient carbon content is necessary to obtain the required level in the mechanical properties and the nickel content is necessary for the fracture toughness properties to be homogeneous between the quarter and the three quarter location through the thickness of the product. The average carbon content is 0.16% for the 16 MND 5 and 0.17% for the 18 MND 5. The average nickel content is 0.73% for the 16 or 18 MND 5 forgings and 0.63% for the 18 MND 5 plates.

3.4 Free Nitrogen Content

The control of the free nitrogen content is performed by vacuum degassing treatment and aluminium killing during the steel making process.

4. MANUFACTURING POINT OF VIEW

4.1 Welding

In parallel with the residual element changes described in paragraph 3, the weldability and weld cracking sensitivity have been permanently checked. During the constant lowering in residual elements, no significant change was observed either in terms of welding cracking sensitivity or in terms of weldability (particularly wettability) during the production of the R.P.V. and S.G.

The heat affected zone formed by the welding operation is also kept under control by assessing the performance of weld production test coupons. In these coupons Charpy-V notch impact tests at 1 mm and 4 mm from the fusion line are performed. The impact test at 1 mm is performed to test an area as close as possible to the fusion line and corresponding to the metal in the most highly cooled condition. The location of the impact test at 4 mm corresponds to the base metal area in the maximum tempered condition due to the successive welding passes. The improvement of the fracture toughness properties in the base metal (see Figs 4 and 5) has been also observed in the heat affected zone.

4.2 Machining

The machining of the heavy forgings was not affected by the lowering of the sulphur content to very low values. In order to check the machinability, particularly to deep drilling, a full scale test of deep drilling on an actual tubesheet (600 mm in thickness) with a sulphur content of 0.0002% has been performed. No significant difference was observed between this tubesheet and the tubesheets with a sulphur content of 0.005%.

5. CONCLUSIONS

The lowering of the residual element contents has become a permanent objective for FRAMATOME and their suppliers, and it has become possible to improve the life-time margins of the components, and also the mechanical properties of the 16 or 18 MND 5 low alloy steels (SA 508 Class 3/SA 533 Type B Class 1). In parallel with the lowering of the residual element contents there have been no significant consequences for the implementation of the manufacturing processes.

REFERENCES

1. RCC-M: French Code issued by AFCEN. Règles de Conception et de Construction des Matériels Mécaniques des ilôts nucléaire REP. Design and Construction Rules for Mechanical Components of PWR Nuclear Islands.
2. RCC-M Code appendix ZG 3430.
3. C. Brillaud, F. Hedin and B. Houssin: 'A comparison between French Surveillance Program Results and Prediction of Irradiation Embrittlement', *Influence of Radiation on Material Properties: 13th Int. Symp. (Part II)*, ASTM STP 956, American Society for Testing and Materials, Philadelphia, 1986, 420–447.
4. S. G. Druce, G. Gage and G. Jordan: 'Effect of ageing on properties of pressure vessel steels', *Acta Met*, 1986, **34**(4), 631–652.
5. B. Houssin, G. Slama and P. Moulin: 'Strain ageing sensitivity of Pressure Vessels and Welds of Nuclear Reactor Components', *Assuring Structural Integrity of Steel Reactor Pressure Vessel*, L. E. Steel and K. E. Stahlkopf eds, Berli, Aug 20–21 1979, Appl Sci. Pul. Ltd, London, 1980, 57–67.
6. R. Dumont, A. Cheviet and I. Poitrault: 'Thirty years of experience in the nuclear industry', *11th International Forgemasters Meeting*, SPOLETTO, Italy, June 11–14, 1991.
7. A. Cheviet and J. H. Roux: 'Heavy gauge plates for nuclear components', *2nd symposium special steels and nuclear energy*, December 1976.

Recent Developments in High Pressure Vessels for the Petrochemical Industry: Materials and Products

P. BOCQUET, A. CHEVIET and R. DUMONT

Creusot-Loire Industrie, 56 Clemenceau BP56, 71202 Le Creusot Cedex, France

ABSTRACT

During the life of a pressure vessel, the Cr Mo steels used in petrochemical plants are submitted to conditions likely to reduce their properties (temper and creep embrittlement, hydrogen damages ...).

Improvements which may be obtained by the steelmakers are presented:

extra low P is the response to different kinds of embrittlement
disbonding observed on rings made from conventional ingot finds one solution in the use of hollow ingot.

The evolution of the performance requirements for hydroprocessing reactors materials in the oil industry is driven by the needs of the processes which are going to higher pressures (total pressure and hydrogen partial pressure) and higher temperatures.

New materials are used, or under development, to reach these objectives.

They are compared to the standard 2.25 Cr 1 Mo steel grade: enhanced 2.25 Cr 1 Mo (ASME case 1960), 2.25 Cr 1 Mo V (Case 2098) 3 Cr 1 Mo V (Case 1961) and grade 91 steel.

The mechanical properties of forgings made with these new materials have been investigated.

Their behaviour in hydrogen environment has been checked through different tests: the resistance to hydrogen attack at operating temperature, the sensitivity to hydrogen embrittlement, and the resistance of the clad materials to disbonding when cooling at room temperature.

Weldability aspects are also considered.

1. INTRODUCTION

The various high pressure vessels commonly used in the refining industry hydrotreaters, hydrosulphurisers, hydrocrackers running usually at high temperature, high pressure as well as high partial hydrogen pressure have to face during their in service life several types of degradation.

Made usually from the Cr Mo materials (1.25 Cr–0.5 Mo, 2.25 Cr 1 Mo) those pressure vessels working in the range of 450°C are susceptible to temper embrittlement. To estimate this susceptibility to temper embrittlement the mechanical properties of the material are tested in the as treated and post weld heat treated conditions as well as after step cooling treatment which is supposed to simulate the material after long term exposure to high temperatures.

If locally the temperature happens to rise to a high level the material can also suffer creep embrittlement.

With conversion processes using high hydrogen partial pressures, the materials suffer damage from hydrogen.

The most well known damage mechanism is the hot hydrogen attack occuring at service temperature.

The second damage from hydrogen occurs when cooling down the reactor to room temperature, usually for a change of catalyst. The hydrogen embrittlement has a very detrimental effect on the mechanical properties of the material, specially the fracture toughness of both the base metal and the weld metal.

Finally, because most hydrogen conversion processes contain H_2S, the Cr Mo material must be protected with stainless steel cladding. Also, during the cooling down operation the hydrogen content grows at the interface between base metal and the overlay and is a source of cracks which creates the disbonding; decohesion of the overlay from the base metal.

2. IMPROVEMENTS BROUGHT TO STANDARD MATERIAL

Regarding the temper embrittlement susceptibility of the Cr Mo steels attention has been focused for a long time on the highly detrimental effect of the impurities, phosphorus, tin, antimony, arsenic among which P is playing the most important role. The J factor calculated through the formula $(P + Sn) \times (Si + Mn)10^4$ is known as the best way to evaluate quickly through the chemistry the susceptibility of the steel to temper embrittlement. Using conventional silicon content of 0.18% CLI guarantee a J factor lower than 80 for a steel which is not susceptible to temper embrittlement.

CLI has identified the parameter (P + Sn) as the most important of the formula for the long term temper embrittlement. Figure 1 shows that with P + Sn lower than 0.010% no shift in transition curve is to be expected.

When using a low silicon content material (0.020%), to expect a similar result, the J factor must be lower than 50.

Low impurities are also important to maintain a good quality of the HAZ, see Fig. 2.

Working in the condition of creep embrittlement, mainly seen on 1.25 Cr 0.5 Mo material, once again an extra low level of phosphorus is the answer. Figure 3 shows the decrease of creep ductility for two levels of P.

CLI is now proposing steel in the grade with P lower than 0.002% which

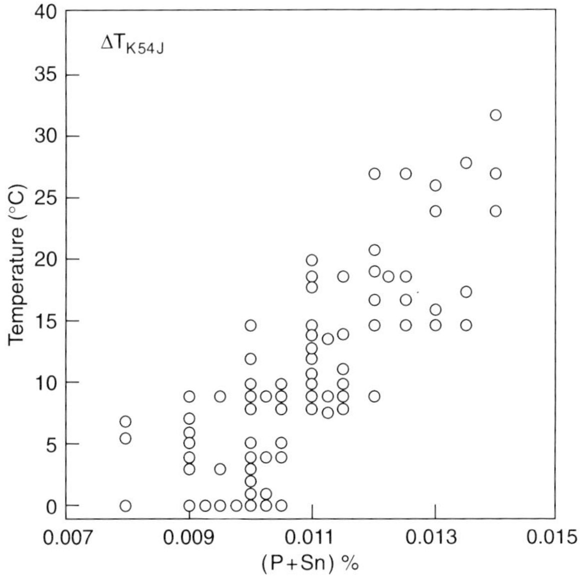

Fig. 1 Influence of P + Sn Parameter on Temper Embrittlement

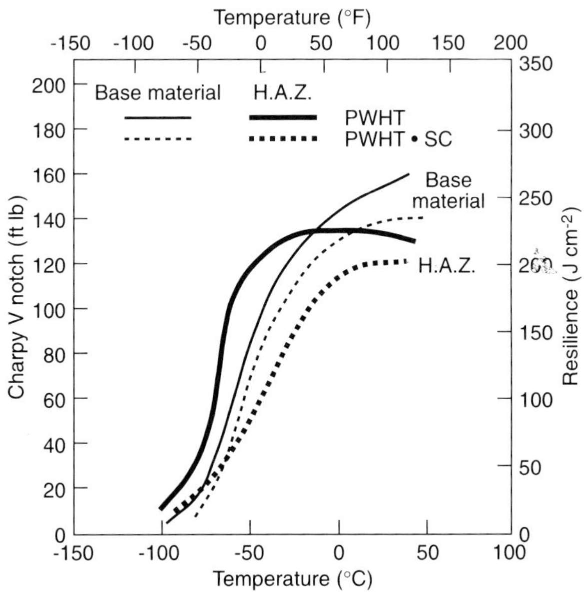

Fig. 2 Influence of Temperature on Charpy V Notch impact energy

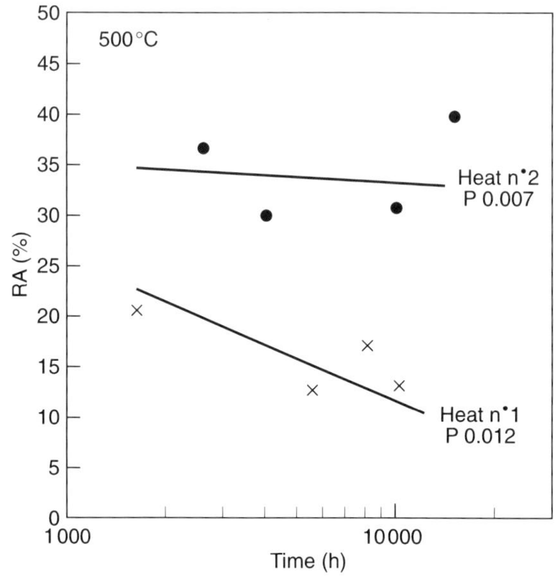

Fig. 3 Creep Ductility SA 387 g 11

must be associated with an extra low level of sulphur to obtain optimum mechanical properties in the HAZ. Today CLI is proposing $P + S \leq 0.005$.

Concerning disbonding phenomena extensive investigations had been conducted in CLI's Metallurgical Research Center.

Disbonding occurs mainly in the overlay close to the interface and the microstructure of the overlay at the interface is directly related to dilution during welding, and to carbon diffusion during post weld heat treatment. It had been demonstrated that the lower the carbon content of the base metal at the overlaid surface, the lower is the disbonding sensitivity. On Fig. 4 disbonding levels are compared for different levels of base material carbon contents. Two welding processes have been used with different dilution levels (5% for ESW, 19% for SAW). It is clearly shown that disbonding grows when the C content increases in the overlay. This may come from the carbon content of the base material or by a dilution effect. One solution is to be found by decreasing the carbon content of the base metal surface to be overlaid.

This is quite impossible to achieve when using a conventional ingot to manufacture the shell (except having heat analysis with extra low levels of C, as the heat analysis will not be compatible with the mechanical properties) because of secondary segregation areas remaining on the surface of the shell after completion of the manufacturing operation. To provide a solution to this problem CLI has developed 10 years ago the hollow ingot. Figure 5 compares the location of areas with increased carbon content on shells made from conventional ingots, hollow ingots and heavy plates.

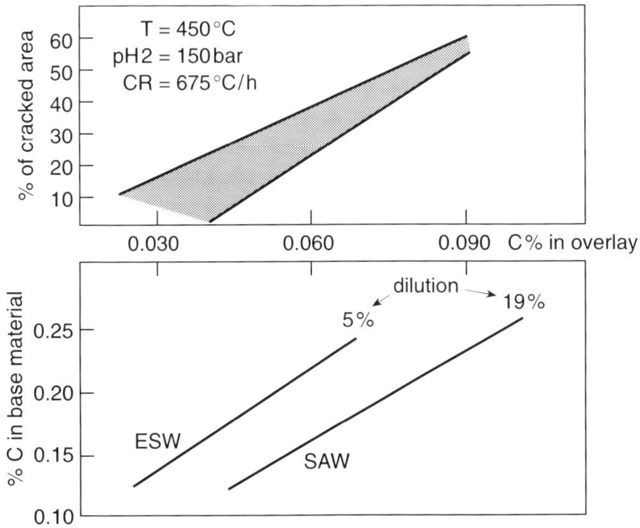

Fig. 4 Influence of C Content and Dilution on Disbonding Sensitivity

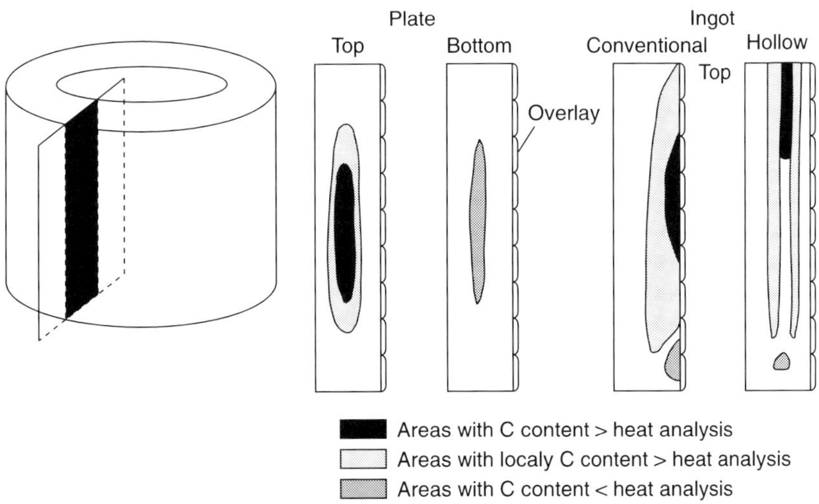

Fig. 5 Comparison of Carbon Distribution in Plates, Conventional Ingots and Hollow Ingots

3. RECENT DEVELOPMENTS IN MATERIALS

The performance requirements for hydroprocessing reactors are to go to higher pressures (total pressure as well as hydrogen partial pressure), higher production capacity and then to larger sizes for the reactors which with higher temperatures have brought other important changes in the material requirements.

Table 1 shows the different types of materials already industrially used or under development.

Table 1 Steels Proposed to be Used

	Chemical composition	Tempering PWHT	Mechanical characteristics YS	UTS
A 387 gde22	2.25Cr–1Mo	>675°C (1245°F)	>310 MPa (45 Ksi)	515/690 (75/100)
CASE 1960		620/663°C (1150/1225°F)	>380 (55)	585/760 85/110
CASE 1961	3Cr–1Mo–0.25V	>675°C (1245°F)	>415 (60)	585/760 85/110
CASE 2098	2.25Cr–1Mo–0.25V	>675°C (1245°F)	>415 (60)	585/760 85/110
A 387 gde 91 cl2	9Cr–1Mo–V–nb	730°C (1345°F)	>415 (60)	585/760 85/110

CLI has manufactured a large number of shells for heavy pressure vessels made from these materials. Shells with outside diameters of up to 5100 mm and wall thickness 220 to 240 mm had been manufactured. Table 2 gives the main mechanical properties obtained on these forgings.

3.1 Enhanced 2.25 Cr 1 Mo

First of all, enhanced 2.25 Cr Mo Code case 1960 allows for service temperatures staying lower than 454°C, the use of higher allowable stresses and then an interesting reduction of the necessary wall-thickness.

In this way the weight of the reactor can be decreased and this has a beneficial effect on the total cost of the installation.

But this material cannot be used in more severe hydrogen conditions than the standard material and a critical point in the manufacturing of this material is to get the balance between the minimum level of hardness in the base metal and the maximum hardness in the heat affected zone taking into account the temperature/time of the PWHT permitted.

Table 2 Mechanical Properties of Enhanced 2.25 Cr 1 Mo Steel

Part	Loc.	Tensile tests T * T/2			Impact tests Trans T * T/2				
		Room Temperature Y.S. >= 385	U.T.S. 615/750	El% >= 20	KCV −29°C >= 54J Average	T[54J] °C	KCV −29°C >= 54J Average	T[54J] °C	T[54J] − 3DT[54J] °C
					Temp. 640°C − Holding 30' per "				
R 6902		PWHT Max 655°C − 38 h					PWHT Max + Step C		
Shell No 1	0°	600	721	22.0	187	−85	167	−85	−85
	180°	564	684	21.0	165	−85	184	−85	−85
Shell No 2	0°	579	702	22.4	204	−85	179	−75	−55
	180°	559	681	20.6	252	−80	187	−70	−50
Shell No 3	o°	548	673	21.6	175	−80	198	−80	−80
	180°	556	679	22.1	179	−75	177	−75	−75
Shell No 4	0°	529	660	21.0	184	−75	183	−60	−30
	180°	536	668	23.2	225	−80	207	−67	−41
Min		529	660	20.6	165	−75	167	−60	−30
Max		600	721	23.2	252	−85	207	−85	−85
Aver.		559	684	21.7	196	−81	185	−75	−63

3.2 *Development of V/Nb Modified Cr Mo Steels*

3.2.1 *2.25 Cr 1 Mo 0.25 V – case 2090 and 3Cr 1 Mo 0.25 V Ti B case 1961*

These two materials which have had only short industrial use are still under development provide allowable stresses similar to enhanced material as shown by Table 2 but because of their good creep resistance properties they can operate at higher temperatures: a 482°C service temperature is already allowed by the ASME code for the 2.25 Cr 1 Mo V. Such a temperature is also allowed with 3 Cr 1 Mo 0.25 V Ti B in European codes.

Furthermore as it will be shown later these steels will tolerate much more severe hydrogen conditions.

In the framework of an EC contract CLI, in collaboration with BELLELI and ATB has manufactured several products from these materials.

One heat of 60 Tonnes was poured into a parallelepipedic ingot to be rolled into heavy plates of 150 and 300 mm thick in 2.25 Cr 1 Mo V steel. A second heat of 123 Tonnes was used for one hollow ingot to be forged in a 300 mm thick wall shell in steel grade 3 Cr 1 Mo 0.25 V Ti B.

The forging had been cut into different rings for complete investigation of the material and also the feasibility of cladding and welding.

Table 3 gives the chemical compositions.

Tables 4 and 5 give the mechanical properties for the two materials.

Table 3 Chemical compositions (wt%)

	Required Case 2098	2.25 1 Mo 0.25 V Heat for plates	Required Case 1961	3 Cr 1 Mo 0.25 V Ti B Heat for forging
C	.11/.15	.135	.10/.15	.147
S	.010 max	.004	.020 max	.001
P	.015 max	.005	.020 max	.005
Si	.10 max	.065	.20 max	.087
Mn	.30/.60	.460	.30/.60	.503
Ni	.25 max	.205	.25 max	.163
Cr	2.00/2.50	2.200	2.75/3.25	3.079
Mo	.90/1.10	1.040	.90/1.10	.996
V	.25/.35	.270	.20/.30	.287
Cu	.20 max	.050	.25 max	.051
Ti	.030 max	–	.015/.035	.024
B	.0020 max	–	.001/.003	.0019
Nb	.07 max	–		.003
Ca	.015 max	–		–
X		7		9
J		37		37

$J = (Si + Mn)(P + Sn)10^4(\%)^2$ $\bar{X} = \frac{10P + 5\ Sb + 4\ Sn + As}{100}$ ppm

Table 4 Tensile Properties at Room Temperature for PWHT: 690°C – 25 h
Material: Forging 3 Cr 1 Mo 0.25 V Ti B 300 mm thick

Test Location		Y.S. (MPa)	UTS (MPa)	El (%)	R.A. %
Required at T/4 Case 1961		⩾415	585/760	⩾18	
BOTTOM T/2	C	481 480	617 615	25 22	76 78
	A	475 488	602 620	24 23	79 79
	R	490 486	619 616	21 21	78 74
BOTTOM T/4	A	493 489	623 622	21 23	78 78
TOP T/2	C	523 521	639 639	22 21	78 78
	A	516 515	637 637	22 22	77 77
	R	519 519	639 638	22 21	77 77
TOP T/4	A	520 518	643 640	22 21	78 78

C: Circumference
A: Axial
R: Radial
} Specimen orientation

Table 6 summarises the Charpy V notch impact test results obtained on the two materials.

It is clear that these materials with adequate cleanliness do not suffer as much from temper embrittlement as the conventional material.

Figure 6 shows the change of tensile properties versus tempering parameter and Fig. 9 shows the already available creep test results of these materials.

3.2.2 9 Cr 1 Mo V/NbN modified steel

Having basic tensile properties similar to the previous material this new material has a much higher creep resistance and may be used up to service temperatures of 520°C.

Table 5 Tensile Properties at Room Temperature for PWHT: 690°C –25 h
Material: Plate 2.25 Cr 1 Mo 0.25 V

Test Location		Y.S. (MPa)	UTS (MPa)	El (%)	R.A. %
Required at T/4 Case 2098		⩾415	585 760	⩾18	
150 mm thick BOTTOM T/2	L	511 512	623 623	21 21	80 79
	T	507 507	619 617	24 21	79 78
	ST	513 511	621 620	19 20	73 74
150 mm thick BOTTOM T/4	T	502 500	615 612	22 23	79 79
300 mm thick T/4	T	521 517	637 632	25 24	N.D.

L: Long
T: Transverse
ST: Short Transverse
} Specimen location

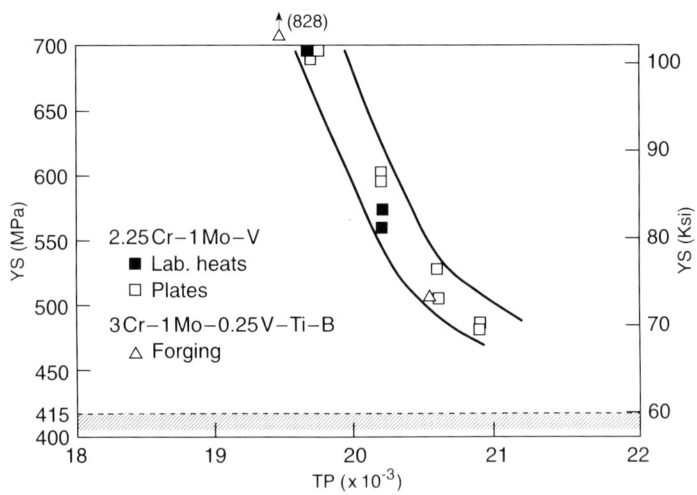

Fig. 6 Change of Tensile Properties at Room Temperature with the Tempering Parameter TP = T(20 + log t)

Table 6 Charpy V Notch Test Results in PWHT: 690°C – 25 h Condition

Test location		KV 1 - (ft.ib)				
	R.T.		−18°C (0°F)		−40°C (−40°F)	
Required T/4 Case 1961 and Case 2098			⩾54 (40) average ⩾47 (35) one specimen			
3Cr 1Mo V Ti B forging 300 mm						
T/2 Axial	223 230 254	236 (174)	100 164 192	152 (112)	12 13 14	23 (17)
Bottom						
T/4 Axial	228 250 262	247 (182)	145 204 211	187 (136)	13 108 17	47 (35)
2.25 Cr 1 Mo V Plate 150 mm						
T/2 Transverse	244 245 248	246 (181)	245 250 255	250 (184)	183 240 253	225 (166)
Bottom						
T/4 Transverse	245 251 252	249 (184)	260 262 264	262 (193)	243 255 260	253 (186)
2.25 Cr 1 Mo V Plate 300 mm						
T/4 Transverse	220 242 237	223 (164)	238 250 240	243 (179)		ND

Up to now main industrial experience is limited to pipes and tubes used in power stations, but the high creep resistance and high resistance to hydrogen makes it one of the basic materials for future development of high pressure vessels for the refining industry.

To investigate the properties of this material on industrial size components, CLI made two heats of 100 Tonnes each. The chemical compositions are given in Table 7.

These two heats had been used to make plates with different thicknesses (from 40 mm to 300 mm) and one hollow forging with a wall thickness of 150 mm.

Different heat treatment procedures have been used depending upon the thickness:

Table 7 Chemical Composition

	Required	Obtained (wt%) Heat 1	Heat 2
C	0.08–0.12	0.015	0.015
Mn	0.30–0.60	0.38	0.44
P	0.020 MAX	0.009	0.008
S	0.010 MAX	0.0025	0.0016
Si	0.20–0.50	0.43	0.24
Cr	8.95	8.26	8.41
Mo	0.85–1.05	0.95	0.95
Ni	0.40 MAX	0.13	0.08
V	0.18–0.25	0.20	0.25
Nb (Cb)	0.06–0.10	0.075	0.080
N_2	0.030–0.070	0.055	0.033
Al	0.04 MAX	0.024	0.013
Cu		0.080	0.035
As		0.017	0.008
Sn		0.008	0.004

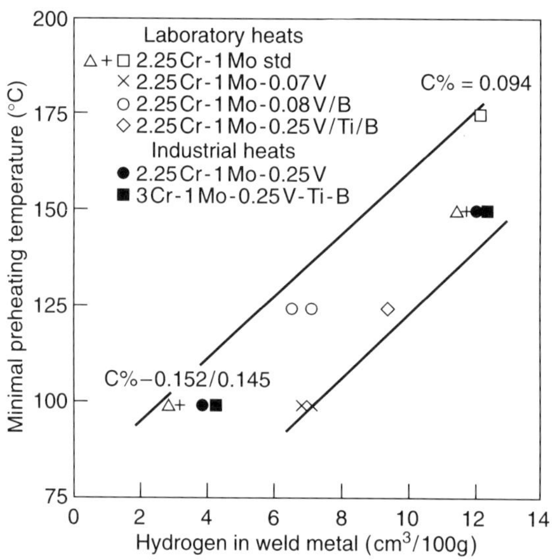

Fig. 7 Results of Implant Cold Cracking Tests. Effect of Weld Metal Hydrogen Content on the Minimal Preheating Temperature

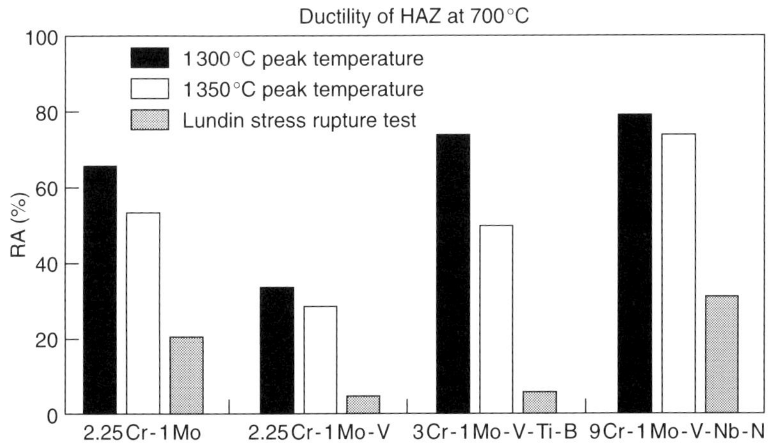

Fig. 8 Reheat Cracking Sensibility Evaluation by Tenstile Testing at PWHT Temperature on Simulated HAZ

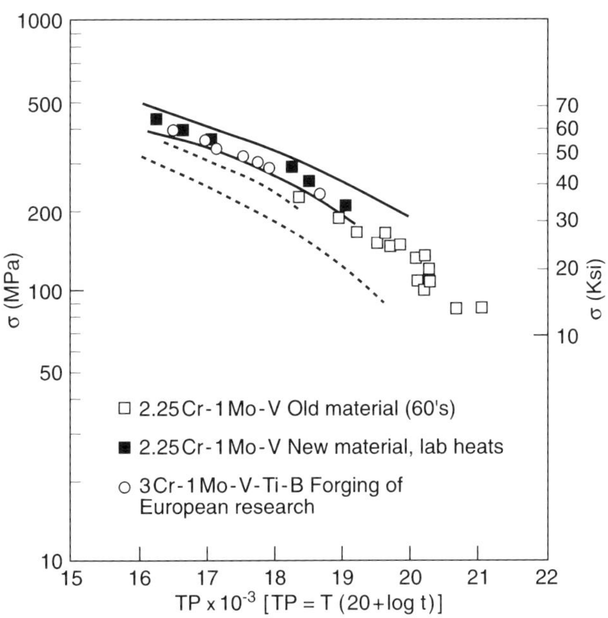

Fig. 9 Creep Test Results

Table 8 Transverse Tensile and Charpy V Notch Properties at 1/4 Thickness

Heat N°	Material thickness mm	Heat Treatment	Tensile YS. Mpa	UTS MPa	E1 %	RA %	KCV (J) at −20°C
1	20	N–T	509	690	23	69	85
1	40	N–T	502	669	24	70	87
1	40	N–T	491	664	24	70	160
1	100	D.Q–T	519	703	21	67	51
2	150(1)	D.Q–T	575	709	21	70	54
1	300	D.Q–T	484	657	20	68	70
2	300	D.Q–T	473	632	27	75	62

(1) forging

1. to get a grain size as fine as possible
2. to get a fully martensitic structure.

Mechanical properties obtained on the different products are given in Table 8.

Figure 10 shows the creep rupture value on tests carried out from different thicknesses (40 to 300 mm).

No significant difference had been found in the creep rupture values.

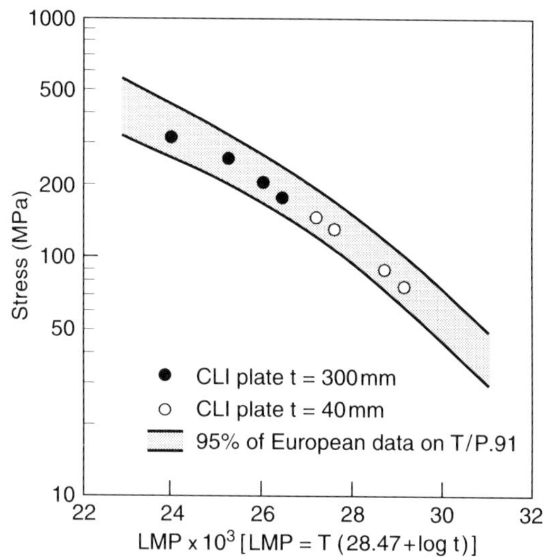

Fig. 10 Creep Rupture Test Results showing Comparison of CLI Data with Computer Optimised Larson-Miller Formula for all Available Data

4. COMPARISON OF WELDABILITY AND HYDROGEN RESISTANCE OF THESE DIFFERENT MATERIALS

4.1 Weldability (Cracking Sensitivity)

4.1.1 Cold cracking sensitivity

Determined by the implant test method the sensitivity of all modified steels has been compared to standard 2.25 Cr 1 Mo material.

Figure 7 shows the relationship between the minimum preheating temperature to avoid cracking when applying a stress of 500 MPa to the implant test and the total hydrogen content of the weld metal for 2.25 Cr and 3 Cr 1 Mo 0.25 V.

With 9 Cr 1 Mo 0.25 V Nb N, results of implant tests show that a minimum preheating temperature of 200°C is required.

4.1.2 Reheat cracking sensitivity

The reheat cracking sensitivity had been evaluated by measurement of HAZ ductility at PWHT temperatures.

The HAZ coarse grain area is simulated on a tensile specimen with the Gleeble thermomechanical simulator machine.

The tensile test is performed at PWHT temperature with a rate of $10^{-1}\% . s^{-1}$.

Figure 8 shows the results of HAZ ductility (RA%) for the different materials at 700°C.

The effect of V addition is detrimental, but it is clear that increasing Cr in these steels is providing a significant improvement of the ductility: 9 Cr 1 Mo 0.25 V is completely insensitive to reheat cracking.

4.2 Hydrogen Resistance

4.2.1 Hydrogen attack resistance

Hydrogen attack happens during exposure of the steel to high hydrogen pressures at high temperatures. It manifests itself by the formation and growth of methane bubbles within the steel. A mechanism has been proposed by F. M. Vitovec.

The resistance to hydrogen attack of the different steels has been performed on tensile test specimens charged with hydrogen at a high temperature.

a – Exposure at 600°C for 720 h under 150 bars H_2

b – Exposure at 600°C for 1000 h under 230 bar H_2.

The mechanical properties are compared to the initial properties of the steels.

Dissolved hydrogen is first removed from the tensile sample by degassing to check only the effect of H_2 transformed into methane.

The results are given in Fig. 11 for RA%.

These results show clearly that a V addition increases the resistance to

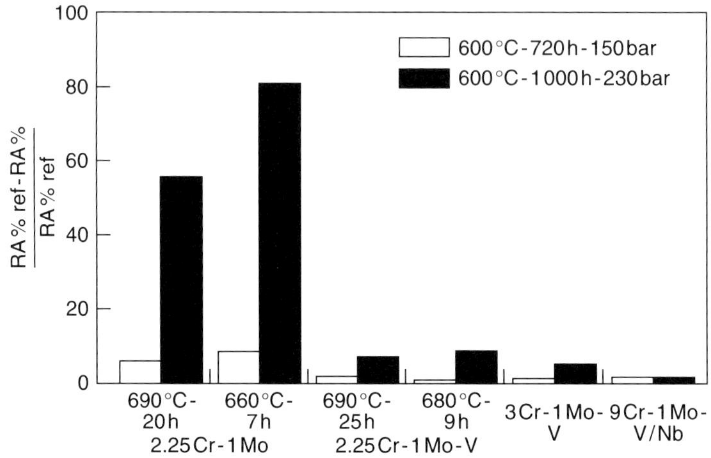

Fig. 11 Influence of Hydrogen Attack on RA%

hydrogen attack. 2.25 Cr 1 Mo 0.25 V has the highest hydrogen attack resistance.

4.2.2 Hydrogen embrittlement resistance

As outlined above the hydrogen embrittlement of Cr Mo steels occurs when the reactor is cooled down to room temperature. A hydrogen content of 3/4 ppm is expected. Hydrogen embrittlement can result in a loss of steel ductility and in hydrogen assisted crack growth.

We have evaluated mainly embrittlement sensitivity by the F% test.

The reduction of areas of tensile tests had been checked on precharged with hydrogen and uncharged specimens.

$$F = \frac{RA_{uc} - RA_c}{RA_{uc}} \times 100$$

Different charging methods can be used in order to obtain different concentrations of hydrogen in the steels. Results are shown in Fig. 12.

It has been shown that steels having the higher trapping capacities (traps being locations in the steel from which H_2 cannot be released without an high activation energy), give the best hydrogen embrittlement resistance. Main traps in V modified Cr Mo steels are the Vanadium carbides.

4.2.3 Resistance to disbonding

During the cooling of the reactor, an increase in hydrogen concentration is expected in the stainless steel just near the boundary where sensitive microstructures can be found: a thin martensitic layer resulting from carburisation during post weld heat treatment.

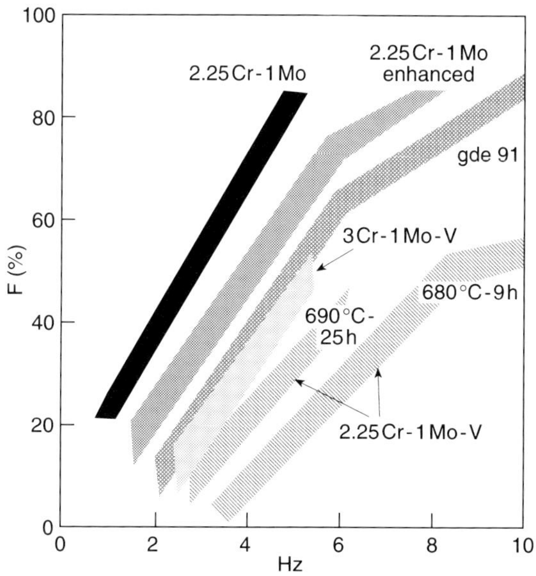

Fig. 12 Results of F% Tests

It had been explained the interest to have the lowest carbon content at this boundary, but carbon diffusivity in the base material depends on carbide stability. If carbides are stable at high temperature, carbon diffusion is less likely to occur. Vanadium carbides are stable, this is why V modified Cr Mo steels are less sensitive to disbonding risks. However, the satisfactory behaviour of these steels is also due to a lower hydrogen diffusivity, and hydrogen trapping capacity, when compared with standard 2.25 Cr 1 Mo.

Calculations have been performed to determine the hydrogen concentration in the cladding after cooling (100 hours): for standard 2.25 Cr 1 Mo H $\simeq$ 5000 mol H/m^3, for 9 Cr 1 Mo V Nb H $\simeq$ 2500 mol H/m^3 and for 3 Cr 1 Mo 0.25 V H = 300 mol H/m^3.

5. CONCLUSIONS

All these considerations can be summarised as shown in Table 9. 2.25 Cr 1 Mo is a material having been widely used will stay for a long time as the main reference material. All the improvements brought to this material, low impurities, cleanliness as well to manufacturing parameters will ensure that it will withstand the test of time. V modified 2.25 Cr and 3 Cr 1 Mo steels are bringing some interesting improvements in the service parameters, as they specially allow higher service temperature and at the same time bring some substantial advantages for the in-service behaviour: better resistance to

Table 9 Steel Grades Evolution for Petrochemical Applications

	2.25 Cr 1 Mo	2.25 Cr 1 Mo Low Temp. PWHT	2.25 Cr or 3 Cr 1 Mo V (Case 1961 Case 2098)	9 Cr 1 Mo V Nb
Mechanical properties at room temperature Rp0.2 ⩾ (MPa)	310	380	415	415
Estimated maximum in operation temperature (0°C)	450	450	480	520
Hydrogen attack resistance	0	–	++	+++
Hydrogen embrittlement resistance	0	)	++	++
Sensitivity to disbonding	0	+	+++	++
Weldability cold cracking	0	0	0	–
Weldability reheat cracking	0	0	–	0
HAZ hardness max. Req. ≤ 250 HV	220/240	235/255	250/265	270/300

0 behaviour equivalent to standard 2.25 Cr 1 Mo
\+ better behaviour than standard 2.25 Cr 1 Mo
– worse behaviour than standard 2.25 Cr 1 Mo

hydrogen attack and hydrogen embrittlement, less disbonding without change of the manufacturing parameters. They are the materials for tomorrows heavy pressure vessels.

9 Cr 1 Mo 0.25 V Nb/N steel is the material that will substantially allow an increase in service temperature. Some aspects of the manufacturing of this steel and its behaviour have still to be clarified. It is also the material for the heavy pressure vessels of the future.

REFERENCES

M. J. Humphries, P. F. Ivens, R. C. Brouwer, A. Cheviet, L. Coudreuse, P. Bocquet: *Two days workshop on new generation CrMo and Cr Mo V steels for use in refining reactors in Europe*, Abbaye de Collonges, France, December 9–10, 1992.

Low Residual HY80 Steel for Offshore Applications

G. A. HONEYMAN, M. A. WALSH and M. G. KEARNEY

Forgemasters Steel & Engineering Ltd, P.O. Box 287, Brightside Lane, Sheffield, United Kingdom

1. INTRODUCTION

One of the most demanding markets for forged low alloy steels is in the manufacture of critical components associated with offshore tension leg platforms (TLP). The TLP is a relatively modern type of oil production platform in which a floating structure is tethered by vertical buoyant tubular sections to the template on the seabed. TLP structures have been advanced due to the Oil Industry requiring exploration and extraction in deeper waters than previously found necessary (Fig. 1). Currently, there are four TLP's in service (Table 1) with a further two under construction. In deep sea fields,

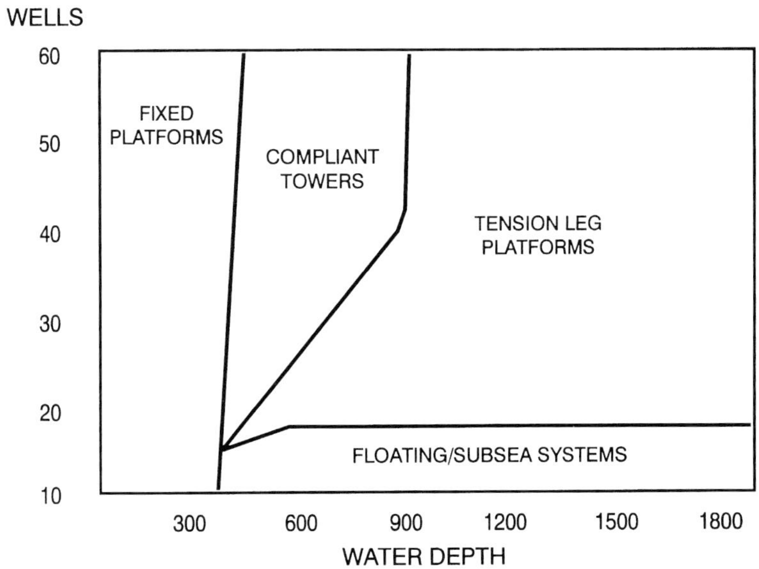

Fig. 1 Platform type with water depth

Table 1 Materials for deep sea offshore platforms

TLP	Location	Designer	Riser Material
Hutton	North Sea	Conoco	3.5 NCMV
Joliet	Gulf of Mexico	Conoco	CMV
Snorre	North Sea	Saga	HY80
Auger	Gulf of Mexico	Shell	A707 MOD
Heidrun	North Sea	Conoco	HY80
Mars	Gulf of Mexico	Shell	A707 MOD

their design offers significant cost reductions and improved reliability over the more conventional 'jacket' or 'semi- submersible' type rigs.

The mooring and riser systems of a TLP require exceptional heavy wall strength, fatigue strength, fracture toughness and resistance to stress corrosion cracking due to the high dynamic and static stresses endured in service.

The content of residuals in steels for other Industrial Sectors is generally important in order to minimise their effects on in-service embrittlement due to components operating at elevated temperatures, however, in offshore structures, which generally operate at low temperatures, the residual content is kept to a minimum to optimise fracture toughness.

2. MATERIALS

In the 1970s and 80s, thick wall forgings for the offshore industry were manufactured from low alloy steels typically NiCrMoV or CrMo types (Table 2). These steels had been developed from the Power Generation and Pressure Vessel Industries and were generally not welded. The Hutton TLP mooring system utilised 3.5% NiCrMoV type steels (Table 1). This design required a yield stress in excess of 850 MPa with excellent cryogenic toughness (Table 3). However, the Offshore Industry preferred forging grade steels to match the weldability of pipeline steels and this brought a gradual decrease in carbon and alloy contents and consequently reduced carbon equivalents and hardenability. The Snorre TLP was the first to use welded tethers and risers instead of conventional mechanical fasteners. These components were manufactured from modified grades of HY80 and 2.25% CrMo to achieve the minimum yield strength of 550 MPa and retain good weldability (Tables 2 and 3). The potential for increasingly 'leaner' steels of this type is limited by their hardenability and the necessity to maintain a fully bainitic microstructure.

An alternative alloy system has been the development of a very low carbon, precipitation hardening steel for the Auger and Mars platforms. This steel is based on the A707 specification and relies on copper precipitation to offset the

Table 2 Typical chemical compostions of steels for offshore applications

Steel	Chemical Compositon (Wt%)												
	C	Si	Mn	P	S	Cr	Mo	Ni	V	Cu	Nb	Al	CE
3.5% Ni–Cr–Mo–V	.27	.25	.40	.010	.010	1.6	.45	3.5	.12	–	–	–	1.0
3% Ni–Cr–Mo (HY80)	.18	.25	.30	.010	.010	1.6	.45	3.0	–	–	–	.030	.85
3% Ni–Cr–Mo Modified HY80	.10	.08	1.0	.005	.005	1.0	.45	2.8	–	–	–	.030	.75
3% Ni–Mo	.07	.08	1.0	.005	.005	–	.45	2.8	–	–	–	.030	.52
2.25% Cr–Mo	.12	.05	.40	.005	.005	2.25	1.0	–	–	–	–	–	.85
MnNiCrMoCuNb A707 Modified	.04	.20	1.30	.010	.005	.70	.45	1.6	–	1.1	.045	.02	.70
CMnNb	.12	.20	1.50	.010	.005	–	–	–	–	–	.045	.010	.40

Table 3 Typical mechanical properties of steels for offshore applications

Steel	Mechanical Properties		CVN Energy
	Yield Strength (MPa)	Tensile Strength (MPa)	FATT °C
3.5% Ni–Cr–Mo–V	850	950	−100
3% Ni–Cr–Mo (HY80)	600	730	−50
3% Ni–Cr–Mo Modified HY80	500	620	−50
3% Ni–Mo	450	600	−40
2.25% Cr–Mo	400	550	−70
MnNiCrMoCuNb A707 Modified	600	700	−70
CMnNb	320	450	−30

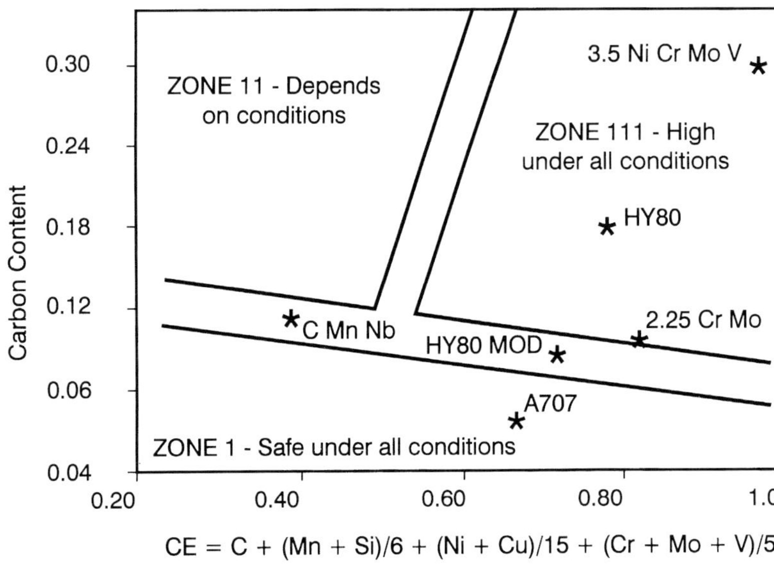

Fig. 2 Weldability of steel as a function of carbon and carbon equivalent from Graville

reduction in carbon content (Tables 1, 2 and 3). The development and properties of this steel were recently published at the International Forgemasters Meeting in September 1994.

The relative weldability of various alloy chemistries can be represented graphically on the Graville Diagram[2] as shown in Fig. 2. It can be seen that, despite their relatively high carbon equivalents, the low carbon contents of modified HY80 and modified A707 result in a low propensity for weld cracking during fabrication. The higher alloy content, however, is used to obtain a high strength bainitic microstructure suitable for thick wall sections.

3. RESIDUAL CONTENT IN HY80 MODIFIED

Forgemasters Steel and Engineering Limited (FSEL) have manufactured a number of components for the Heidrun field in the North Sea including the critical Taper Stress Joints and the Tensioner Spool Joints. The selected material was modified HY80 and typical cast analyses are shown in Table 4.

The Taper Stress Joint has a thick section flanged end which is joined to a Tie Back Connector during assembly. Due to the flange section, the silicon content was maintained at approximately 0.25% to ensure through hardenability by solid solution strengthening. It is not possible to increase the carbon content due to weldability considerations. The Tensioner Spool Joint, however, is not flanged and the maximum section thickness is significantly

Table 4 Ladle analyses with J Factors for HY80 modified grades (wt%)

Component	C	Si	Mn	P	S	Cr	Mo	Ni	Al	Cu	Sn	V	B	Sb	Pb	H_2	O_2	N_2	J
Taper Stress Joint	.12	.25	.86	.007	.003	.85	.46	2.77	.022	.03	.005	.02	.0014	.002	.0001	1.6 ppm	.002	.008	133
Tensioner Spool Joint	.12	.06	.90	.004	.002	.78	.47	2.67	.025	.02	.006	.02	.0007	.001	.0001	1.6 ppm	.002	.006	96

Table 5 Typical mechanical properties for modified grades of HY80

Component	Rp0.2 MPa	Rm MPa	A %	Z %
Taper Stress Joint	585	693	25	75
Tensioner Spool Joint	575	680	25	78

smaller than the Taper Stress Joint. It was, therefore, possible to utilise a low silicon grade of modified HY80 without affecting the through hardenability. This was achieved by VCD practice during secondary steelmaking. Typical mechanical properties for each component is shown in Table 5.

In order to evaluate fully the effects of silicon on the toughness properties of modified HY80, laboratory trials were conducted particularly with respect to cooling rate after tempering to ascertain the susceptibility of the material to embrittlement during processing. It was considered that the lower residual modified HY80 as expressed by the J factor:

$$J = (Mn+Si)(P+Sn) \times 10^4$$

would show superior resistance to embrittlement for a given cooling rate. The J values for the Taper Stress Joint and Tensioner Spool Joint are shown in Table 4.

4. RESULTS

The variation in fracture appearance transition temperature (FATT) with cooling rate after tempering is shown in Fig. 3. All test pieces had been previously austenitised and quenched in water. The cooling rate of 250°C per hour after tempering represents a rate typical of water quenching under which conditions no embrittlement is perceived. SC represents a very slow cool after tempering typically less than 10°C per hour.

The effect of cooling rate after tempering on the charpy impact energy at −20°C is shown in Fig.4.

CTOD tests were conducted on both Taper Stress Joints and Tensioner Spool Joints in the water quenched and air cooled conditions (Table 6).

Table 7 gives the CTOD results in air and seawater of material with J values towards the upper and lower limits of the specified chemistry range. The results include the variation in cooling rate after tempering.

Table 8 demonstrates the high cycle fatigue results of the materials in both air and seawater under cathodic protection.

All results were presented as part of a qualification programme for both the low silicon and high silicon modified HY80 material.

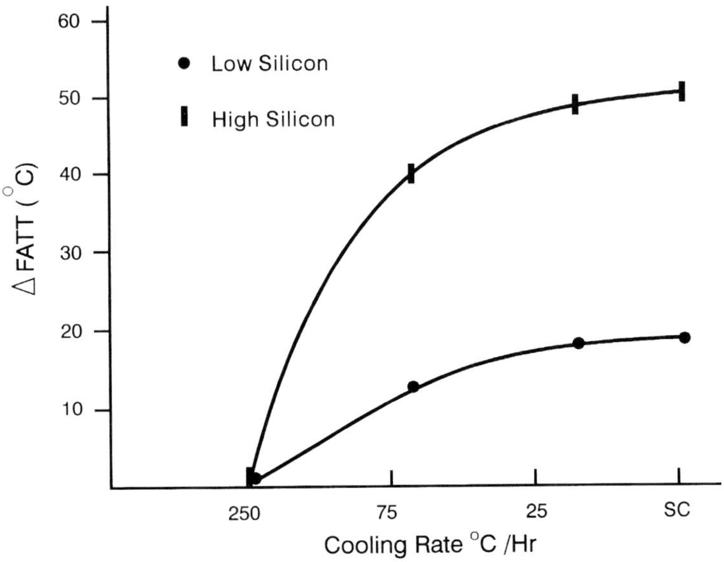

Fig. 3 Variation in FATT with cooling rate after tempering in laboratory trials

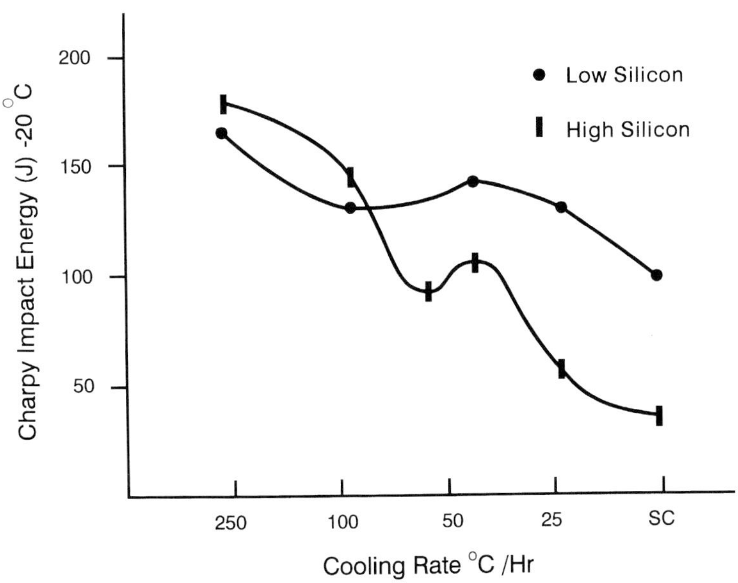

Fig. 4 Variation in impact energy with cooling rate after tempering in laboratory trials

Table 6 Variation of CTOD values with J Factor and cooling rate after tempering

Component	J Average	Cooling Rate	CTOD Average (mm)
Taper Stress Joint	113	WQ	1.126
Taper Stress Joint	90	WQ	1.240
Tensioner Spool Joint	96	AC	1.122

Table 7 Comparison of average CTOD value for different J Factors

Component	J Factor	Cool from Temper	Average CTOD (mm)	Average CTOD SW/CP (mm)
Taper Stress Joint	133	WQ	1.105	1.026
Tensioner Spool Joint	96	AC	1.122	0.977

Table 8 S–N fatigue comparison with different J Factors

					Endurance 100,000 cycles	
Component	J Factor	Cool from Temper	Stress Range	Stress Ratio	Air	SW/CP
Taper Stress Joint	133	WQ	400	0.5	NF	NF
Tensioner Spool Joint	96	AC	400	0.5	NF	NF

SW = Sea water
CP = Cathode protection
NF = No failure

5. DISCUSSION

The results of the FATT measurements for both the high and low silicon grades (Fig. 3) show the substantial effect of silicon in controlling the rate of embrittlement in modified HY80 material. The low silicon version shows only a marginal deterioration in properties between the water quenched and slow cooled conditions (ΔFATT 20°C maximum) whereas the high silicon steel gives a substantial change of approximately 50°C at equivalent cooling rates.

Similarly, the impact energy evaluations at −20°C (Fig. 4) show a significantly higher reduction for the high silicon material as the rate of cooling after tempering is reduced. It is interesting to note that, for both grades of steel, there appears to be a partial recovery of the embrittlement tendency at a cooling rate of approximately 50°C per hour. This can be explained by the influence of the cooling rate on the softening of the microstructure which overcompensates for the negative effect of embrittlement at this particular cooling rate. Thereafter, the effect of a further cooling rate reduction results in a combined mechanism which is predominantly embrittlement rate controlled.

The influence of residual silicon on the absolute fracture toughness of the material is certainly less pronounced for crack opening displacement measurements (Tables 6 and 7) whether the tests are conducted in air or in seawater under cathodic protection. The results, however, indicate that optimum properties are achieved with low residual HY80 water quenched from temper.

The effect of composition and tempering treatment variations have no influence on the high cycle fatigue properties of modified HY80 tested in either air or seawater under the stress range, stress ratios and fatigue life evaluated (Table 8).

In selecting the optimum chemical composition for TLP components, a number of important criteria require critical appraisal to ensure a material that is capable of meeting all the mechanical property requirements for the onerous offshore environment. These include the following:

- through thickness tensile properties
- high fracture toughness at low temperatures
- high cycle fatigue resistance
- resistance to stress corrosion cracking/corrosion fatigue
- weldability

Modified HY80 has been developed to achieve these requirements for various types of TLP structures. For thick section Taper Stress Joints, hardenability is obtained from a combination of the major alloying elements of carbon, manganese, silicon, nickel, chromium and molybdenum. Carbon is deliberately kept low to ensure good weldability. In order to achieve a high fracture toughness at low temperatures, quenching after tempering can remove the tendency to embrittle the steel whilst maintaining an excellent resistance to high cycle fatigue, stress corrosion cracking and corrosion fatigue. However, to improve further the embrittlement tendency, the residual elements, in particular silicon, phosphorus and tin, can be reduced on thinner walled components, such as the Tensioner Spool Joints, which also permits the use of air cooling after tempering during the processing route.

6. CONCLUSIONS

1. Modified HY80 has been developed to meet the difficult environmental conditions required by offshore TLP components.
2. For thick walled components, silicon is required for through section hardenability, however, embrittlement can be eliminated by rapid cooling after tempering.
3. For thin walled components, silicon can be reduced to minimise embrittlement during heat treatment and can further enhance the fracture toughness of the steel at low temperatures.

ACKNOWLEDGEMENTS

Forgemasters Steel and Engineering Limited wish to thank Conoco Norway Inc, Hydralift AS, and Statoil for permission to publish the data.

REFERENCES

1. G. A. Honeyman and M. G. Kearney: *Copper Bearing Steels for Offshore Applications,* 12th International Forgemasters Meeting, Chicago, September 11–16, 1994.
2. B. A. Graville: *Cold Cracking, Cracking in Welds in HSLA Steels,* Welding of HSLA Structural Steels ASM, Rome, Italy, November 9–12, 1976.

Clean Steel/Superclean Steel Workshop – Questions and Answers

RESIDUAL CONTENTS IN PURCHASED SCRAP FOR USE IN BASIC ELECTRIC ARC TECHNOLOGY

P. A. BRALSFORD, E. HYDES, G. A. HONEYMAN

Question

Have you found that a shortage of scrap results in an increase in the price of scrap and also an increase in the residual elements such as Cu, Sn, As and P of premium grade scrap? We have definitely experienced this trend in the USA.

Response

We have not found that a shortage of scrap results in an increase of price. Similarly we have not found an increase in the residual contents of premium quality scrap. However we do feel that further improvements could be made in the quality of scrap if better methods of scrap segregation were adopted by steel users. However the results we have obtained with high carbon 3.5 Ni–Cr–Mo–V steel where a minimum Mn content of 0.2% is required confirm the interpretation that a sulphur content of 0.003% is satisfactory.

Question

What do you see as the trend towards higher Cu contents?
Preparation of by-product scrap (sinkheads etc.) are increasing due to environmental issues, is this seen in the UK?

Response

In general we do not see an increasing copper content in scrap. However from time to time rogue scrap batches are received in which the copper content is much higher than usual.

Question

How much more expensive is the EAF scrap charge for a superclean heat versus your 'standard' scrap charge?

Response

EAF scrap for superclean production can be up to 5% more expensive than the scrap used for conventional grades. Increases in cost also come about from the use of more ferroalloy additions, and the length of time the charge has to spend in the furnace in order to remove elements such as Mn, S and P. We do not use electrolytic iron but in order to keep the P content of the charge low in superclean steels we use low P superalloys.

MANUFACTURE OF SUPERCLEAN 3–3.5% NiCrMoV STEELS FOR GAS TURBINE COMPONENTS

E. POTTHAST, F. TINCE, K. LANGNER

Question

Your Figure eleven indicates that there are six conventional steels at the lower end of the superclean steel band. Have you investigated these steels to ascertain why they are very similar in 'J' energy levels to the superclean steel? Maybe it is some reason other than just lower residuals making a contribution.

Response

The high level of J energy of these six steels in conventional steel band cannot be attributed to the low level of Si, Mn or residuals elements. It is caused by the favourable test position namely tangential outside.

Question

Have you compared the superclean steel and conventional steels in terms of FATT?

Response

FATT values are not available. But due to the high J energy values the FATT of the superclean steels is even still higher in comparison to the conventional steels.

Question

Your steelmaking method utilises a strong aluminium deoxidation followed by a designed fade to levels less than 0.01% Al. Have you performed any creep testing to determine any detrimental effects possibly caused by this steelmaking method?

Response

We have not performed creep rupture tests. But our Al content is comparable to the Al level or EPRI superclean forgings on which creep results are available.

PRODUCTION AND PROPERTIES OF SUPERCLEAN MONOBLOCK LP TURBINE COMPONENTS
T. FUKUDA, Y. IKEDA, H. YOSHIDA, Y. TANAKA

Question

Have you analysed the oxygen and TAE aluminium contents of the super-clean steels and compared the results with conventional cast steels to see if there is a relationship between these two elements in the improvement of toughness and reduction of embrittlement? Lower O_2 content results in improved cleanliness and higher aluminium results in a finer grain size but more inclusions.

Response

Yes, we have analysed them. Both superclean and conventional steels show extremely low contents of oxygen and aluminium and similar tensile and V-Charpy impact properties. Generally, such low contents of oxygen and aluminium are not thought to cause a reduction of toughness. Finer grain structures may not be expected for these steels because of the lower nitrogen content. This still applies even if these steels contain more than 0.02 wt% nitrogen.

Question

Does JSW use Eccentric Bottom Tapping in addition to reladling or is reladle used instead?

Response

JSW does not use eccentric bottom tapping (EBT). Instead we use the top tapping. In this top tapping, we use the top hole slide valve (THSV) to remove the slag. The removability of slag by THSV is almost the same as that by EBT. On the other hand, the reladle method has formerly been used by JSW.

A NEW SUPERCLEAN HIGH PRESSURE–LOW PRESSURE ROTOR STEEL FOR ADVANCED CYCLE POWER PLANT

M. YAMADA, Y. TSUDA, M. MIYAZAKI, Y. TANAKA, T. AZUMA, Y. IKEDA

Question

Have you measured the upper shelf energies of the UCD and ESR materials?

Response

Yes the upper shelf energies were measured. The value of VCD material is about 18 kgfm. The value of ESR material is nearly the same.

Question

To quote the duration range at temperatures used for creep tests and to give information on rupture ductility valves – ie parameter plots are not sufficient to indicate real extent of the data.

Response

The maximum duration of creep rupture tests which have been carried out so far is about 5000 hours. Although the test points are not extensive, all the results indicate that the rupture elongations are over 20% and the reductions of area are over 80%. The longer creep tests are now on going.

Question

Carbon distribution was measured up the length of the ingot. Did you measure aluminium? If so, what was the top to bottom variation?

Response

Aluminium distribution was measured. The level of the content of aluminium was about 0.005% and the difference between the top and the bottom of the ingot was negligible.

DEVELOPMENT OF SUPERCLEAN 12 Cr STEEL DISC FOR HEAVY DUTY GAS TURBINES

Y. FUKUI, M. SHIGA, S. NAKAMURA, K. HIDAKA

Question

You conclude that improvement in creep strength of HGTD2S is due to higher Mo content, but if one compares the composition of HGTD2S and HGTDISC

(Table 5), the former also has a higher Nb. How did you separate the effect of Nb to conclude that improvement is only due to Mo?

Response

Mo addition in our 12% Cr steel brings about the improvement of creep rupture strength. Nb addition is effective in the improvement of toughness and ageing embrittlement. We consider that the improvement of creep rupture strength due to a Nb addition is subsidiary because a Nb addition above 0.15 wt% in the 12% Cr steel degraded the creep rupture strength but gave improved toughness. Thus, we may conclude that an increase of creep rupture strength in HGTD2S is caused by the higher Mo addition. However the Mo addition caused a degradation of toughness thus a Nb addition of about 0.11 wt% is necessary for HGTD2S.

Question

To quote the duration range at temperatures used for creep tests and to give information on rupture ductility valves – ie parameter plots are not sufficient to indicate real extent of the data.

Response

The creep rupture examination was performed at 450, 500, 550°C, respectively. The tests have been carried out for up to 20,000 hours. The Larson-Miller Plot is used because it is a simple and easy way to evaluate creep rupture strength. We believe the extrapolated strength data to be correct because the % elongation (20%) and the reduction of area values (60–70%) of steel HGTD2S obtained after creep rupture testing did not change when the time of testing was increased beyond 1500 hours.

SUSCEPTIBILITY TO EMBRITTLEMENT OF 4% Ni–Cr–Mo HIGH STRENGTH STEELS AND 11% Cr–NiMoV HEAT-RESISTANT STEEL-ROLES OF MELTING METHODS AND RESIDUAL ELEMENT

H. SCHAFF, J. BOURRAT

Question

Do you attribute the lower FATT of melt D over that of melt B to the lower manganese of the former or is there another reason?

Response

The main reason for the improvement of the FATT of D melt (compared to that of B melt) is the remelting. Improvement given by remelting is evident in these results, because the tests are in the *transverse direction* of large diameter bars, and therefore are very sensitive to inclusion cleanliness and to homogeneity of the microstructure.

Question

Please explain why when heat A was vacuum arc remelted to give heat C there was an increase in % Mn from 0.42 to 0.45.

Response

We have checked the analysis sheets of this program. There is a typing mistake in Table 1: 0.45 is for heat A and 0.42 for heat C. Apologies.

Question

Can you comment on fracture mode regarding the differences you show?

Response

The fracture modes of all the specimens in Figure 1 are classical modes in transition curves. Cleavage, mixed and ductile (dimpling). In Figure 2 (after embrittlement) for heats A, B and C the fracture mode is mainly intergranular. It is mixed for heat D, with some intergranular zones still visible.

Question

For the remelted steels C and D, the total oxygen contents are halved compared with steels A and B indicating a reduction in the levels of non-metallic inclusions. Do you think this might influence the changes in Charpy values? Have you examined the changes in inclusion populations?

Response

Yes, there is an important change in the inclusion populations. The remelted steels C and D are very clean and exhibit only 'D type' inclusions (dispersed very small oxides). This has an effect on the Charpy values in the transverse direction.

100,000 h ISOTHERMAL AGEING TEST RESULTS OF NiCrMoV STEELS FOR LOW PRESSURE STEAM TURBINE
Y. TANAKA, T. AZUMA, N. YAEGASHI, Y. IKEDA

Question

It is important to take into account the starting toughness as well as the shift in the toughness as a result of the exposure to the elevated temperature.

Response

Starting toughness (the toughness of the as-heat treated condition) is also an important factor to assess the reliability of turbines. When a forging has a very low temperature FATT, the increase of FATT due to ageing can be allowed to a certain extent. This margin depends upon the design concept of the turbine and should be assigned by the turbine designer. From the standpoint of the material, steels with more residual contents may suffer from larger temper embrittlement during production which leads to a worse starting toughness. Therefore, clean and superclean steels have the advantage both in starting toughness and in ageing embrittlement.

ZERO INCLUSION STEELS
A. MITCHELL

Question

Similar methods to EBCHM give rise to problems when applied to nickel based super alloys because of the wetting inclusions. Is there the same problem with steels?

Response

The wetting problems which influence the agglomeration of oxide inclusions on the surface of the melt in EB techniques appear to be associated with the sulphur content of the alloy in the case of nickel based super alloy melting. We have not seen any evidence of the same effects in the alloy steels presently under investigation.

Question

What happens to Exogenous inclusions?

Response

As the melting of the feedstock proceeds, the exogenous inclusions are rejected to the surface of the forming liquid by interfacial forces. When the liquid falls to the EBCHM hearth, the inclusions stay on the surface of the melt. By manipulating the electron beam scanning so as to create an electrothermal barrier on the melt surface, the inclusions may be held at the incoming end of the hearth and prevented from passing into the ingot. Since the incoming alloy is already quite clean, the total inclusion raft formed in this way is not large enough to require an arrangement for periodic removal during the course of the melt.

TEMPER EMBRITTLEMENT IN HIGH STRENGTH POLYETHYLENE MANUFACTURE
G. A. HONEYMAN, J. E. HEESON

Question

Your Figure 7, plotting (S & Mn) against upper shelf energy indicated a drop off at about 0.2. Can this be interpreted such that a Mn content of 0.2% and a S content about 0.003% is satisfactory?

Response

The results we have obtained with high carbon 3.5 Ni–Cr–Mo–V steel where a minimum Mn content of 0.2% is required confirm the interpretation that a sulphur content of 0.003% is satisfactory.

IMPROVEMENTS OF MANUFACTURING AND IN SERVICE MECHANICAL PROPERTIES OF NUCLEAR COMPONENTS IN Mn Ni Mo STEEL THROUGH THE LOWERING OF CONTENTS IN RESIDUAL ELEMENTS
L. DUNAND-ROUX, T. BERGER

Question

Have you been able to purchase, without any difficulty or increase in price, low residual welding consumables to match the low residual steels shown in Table III?

Response

Framatome purchases and uses without any technical difficulties welding consumables (wire, flux for SAW process and coated electrodes for the SMAW process) having the same residual element contents as those described in Table III of our presentation. However, these welding products with low residual contents are used only for the manufacturing of reactor pressure vessels and the primary side of steam generators. They are more expensive than those used for the secondary side of steam generators for which the welding products are purchased without special requirements on the residual elements.

CLEAN STEELS IN PETROCHEMICAL INDUSTRY
R. DUMONT, P. BOCQUET, A. CHEVIET

Question

Is there any influence of microstructure eg. martensite or bainite being present in steel 91 which could account for its overall better performance than the other materials?

Response

In relation to hydrogen attack the most important parameter for good service performance is the stability of the carbides. With a 9% Cr steel containing vanadium which has been tempered at a high temperature the carbides are really very stable. In these circumstances the microstructure of the matrix is not of prime importance.

However in relation to hydrogen embrittlement it is well known that the most resistant microstructure is that of tempered martensite. This is the structure that is generally present in a 9% Cr Mo V steel.

Question

What size of hollow ingot is produced (outside diameter-inside diameter-length)?

Response

We have two kinds of hollow ingots: **'small ones'** outside diameter 2380 mm inside diameter 1000 mm. Length depends on the weight which is ranging between 75 T to 125 T. **'big ones'** outside diameter 3500 mm inside diameter from 1300 mm to 1900 mm. Weight from 125 T to 195 T.

Question

What device is used to make the hollow in the hollow ingot?

Response

The hollows are made by a mandrel being applied to the hollow ingot. After enlarging the hollow the ingot is air cooled.

Question

Approximately how many tons of hollow ingots are produced per year?

Response

Each year we make on average 7,500 tonnes.

Question

The comparison of the hollow ingot vs conventional presumes the conventional ingot is solid punched as opposed to hot trepan coring?

Response

In relation to the removal of primary carbon segregation hot trepan coring is much more efficient than solid punching, but the efficiency depends upon the diameter of the hot trepanning. The main advantage of the hollow ingot is associated with the fact that the inside surface of the forging is free from any kind of segregation; in particular it is free from 'A' segregations.

LOW RESIDUAL HY80 STEEL FOR OFFSHORE APPLICATIONS
M. A. WALSH, G. A. HONEYMAN, M. G. KEARNEY

Question

Could you give a few more details on the CTOD testing, specifically the test temperature, and information on how the testing in conjunction with cathodic protection was carried out?

Response

The details requested for CTOD testing are as follows:
Transverse, full thickness CTOD test pieces ($B \times 2B$) where $B = 25\,mm$ with through thickness (TT) notch shall be prepared from the samples allocated for this testing. Conventional CTOD tests were conducted in air at 0°C in accordance with BS 7448–91. For special CTOD testing the test pieces shall be exposed to cathodic protection in artificial sea water (ASTM D1141) prior to testing.

To simulate realistic conditions, the CTOD test pieces shall be machine finished and pre-cracked, and then loaded in bending to a stress intensity factor equal to 1500 $Nmm^{-3/2}$ by means of fixture.

The pre-loaded test pieces shall be immersed in aerated artificial sea water at 4°C, under a cathodic protection potential of minus 1050 ± 10 mV (Ag/AgCl).

After six weeks exposure time in this environment the test pieces are ready for testing.

In order to avoid possible springback effects in the crack tip region, unloading of the test pieces shall be avoided, for instance by inserting a wedge into the notch prior to removal from the fixture.

It is essential that as little hydrogen as possible is lost from the crack tip region prior to testing. The time between removal from the sea water bath and the actual testing shall therefore be minimised, and shall not exceed 15 minutes.

For practical reasons the testing temperature shall be the same as during exposure in sea water, that is +4°C.

The testing shall be carried out under load control and in accordance with BS 7448–91.

Report on the Round Table Discussion

J. NUTTING

The final session of the Workshop was devoted to a round table discussion which lasted for about one and a half hours. Throughout there was a lively exchange of views.

Prior to the discussion, members of the committee and a few others drew up a list of topics which were related to issues raised earlier during the meeting and which it was thought should be given a further airing.

Whilst the final period was devoted to a consideration of where the focus of new developments should be, and if there should be further workshops devoted to these developments.

The topics considered are presented below in the order they were actually discussed at the round table.

WELDING

The welding of superclean steels was not thoroughly discussed at the meeting, but questions had been raised about the ways by which the materials could be welded to give sound welds with satisfactory properties. The general view was that there were no problems, good joints could be produced in superclean steels. This view was also supported by the answers given in the EPRI survey, where no major issues were raised.

A LIMITING 'J' VALUE

An important issue was based upon the argument – is there a critical 'J', or similar value below which no further improvement in temper embrittlement resistance occurred. If this were so, it would provide a guide limit to steelmakers. It generally seemed to be accepted that, with a 2.5% Cr–Mo steels, a 'J' value of 50×10^{-4} was the lower limit, but with 3.5% Ni–Cr–Mo–V, it seemed to be a case of, the lower, the better.

The views of some power plant makers in relation to rotor design was that perhaps a different approach should be adopted. A possible requirement was that at the end of the service life of a rotor, the FATT should not exceed, say +20°C. The initial FATT would clearly be lower, but the change (ΔFATT) did not matter, provided the upper limit were not exceeded. This implies that ΔFATT can be large, provided that the initial FATT is very low. But, how can the initial FATT be lowered? No satisfactory answers were forthcoming, although fundamentally, the answer is clear – get a better microstructure.

Not all the power plant makers accepted the FATT criterion – an obvious problem was, where should the samples be taken from? Perhaps a more sophisticated requirement was to design a critical crack size, and then assess how K_{1C} varied with time, temperature, and composition – a mammoth undertaking.

A plea was made for extending the knowledge base, so that the plant makers could then use the appropriate data in relation to their own designs, rather than asking the steel makers to meet an appropriate FATT specification after a simulated service life.

RELAXING THE EPRI GUIDELINES

Discussions of 'J' values led naturally to the possibility of relaxing the EPRI composition, so making life somewhat easier for the steel makers and possibly lowering the initial cost of rotors. Many were at pains to point out that the use of superclean steels was not solely as a means of alleviating problems associated with temper brittleness, but these steels had great advantages in overcoming problems associated with stress corrosion cracking.

From Japanese experience, it appears that many orders for superclean materials were received, specifically because of the improved stress corrosion properties. Yet it was not known which elements, or combination of elements, were specifically responsible for this improvement.

It was also pointed out that improved creep ductility could be linked to low sulphur–low manganese levels and that good creep ductility was a vital in-service requirement.

As it appeared that, internationally, some seven steel makers could meet the EPRI guidelines, a considered view was not to tinker with minor compositional variations. In the final analysis, the degree of adherence to the EPRI guidelines should be a matter of choice to be negotiable between the equipment supplier and the user.

ANALYTICAL VARIATIONS

Most of the steel makers agreed that the element for which they had most difficulty in achieving consistent results in relation to the EPRI composition was phosphorus.

A lengthy discussion then followed on the wisdom of organising a 'round robin' amongst the major steel makers to check analytical results, since it was felt important that the purchasers of superclean steels should have confidence that they were getting what they had paid for.

A tentative proposal was made that a sample of a superclean steel, supplied previously to a plant maker, should be circulated to the major steel makers for analysis, with possibly EPRI assessing the results. This discussion led naturally to the next topic.

SUPERCLEAN STEEL – EPRI GUIDELINES OR A PREFERRED COMPOSITION

Since drawing up the EPRI compositional limits for 3.5% Ni–Cr–Mo–V steel, to meet a superclean requirement, many suppliers and users had accepted the values quoted as a specification. EPRI were at pains to point out that they were not a specifying body, and that the limits proposed were a preferred composition, not a legally defined specification. From the steel makers point of view, this approach was beneficial, since easements in relation to the preferred composition could be negotiated if there were minor differences in a cast analysis. No firm conclusions were reached as to the action to be taken, but it was made very clear that the present composition limits were not a legally defined specification.

The view was expressed that, if the knowledge base with regard to 3.5% Ni–Cr–Mo–V steels were improved, then it would be possible for the plant maker, knowing the specific service requirements of the plant user, to order steel from the steel maker so that the property requirements could be met. Whilst the use of 'J' values and similar parameters might give some indication of suitability, they were not really a substitute for a full understanding of compositional interactions.

FUTURE DEVELOPMENTS

Three topics were raised as possible areas for future developments and in which EPRI might take an interest.

1. To improve 3.5% Ni–Cr–Mo–V steel.
2. To develop further a superclean steel suitable for a single shaft HP/LP rotor.
3. To encourage the development of improved 9–12% Cr steels.

Improved 3.5% Ni–Cr–Mo–V Steel

It was generally felt that, in view of the pioneering work of EPRI on these steels, and that they were now being used extensively, it was no longer necessary for EPRI to support general research programmes on such steels. However, there was a need to encourage steel makers and steel users, to improve the knowledge base. There may be specific topics which need clarification which EPRI could support, but no topics of this type were put forward.

Single Shaft HP/LP Rotor Steel

Many comments were made about the need to develop improved ferritic Cr–Ni–Mo–V steels to meet the new service requirements. A major requirement was improved fracture toughness, but further fundamental work was needed on the factors influencing creep resistance, and both low and high temperature strength.

There were interesting comments made on the interaction between markets, as driven by the plant users, and materials availability, all linked with the view that further pump priming money for steel development could lead to much wider use of single shaft designs.

This would seem to be an area that EPRI could consider supporting.

9–12% Chromium Steels

The proceedings of the Workshop had shown clearly the advantages to be gained from using superclean steels in the 9–12% chromium range. It has now become recognised that such steels embrittle in service and this embrittlement could be greatly reduced by moving towards a superclean composition.

One of the major difficulties was associated with the requirement to keep the phosphorus levels low. The additions of conventional ferro-chromium were not satisfactory as a means of raising the chromium level because of the high associated phosphorus content and the difficulty of then removing the phosphorus during refining. A possible solution was to use electrolytic-chromium, but there was a need for further discussion on these issues.

During the meeting a questionnaire had been circulated seeking views on the topics to be discussed at the next EPRI Superclean Conference. The results from the questionnaire and the subsequent discussion were almost unanimously in favour of the next meeting being organised to deal with 9–12% chromium steels.

RECOMMENDATIONS FOR FUTURE WORK

Specific recommendations on what EPRI should do.

1. Assist in the organisation of a Round Robin to assess compositional accuracy of analyses of steels made to an EPRI preferred composition, with particular emphasis being made to the phosphorus content.
2. Apart from the above, no further development work on 3.5% Ni–Cr–Mo–V should be supported by EPRI.
3. Encouragement should be given through financial support to the

development of single shaft rotor steels, with a view to producing an EPRI preferred composition of a superclean type.

4. The next EPRI superclean conference should be organised primarily around the development and use of 9–12% chromium steels, with one of the major topics to be discussed being the steel making methods to be used for producing low phosphorus contents.

AUTHOR INDEX

Azuma, T. 109–122, 181–193, 298, 301
Bauer, R. 89–100
Berger, C. 195–211
Berger, T. 255–264, 302–303
Bocquet, P. 265–283, 303–304
Bourrat, J. 157–165, 299–300
Bralsford, P. A. 53–58, 295–296
Cheviet, A. 265–283, 303–304
Denk, J. 167–180
Dumont, R. 265–283, 303–304
Dunand-Roux, L. 255–264, 302–303
Everson, H. 213–221
Fukuda, T. 71–87, 297
Fukui, Y. 123–131, 298–299
Gelles, D. S. 233–241
Heeson, J. E. 243–253, 302
Hidaka, K. 123–131, 298–299
Holdsworth, S. R. 133–155
Honeyman, G. A. 53–58, 243–253, 295–296, 302, 304–305
Houssin, B. 255–264
Hydes, E. 53–58, 295–296
Ikeda, Y. 71–87, 102–122, 181–193, 297, 298, 301
Kearney, M. G. 285–296, 304–305
Langer, K. 59–70, 296–297
Lowe, J. H. 223–232
Mayer, K. H. 195–211
Meyer, W. 89–100
Mitchell, A. 223–232, 301–302
Miyakawa, M. 101–108
Miyazaki, M. 102–122, 298
Murphy, P. 101–108
Nakamura, S. 123–131, 298–299
Nakashima, H. 101–108
Nougaret, M. 133–155
Nutting, J. 33–51
Okamura, M. 101–108
Orr, J. 213–221
Potthast, E. 59–70, 296–297
Schaff, H. 157–165, 299–300
Schwant, R. C. 101–108
Shiga, M. 123–131, 298–299
Tanaka, Y. 71–87, 102–122, 181–193, 297–298, 301
Tince, F. 59–70, 296–297
Tsuchiyama, T. 101–108
Tsuda, Y. 102–122, 298
Viswanathan, R. 1–31
Vitteman, B. 133–155
Walsh, M. A. 285–296, 304–305
Yaegashi, N. 181–193, 301
Yamada, M. 102–122, 298
Yoshida, H. 71–87, 297
Zeiler, G. 89–100

INDEX

A
ABB Power Generation Ltd 167
aerospace applications 227, 230–231, 238
Afnor steels 158, 159, 161, 162, 163–164
ageing
  and embrittlement 184–185
  and FATT 127–128, 184, 185, 301
  grain boundary effects 258
  in-service 257, 258
  long-term results 181
alloying elements 13, 40, 123
alumina 89–90, 228
aluminium
  in clean/superclean steel 59, 59–61, 296–297
  detrimental effects 1, 206
  inclusions 1, 168
  pickup 89
  in rotor steels 59–61, 298
  and rupture ductility 214
  in stainless steels 230
  and strain ageing 259
aluminium deoxidation 2, 60, 89, 296–297
aluminium killing 90, 263
analytical tolerances 38, 39, 50, 308–309
antimony
  analysis 40
  in clean/superclean steel 59, 62–65
  economic considerations 51
  Electric Power Research Institute specification 37
  and embrittlement 1, 157, 185, 200, 252, 266
  grain boundary segregation 34
  and R-factor 216
  scrap selection 2, 54, 57, 62–63
  synergism with other impurities 27
AOD treatment 205
arsenic
  analysis 40
  in clean/superclean steel 54, 157
  economic considerations 51
  Electric Power Research Institute specifications 37, 57, 59
  and embrittlement 1, 62–65, 185, 200, 215, 252, 266
  grain boundary segregation 34, 258
  scrap selection 2, 37, 54, 57, 59, 62–65
ASEA Brown Boveri 7, 8, 20
ASEA-SKF system 36
Aubert et Duval 157
Auger spectroscopy 34, 189, 200, 205, 220
austenitising process 5, 102, 111, 116, 138, 159, 163, 206, 262–263
Axel Johnson Metals Corporation 223, 231

B
bainite structure 103, 112, 137, 200, 286
basic oxygen furnace steel scrap 2, 55, 57, 246
Bethlehem Steel 6, 7
blade steel 11, 206
Bohler Edelstahl GmbH 89
boilermakers 48
bolting steel 206, 213–221
boron 214
breakdown potential 170
British Steel Technical 213
brittle fracture 2
Bruscato factor *see* X-factor

C
calcium aluminate slag 89–90
calcium carbide method 98
calcium fluoride 62
calcium oxide 62, 228–229
carbon
  chemical segregation 112, 268–269, 304
  diffusivity 281
  and disbonding 268–269, 280–281
  and ductility 252
  and hardenability 286
  and mechanical properties 263
  in pressure vessel steel 243
  and strain ageing 259
  and weldability 288, 293
cathodic protection 290, 304
Central Electricity Generating Board (UK) 214, 215
Charpy impact data
  Afnor steels 159, 160
  aged specimens 182, 259–260
  clean/superclean steel 129–130, 163
  and cooling rate 290, 291, 293
  12steel 129–130
  CrMoV and CrMoVTi 273

Charpy impact data – *continued*
CrNiMoVNbW 112–114
and FATT 247
NiCrMoV 201
and niobium 125
rotor steels 82–83
temperature effect 267
transverse direction 300
weldments 263
chemical composition
Afnor steels 158, 161
bolting steel 217
clean/superclean steel 14, 63, 72–73, 74, 80–82, 127, 135, 138, 158, 162, 168
12steel 124, 129
CrMoV 10, 272
CrMoVTiB 217
CrNiMoVNbW 111
electroslag ingot 116
low activation steels 238–239
MnNiMo steel 257
NiCrMoV steels 19, 53, 63, 74, 135, 182, 183
nuclear reactor steels 257, 262
offshore applications 287
pressure vessel steels 247–248, 270
scrap 57–58
turbine steels 80–82, 103
zero-inclusion steel 228
chemical homogeneity 80, 82, 92–94, 107, 112–116
chemical industry 37, 38, 53, 230, 243–253, 265–283
chemical segregation
carbon 268–269
clean/superclean steel 20, 118
and electroslag refining 117
residual elements 209
and toughness 121
chromium 49, 56, 57, 94
Chubu Electric Company 14
clean/superclean steel
ageing embrittlement 190, 301
aluminium content 296–297
analytical tolerances 38, 39
bolting steels 213
Charpy impact data 128, 163
chemical composition 63, 72–74, 127, 135, 138, 158, 162, 168
chemical homogeneity 92–94
chemical industry 53
chromium steels 123–131, 310
creep rupture strength 15, 128, 150, 203
definition 1, 27, 59
economic aspects 3–5, 4, 24, 28, 37, 50
Electric Power Research Institute survey 33–51
embrittlement 157, 164–165, 190–191, 246
embrittlement factors 65–69
FATT 296–297
forgeability 5
fracture toughness 16
fusion energy production 233–241
hardness 138
heat treatment 127–128, 138, 163
HP/LP combined rotors 89, 109, 298, 309–310
inclusions 134, 226
J-factor 186
K-factor 65–69
long-term data 190, 203, 301
LP rotors 13–23, 27–28, 71–87, 89, 133, 182
manganese content 12–13, 134
mechanical properties 24, 53, 169, 183
molybdenum content 129
NiCrMoV 249
offshore applications 285
operating temperatures 13
oxygen content 297
pitting resistance 154, 178–179
power generation 53
pressure vessels 245
price premium 37–38, 48, 50, 56, 58
production and demand 27–28, 36, 41–47, 50, 58–60, 76, 98
requirements 27, 40
silicon content 129
specifications 7, 23, 35–37, 74, 75, 82, 246, 308, 309
starting toughness 301
stress corrosion cracking 35, 133, 146–150, 154, 175, 308
tensile strength 128, 163
toughness properties 65–69, 203
use by power plant manufacturers 48–49
welding 307
X-factor 65–69, 186

see also turbine steels
cleanness *see* steel, cleanness
cobalt 239
cold cracking sensitivity 279
combined cycle power generation 51, 109, 209, 298, 309–310
Conoco Norway Inc 294
cooling rate 13, 290, 291, 293
copper 57, 215, 234, 258, 286, 295
costs *see* economic considerations
9–12clean/superclean steel 310, 311
12clean/superclean steel
  alloying elements 123, 129–131, 236
  bolt and blade steels 206
  Charpy impact data 129, 129–130
  chemical composition 124, 129
  chemical segregation 118
  electroslag remelting 123–125
  embrittlement 126
  FATT 125–126, 129–130
  fusion reactor steels 236–238
  high temperature properties 8, 123
  ladle refining 128–130
  nitrogen addition 236
  production and properties 98–99
  silicon content 129
  tensile strength 129
  turbine steels 5, 13, 27, 123–131
crack initiation 7, 17, 18, 143–153, 173–177, 225
  and inclusions 133, 207–208, 224
cracks, tranverse 300
creep ductility 1, 150, 308
creep embrittlement 266
creep rupture strength
  clean/superclean steel 2, 10, 15, 128, 203, 205
  CrMoV 7, 277
  CrMoVTiB 277
  CrNiMoVNbW 113–114
  and heat treatment 102
  and inclusions 209
  long-term data 298, 299
  and molybdenum 84, 123, 124, 298–299
  NiCrMoV 201–206, 203, 204
  and niobium 125
  pressure vessel steel 278
  and toughness 2
  turbine steels 3, 83, 104, 107, 109, 195, 197
  and vacuum carbon deoxidation (VCD) 204
Creusot-Loire Industrie 265
critical defect length, crack initiation 225
CrMo 205, 206, 266, 271, 286
CrMoNiWV 8, 21–22
CrMoV
  chemical composition 10, 272
  cold cracking sensitivity 279
  creep strength 7, 277
  Durahete 213–214
  electroslag remelting 10
  forgeability 101
  impurity elevels 5, 9
  martensitic structure 303
  rotor forgings 101–109
  superclean specification 40
  temperature resistance 272
  tensile properties 274
  vacuum deoxidation 10
CrMoVNbN 273–275, 279, 282
CrMoVTiB 272, 273, 277
CrNiMoV 157–165, 310
CrNiMoVNbW 109–122
crossover temperature *see* operating temperature
CTOD testing 290–291, 304–305

D
Daedalus Associates 7
Daido Steel Corporation 231
degassing
  ladle 59, 60
  see also vacuum degassing
dehydrogenation 2, 246
dendrite arm spacing 94
denitrogenation 2
deoxidants, adverse effects on steel 1
deoxidation
  aluminium deoxidation 2, 60, 89, 296–297
  diffusion 89
  manganese deoxidation 89, 163
  silicon deoxidation 89, 163
dephosphorisation 10, 54, 60–62, 74, 98, 219, 260–262
desulphurisation 1, 2, 10, 54, 59–60, 62, 74, 228, 230, 260–262
diffusion deoxidation 89
disbonding 268–269, 280–281

dolomite refractory 246
ductile–brittle fracture appearance
  transition temperature *see* FATT
ductility 10, 11, 214, 215, 248, 252
Duke Power Company 15
Durahete bolting steel 213–214

E
eccentric bottom tapping 59, 60, 61, 297
economic considerations
  clean/superclean steels 3–5, 24, 28, 37–38, 50–51
  energy costs 56
  HP/LP combined rotors 3, 20
  operation/inspection costs 2–3
  scrap 55–56, 295–296
  turbine steels 20
electric arc furnace scrap 295–296
electric arc furnace steelmaking 2, 35–36, 53–59, 61–62, 245–246
electric arc melting, vacuum degassing 158
Electric Power Research Institute 1
  research programme 3, 5–8, 12, 14, 20, 34, 35, 71, 89
  superclean steels
    guide 17, 19
    specifications 36–37, 39, 48–49, 74, 75, 82, 246, 252, 308, 309
    survey 33–51
electrochemistry of pitting 170–173
electron beam button melting 227, 228, 229, 230
electron beam cold-hearth refining 229, 230, 301–302
electronics industry applications 229
electroslag ingot, chemical composition 116
electroslag refining 36, 117
electroslag remelting (ESR) process 10, 21, 101, 123–125, 204, 228
electroslag topping (BEST) process 90–94, 98, 100
embrittlement
  Afnor steel 163–164
  and ageing 184–185
  and antimony 1, 157, 185, 252, 266
  and arsenic 185, 215, 252
  clean/superclean steel 15, 126, 157, 164–165, 190–191, 246
  and copper 215
  CrMo 266
  CrNiMoV 157–165
  and grain size 259
  and hydrogen 150, 280, 281
  and impurities 2, 6, 157–165, 185–189, 273
  in-service 40, 286
  and J-factor 266
  long-term data 40, 59, 181, 198–201
  and manganese 37, 115, 186, 195
  and melting methods 157–165
  and neutron irradiation 53
  and nickel 186, 192
  NiCrMo 157–165
  NiCrMoV 71, 82, 89, 181, 186–191, 246, 299–300
  and niobium 131
  and phosphorus 185, 215, 252, 293
  power generation 34, 53, 82
  pressure vessels 243–253, 266–267
  research 1
  and residual elements 157–165, 182
  rotor forgings 13, 34, 249
  and silicon 186, 195, 292–294
  and step cooling 189
  testing 157, 165
  and tin 185, 215, 252, 293
  and toughness 159–161
embrittlement factors 62, 64–65
  clean/superclean steel 65–69, 69
  turbine steels 65–69
see also J-factor, K-factor, X-factor
Engineering Materials and Processes 6, 8
environmental effects, stress corrosion cracking 133, 140, 167, 171–177
European Research Programme 134, 168, 272
expansion characteristics of bolting steel 214

F
fatigue behaviour, thermal cycle 206–207
FATT
  and ageing 127–128, 184, 185, 301
  clean/superclean steel 82, 95–96, 125–126, 129–130, 296–297
  and cooling rate 290, 291
  12steel 125–126, 129–130
  CrNiMoVNbW 113
  Δ FATT
    and J-factor 84, 187

and manganese 200, 202
NiCrMoV 16, 184
and phosphorus 200
and step cooling 189–190
and X-factor 187, 202
and impact testing 247
improved 260–262
and impurities 1
and J-factor 187, 249–252
and manganese 299–300
NiCrMoV 8, 15
and operating temperature 2
and proof strength 120
and remelting 300
and silicon 125–127
and sulphur 261–262
tungsten-stabilised martensitic steel 238
turbine steels 13, 82, 95–96, 104, 157, 307–308
ferritic steels 236, 310
ferroalloy 296
ferrochrome, phosphorus content 58, 310
ferromolybdenum 56
flaw detectability 78, 80, 224
floating out of inclusions 60, 91, 301–302
forgeability 5, 6, 34, 101
Forgemasters Steel & Engineering Ltd 53, 243, 285, 294
forging procedures 74, 101
forgings
hollow 268–269, 275, 303–304
machineability 264
MnNiMo steel 256
monoblock rotors 77–79
offshore applications 286
pressure vessels 246
fossil fuel *see* power generation
fracture appearance transition temperature *see* FATT
fracture modes 300
fracture toughness (KIC)
clean/superclean steel 2, 10, 16, 157
and ductility 11
effect of impurities 1
initial quality 157–158
and nickel 263
rotor forgings 104, 181
temperature effect 2
testing 159
variation 308
Framatome 255, 259, 260, 264, 302–303
fusion reactor steels 236–238

G
Gallatin HP/IP rotor failure 10
gas turbines *see* turbines
General Electric Company 7, 8, 20, 101, 133
ghosts 103
Gleeble thermomechanical simulator 279
grain boundaries
cavity nucleation 1, 35
effects of ageing 258
surface breaking 133, 146
and toughness 252
grain boundary segregation
antimony 34
manganese sulphide inclusions 17, 51
phosphorus and tin 199
residual elements 1, 34, 165, 168, 185, 189, 205
grain size
and embrittlement 259
and forging 103, 106
and heat treatment 74, 116, 124, 262–263
and niobium 115–117, 119–121, 123, 124
and welding 137

H
hardenability
and carbon 286
and nickel 34–35, 168
offshore applications 293
pressure vessels 243
and silicon 288
hardness
clean/superclean steel 138
heat affected zone (HAZ) 270
heat affected zone (HAZ)
hardness 270
and impurities 266
microstructure 136, 137
nuclear reactor steel 263
and stress corrosion cracking 133
tensile testing 277
vulnerability 151
heat treatment
clean/superclean steel 127–128, 138, 163
differential 101–103, 111
and hydrogen cracking 276
monoblock rotor shaft 78

heat treatment – *continued*
pressure vessels 246
see also step cooling
Hitachi Ltd 123
holding times 74, 76, 116
hot shortness 34
HP rotors
clean/superclean steel 5–13
12steel 27
creep stress 195
operating temperature 5
specifications 27
temper embrittlement 34
HP/IP rotors 3, 5, 10
HP/LP rotors
chemical composition 103
clean/superclean steel 7, 20–23, 89, 298
cost effectiveness 3, 20
creep rupture strength 104, 107, 109
CrMoNiWV 21–22
CrMoV 101–108
FATT 104
fracture toughness 104
manufacture 101–103
mechanical properties 20, 103–104
NiCrMoV 8, 13
proof stress 103–104
standard specification 40
summary of research 24–26
superclean NiCrMoV 17, 20
tensile strength 103–104
HY80 low residual steel 285–296
Hydralift AS 294
hydrogen
analytic methods 38
and carbide stability 303
and CrMoVNbN 275
and embrittlement 150, 200, 280, 281
and heat treatment 276
high partial pressure 266, 270, 281
in pressure vessels 265
and vanadium 279–280
hydrogen sulphide 266

I
image analysis 229
impact properties
clean/superclean steel turbine steels 94–95
monoblock LP rotor 85
pressure vessel steels 244, 248
see also Charpy impact testing
Imperial Chemical Industries 243, 244, 245, 246, 252, 253
Imperial College London 245, 252
impurities
adverse effects on steel 1, 3
commercial levels 24, 111
CrMoV steel 5, 9
and embrittlement 6, 157–165, 185–189
and heat affected zone (HAZ) 266
and J-factor 83
removal 77
synergistic effects 2, 27, 40
inclusions
aluminium silicate 168
aluminium–magnesium oxide 168
clean/superclean steel 134
and creep stress 209
detrimental effects 1, 197, 224
floating out 60, 91, 301–302
and life expectancy 209
manganese sulphide 35, 133, 168, 169
non-metallic 1, 2, 5, 34, 91, 117, 168, 169, 197, 199
and oxygen content 227
and pitting 167, 229–230
in rotor steels 10, 169
size 224, 226
solubility 227–229
and stress corrosion cracking 133, 168, 207–208, 224
and sulphur 220–221
surface breaking 224
titanium 230
zero level 223–232
induced radioactivity, guidelines 237
induction levitation melting 227
industrial production
clean/superclean steel 98
clean/superclean turbine steel 121
zero-inclusion steel 228
inert gas 2, 36, 59, 60
ingot size 20, 24, 36, 69, 71, 74, 94, 111, 230, 246, 303–304
ingots, hollow 268–269, 303–304
International Thermonuclear Experimental Reactor (ITER) 233–241
IP rotors 5, 195

irradiation embrittlement 53, 257–258
irradiation induced swelling 236

J
J-energy levels 296–297
J-factor
clean/superclean steel 19, 65–69, 69, 82, 186
CrMoV and CrMoVTi 272
definition 35, 64–65
J′-factor, definition 64–65
J-factor
and embrittlement 192
and FATT 84, 187, 249–252
HY80 289
and impurities 83
limiting value 252, 307–308
NiCrMoV 19
pressure vessel steels 245, 247–248
and silicon 266
and temper embrittlement 266
turbine steels 82, 97
Jaffee, R.I. vii, 20, 28–29, 34, 51
Japan Casting and Forging 14–15
Japan Steel Works Ltd 6–8, 14, 20, 34–35, 71, 109, 181–182, 297
joining problems 37
see also welding

K
K-factor 35, 64–69
KIC *see* fracture toughness
Kobe Steel Ltd 14, 15, 101
KWU-Siemens 8, 20

L
ladle refining
calcium oxide 228–229
12clean/superclean steel 128–130
desulphurisation 2, 10, 60, 76
diffusion deoxidation 89
process 74–76
refining schedule 71, 74
residual elements 260–262
vacuum degassing 1, 36, 54, 59, 76, 111
Larson–Miller creep rupture data 22, 23, 104, 128, 201, 278, 299
lead 27
lime alumina slag 89
load cycles, fossil fuel plants 196, 197
long-term ageing, NiCrMoV 181–193
long-term creep rupture data 298, 299
bolting steel 216
CrMoNiWV 22
long-term embrittlement
clean/superclean steel 190
NiCrMoV 198–201
and phosphorus 201
see also embrittlement, in-service
long-term testing, alumina activity 90
long-term toughness
clean/superclean steel 203
NiCrMoV 203
low residual steel *see* clean/superclean steel
low residual welding consumables 302–303
LP rotors
clean/superclean steel 13–23, 51, 71–87, 89, 133, 297
embrittlement 13, 249
FATT 13
fracture toughness 181
ingot size required 74
mechanical properties 14, 168
monoblock *see* monoblock rotors
NiCrMoV 13–14, 51, 57, 181–193
non-metallic inclusions 169
operating conditions 175, 179
operating temperatures 3, 13, 15, 181
specifications 27
stress corrosion cracking 167–180

M
machinability, and sulphur 220–221, 264
macrocracks 141, 146
macrostructure 112–114, 117, 119
magnesia 168, 228
magnetic particle inspection 112
magnetite surface film 177
MAN Energie GmbH 8, 20, 195
manganese
acceptable levels 40, 51, 54, 165, 252
analysis 40
in clean/superclean steel 12–13, 59, 62, 157, 161
and creep rupture strength 84, 308
effect on Δ FATT 200
and embrittlement 37, 157, 186, 195, 200
and FATT 299–300

manganese – *continued*
and Δ FATT 202
and grain boundary segregation 34
and hardenability 34
pickup 62, 74, 76
in pressure vessel steels 248
reduction 13, 54, 60, 61, 74, 77, 128
in rotor forgings 111
in scrap 296
specifications 27, 37
and steel price 38
and sulphur 1, 3
and tempering temperature 163
and temper embrittlement 115
and toughness 250
and upper shelf energy 251, 302
manganese deoxidation 89, 163
manganese sulphide inclusions 17, 35, 51, 133, 134, 168, 169, 250
maraging steel, aerospace applications 230–231
martensitic steels 235–237
martensitic structure 137, 200, 201, 248, 278, 303
mathematical modelling 77, 165
mechanical properties
and carbon 263
clean/superclean steel 15, 24, 53, 94–98, 169, 183
CrMo 271
CrNiMoVNbW 112–114
HY80 289
improvement 1, 2, 10–11
NiCrMoV 14, 15, 94–98
nuclear power components 255–264
offshore applications 287
pressure vessel steels 247–248
rotor forgings 14, 20, 82–85, 103–104, 168
testing 48–49
melting methods 157–165, 209, 299–300
metallographic investigations 168, 170, 176, 225, 231, 248
microcracks 141, 145–146, 148, 149, 153, 174
microsegregation, clean/superclean steel turbine steels 94–98
microstructure
and cooling rate 293
effect of ageing 189
long-term changes 192
MnNiMo low residual steel 255–264
molybdenum
in clean/superclean steel 129
in 12steel 129–131, 236
and creep rupture strength 123, 124, 298–299
demand 56
microsegregation 94
radioactive half life 234, 235
scrap selection 57
monoblock rotors
creep rupture strength 83
forging 77–79
heat treatment 78
impact properties 85
joining problems 37
mechanical properties 82–84, 85
production and properties 71–87, 297
tensile strength 83
transverse toughness 86

N
nearest neighbour interactions 227
neutron irradiation embrittlement 53, 257, 257–258
nickel
addition, 12steel 236
in chemical process reactor steel 230
and embrittlement 186, 192
and fracture toughness 263
and hardenability 34–35, 168
microsegregation 94
price 56
radioactive half life 234
reduction 49
in scrap 56, 57
and strength 188–189
in turbine steels 7, 123
Nickel Development Institute 231
NiCrMo, embrittlement 157–165
NiCrMoV
Charpy impact data 201
chemical composition 53, 63, 74, 135, 182, 183
crack initiation 18
creep behaviour 201–206
Electric Power Research Institute specifications 309
embrittlement 71, 82, 89, 181, 186–191, 246, 299–300

FATT 8, 15
fracture toughness 16
improved 309–310
joining 37, 133–155
long-term ageing 181–193
long-term embrittlement 199–201
long-term toughness 201, 203
mechanical properties 14
offshore applications 286
operating temperatures 198–199
pressure vessel 243
rotor forgings 8, 13, 13–14, 57, 109, 181–193
scrap mix 57
stress corrosion 17, 18, 133–155
superclean 6, 7, 14–17, 16–17, 19–21, 59–70, 209, 249, 296–297
toughness 7
wide use 36, 49
X-factor 201
NiMoV rotor steels 5
niobium
and Charpy impact data 125
in 12steel 236
and creep rupture 21, 49, 125
and embrittlement 131
eutectics 110–112, 117
in fusion reactor steels 236, 240
and grain size 115, 116–117, 120, 121, 123, 124
in low activation steels 238–239
radioactive half life 234, 235
raw material source 239
and toughness 7, 299
Nippon Kokan Corporation 238
nitrogen
in 12steel 236
dissolved 227
and embrittlement 200
radioactive half life 234, 235
removal 229, 263
and strain ageing 259
non-destructive testing 79, 224
non-metallic inclusions *see* inclusions
normalisation 78, 119
notch properties
bolting steel 215, 216, 219
rotor steels 201
and stress corrosion 172, 173, 178
superclean NiCrMoV 21
see also Charpy impact data
nuclear fusion 233–241
nuclear power *see* power generation, nuclear

O
offshore applications 38, 53, 285, 287, 293
oil hardening 159
oil quenching 11, 12, 244, 246
operating conditions
LP rotors 175, 179
nuclear power generation 255
pressure vessels 265, 270
steam turbine 195–196
operating temperatures
combined rotors 3, 209
CrMoVNbN 273–275, 282
fossil fuel power generation 2, 123, 130, 209, 213
fusion reactors 236
HP rotors 5
IP rotors 5
LP rotors 3, 13, 15, 27, 49–50, 59, 71, 181, 182, 248–249
NiCrMov 198–199
nuclear power generation 258
pressure vessels 243, 250, 265, 270, 282
superclean NiCrMoV 13, 17
oxygen 1, 40, 200, 227, 230, 297, 300
removal *see* deoxidation

P
Pacific Northwest Laboratory 233
Parsons Inc 8, 20
passivation 170, 174, 177, 178
pearlite structure 111, 119
petrochemicals *see* chemical industry
phosphorus 62, 157
analysis 40, 308–309
detrimental effects on steel 1, 266
and embrittlement 185, 200, 201, 215, 252, 293
in ferrochrome 58, 310
and neutron irradiation embrittlement 258
in nuclear reactor steels 260–262
pickup 60, 62, 74, 76
removal *see* dephosphorisation
in rotor steels 27, 34, 59
in scrap 260, 296

phosphorus – *continued*
segregation 34, 189, 199
and toughness 200
pipeline steels, weldability 286
pipework and tubing 37, 48, 49, 51, 229, 275
pitting
electrochemistry 170–173
and inclusions 167, 229–230
resistance, clean/superclean steel 154, 178–179
turbine rotor steel 167–180
in welded joints 141–143
polyethylene manufacture 243–253
power generation
combined cycle plants 51, 109, 209, 298, 309–310
fossil fuel plants 2, 3, 123, 196, 197
fusion energy production 233–241
nuclear 27, 51, 53, 255–264
operating temperatures 2, 27, 49, 50, 59, 71, 123, 130, 182, 209, 213
temper embrittlement 34, 82
power generation products
clean/superclean steel 48–49, 53, 233–241
production volume 36
see also turbine components
power generation utility companies 49–50, 214, 215
pressure vessel steels 79, 245, 247–248, 270
pressure vessels
chemical industry 38, 265–283
forging 246
hardenability 243
heat treatment 246
impact properties 244
nuclear industry 255–264
oil quenching 244
operating conditions 265, 270
operating temperatures 243, 250, 265, 270, 282
reconditioning 243–244
temper embrittlement 243–253, 266–267
prices *see* economic considerations
proof stress, rotor forgings 103–104, 110, 120, 207
pyrometry, imaging 229

Q
quenching 20, 22, 78, 111, 119, 120, 124, 293
questionnaire, clean/superclean steel 41–47

R
R-factor
and antimony 216
and ductility 218
improvement 219–220
and long-term strength 218
radioactivity and radioactive waste 233–241
raw material selection 260
see also scrap selection
reactive gas transport 229
refining schedule 71, 74
refractories 56, 228, 246
reheat cracking sensitivity 279
remelting, and FATT 300
residual elements
acceptable levels 165, 191, 192
in bolting steels 213–221
cost 295–296
detrimental effects 197
and embrittlement 182
grain boundary segregation 165, 209
in HY80 288–290
ladle refining process 260–262
in MnNiMo steels 255–264
and rupture ductility 215
see also impurities; R-factor
residual steel pool method 60
rinsing for cleanliness 60
rotor forgings
clean/superclean steel 35, 71
CrMoV 101–109, 109
CrNiMoVNbW 109–122
defects 10, 199
embrittlement 13, 34, 249
fracture toughness (KIC) 104, 181
life expectancy 207
manganese 111
mechanical properties 14, 20, 82–85, 103–104, 168
NiCrMoV 8, 13–14, 35, 57, 109, 181–193
proof stress 103–104, 110, 120, 207
scrapping rate 197
tensile strength 83, 103–104
transverse toughness 86
rotor steels
Charpy impact data 82
FATT 307–308
single shaft 309–310
strength and toughness 198

rupture ductility *see* ductility
rupture strength, of bolting steel 216

S
Saarschmeide GmbH 59
Saarstahl 8, 20
Sb *see* antimony
scrap
  chemical composition 57–58
  economic considerations 55–56, 295–296
  residual contents 53–58, 295–296
  selection 2, 37, 57–58, 59, 62, 63, 219, 246
seawater 290, 304, 305
segregation *see* chemical homogeneity; chemical segregation; grain boundary segregation
sheet steel, electronics applications 229
Siemens Power Generation (KWU) 195
silica 228
silicon
  acceptable levels 165
  adverse effects on steel 1
  analysis 40
  in clean/superclean steel 6, 54, 77, 129, 157
  and embrittlement 186, 195, 200, 292–293, 293, 294
  and FATT 125–127
  and hardenability 288
  and J-factor 266
  in pressure vessel steels 248
  and price 38
  reduction/removal in steelmaking 2, 11, 13, 54, 59, 60, 62, 128, 161, 293
  and segregation 34
  and strain ageing 259
  and tempering temperature 163
  and toughness 290
silicon carbide 235
silicon deoxidation 89, 163, 204
silicon killing 263
silver 239
slag
  acid 230
  basic 60
  calcium oxide 228
  carryover 60, 61
  changing 59, 60
  composition 89
  low iron oxide activity 89
  oxidising 2, 54
  removal 297
  synthetic reducing 246
slag-free tapping 246
Sn *see* tin
specifications
  bolting steel 217
  clean/superclean steel 7, 23, 35–37, 39–40, 48–49, 74–75, 82, 246, 252, 308–309
  Electric Power Research Institute 36–37, 39, 48–49, 74–75, 82, 246, 252, 308–309
  HP and LP rotors 27, 40
  manganese 27, 37
  NiCrMoV 309
  nuclear reactor steels 256
  pressure vessels 244–245
stainless steels 229–230, 234–235
starting toughness, turbine steels 301
Statoil 294
steam turbines *see* turbines
steel
  adverse effects ofimpurities 1
  cleanness
    assessment 224–225, 231
    effect on turbine operational properties 195–211
    and stress corrosion cracking 197
    and temper embrittlement 273
  heat resistant 162–163
  low activation 238–239
  oxygen content and inclusions 227
  toughness, and cost effectiveness 3
  zero-inclusion 223–232
steelmaking
  bolting steels 219
  clean/superclean steel 76
  control of residual elements 2, 10–11, 13, 54, 59–62, 74, 128, 161, 260–262, 293, 295–296
  and creep-rupture strength 10
  and ductility 10
  and fracture toughness 10
  and mechanical properties 10
  methods 1, 2, 9, 10, 35–36, 53–59, 61, 62, 245–246
  production capability 28
  and rupture ductility 215
step cooling 90, 159, 160, 161, 163, 164, 184, 189–190, 266

strain ageing 259
stress corrosion cracking
  clean/superclean steel 7, 17, 35, 133, 146–150, 154, 175, 197, 308
  crack initiation *see* crack initiation
  environmental effects 133, 140, 167, 171–177
  and inclusions 168, 224
  LP rotor steel 27, 167–180
  NiCrMoV 17, 18, 133–155
  resistance 2, 17, 146–154
  test matrix 139
  welded joints 133–155
stress rupture, NiCrMoV superclean steel 17
sulphur
  acceptable levels 27
  adverse effects on steel 1
  analysis 40
  and creep ductility 308
  dissolved 227
  and embrittlement 200
  and FATT 261–262
  and machinability 220–221, 264
  and manganese 1, 3, 250–252, 302
  and price 38
  reduction/removal in steelmaking *see* desulphurisation
  in scrap 296
  and toughness 6
  and upper shelf energy 250–252, 302
  and welds 268
  and X-factor 205
superclean steel
  CrMoNiWV 20
  CrMoV 40
  NiCrMoV 6, 7, 14–17, 19–21, 59–70, 209, 249, 296–297
  specifications *see* specifications, clean/superclean steel
  temper embrittlement 15
  turbine components 14, 17, 20, 35, 51, 59, 297
  see also clean/superclean steel

T
tantalum 236
taper stress joints 288, 290, 293
temper embrittlement *see* embrittlement
temperature effects 2, 267
temperature resistance 272
tempering temperature 163, 274
tensile properties
  Afnor steels 159, 162
  clean/superclean steel 128, 129, 163
  CrMoV 274
  CrMoVTiB 273
  heat affected zone (HAZ) 277
  pressure vessel steels 248
  rotor forgings 83, 103–104
  testing 159, 163
tensioner spool joints 288, 290, 293
testing
  Charpy impact tests 82, 83, 159
    see also Charpy impact data
  creep rupture tests 128
  CTOD testing 304–305
  embrittlement 157, 165
  fracture toughness tests 159
  non-destructive testing 79, 224
  tensile testing 159, 163
thermal efficiency, power generation 3
thermal fatigue stress 195, 206–207
tie back connector 288
tin
  analysis 40
  detrimental effects on steel 1, 27, 266
  and embrittlement 185, 200, 215, 252, 293
  grain boundary segregation 34, 189, 199, 258
  scrap selection 2, 54, 57, 59, 62–63, 64
  specified levels 37, 51, 157
titanium 214, 230
top tapping 297
torsion leg platforms 285
Toshiba Corporation 7, 8, 14, 20, 109
toughness
  clean/superclean steel 65–69
  and creep strength 2
  cryogenic 286
  and embrittlement 159–161
  and grain boundary segregation 121, 252
  and heat treatment 102
  and manganese sulphide inclusions 250
  NiCrMoV 7
  and niobium 299
  and silicon 290
  and sulphur 6
  turbine steels 65–69, 86
transient stresses, fossil fired plant 2, 196

tungsten 49, 236
tungsten stabilised martensitic steel 235–238
turbine components
  blade length 3, 109–110
  clean/superclean 27–28
  construction 109
  creep stress 197
  electroslag-topping 90–94
  FATT 157
  fracture toughness 157
  life expectancy 209
  manufacturing process 94, 128–130
  materials characteristics 196
  nickel alloys 123
  operational properties 195–211
  operational stresses 195–196
  reliability 157, 165, 181
  superclean NiCrMoV 17, 59
  testing 66–69
turbine condensate, pH values 207
turbine steels
  chemical analysis 80–82
  12Cr 123–131
  embrittlement factors 65–69
  FATT 82, 95–96
  impact values 94–95
  industrial production 121
  J-factor 82, 97
  mechanical properties 94–98
  microsegregation 94–98
  phosphorus removal 98
  pitting 167–180
  stress corrosion cracking 167–180
  toughness properties 65–69, 301
tweaking 2

U
UES Steels 213
ultrasonic testing 78, 80, 112, 197, 208, 209, 224
University of Britsh Columbia 223
University of Leeds 6
University of Pennsylvania 6
upper shelf energies 1, 250–252, 260, 262, 298, 302

V
vacuum arc degassing (VAD) 36, 57, 245–246
vacuum arc remelting 5, 36, 158, 215, 216
vacuum carbon deoxidation (VCD) 1, 2, 10, 11, 13, 20, 101, 111, 128, 182, 204–206, 290
vacuum degassing 1, 54, 158, 260
vacuum induction melting (VIM) 36, 215, 216
vacuum ingot casting 36, 59
vacuum remelting 162
vacuum silicon deoxidation 10
vacuum stream degassing (VSD) 2, 36, 246, 260
vacuum treatment 76
vanadium 259, 279–280, 303
vanadium alloys 235
vanadium carbides 281
Vereinigte Edelstahlwerke 6, 7, 14, 34
void closure 77

W
weldability 79, 288, 293
welded joints 5, 27, 133–155
welding 137, 263, 307
welding consumables, low residual 302–303
Westinghouse Electric 6
wetting problems 301

X
X-factor
  clean/superclean steel 65–69, 186
  definition 35, 64–65
  and Δ FATT 187, 202
  NiCrMov 201
  and sulphur content 205
X-ray analysis 91, 92, 189

Y
yield stress, NiCrMoV 286

Z
zero-inclusion steels 223–232